ILLUSTRATED TEXTBOOK OF

CLINICAL CHEMISTRY

ILLUSTRATED TEXTBOOK OF

CLINICAL CHEMISTRY

WILLIAM J. MARSHALL MA MSc PhD MB BS MRCP MRCPath

Senior Lecturer in Chemical Pathology
King's College School of Medicine and Dentistry
University of London
London, UK

Foreword by:

BARBARA E. CLAYTON DBE PhD MD DSc (Hon) FRCP FRCPEd FRCPath

Honorary Research Professor in Metabolism
University of Southampton
Southampton, UK

J. B. Lippincott Company PHILADELPHIA

Gower Medical Publishing LONDON • NEW YORK

Distributed in USA and Canada by:
J.B. Lippincott Company
East Washington Square
Philadelphia, PA 19105
USA

Distributed in Southeast Asia, Hong Kong, India and Pakistan by:
Harper & Row Publishers (Asia) Pte Ltd
37 Jalan Pemimpin 02-01
Singapore 2057

Distributed in UK and Continental Europe by:
Harper & Row Ltd
Middlesex House
34-42 Cleveland Street
London W1P 5FB, UK

Distributed in Japan by:
Igaku Shoin Ltd
Tokyo International
PO Box 5063
Tokyo, Japan

Distributed in Australia and New Zealand by:
Harper & Row (Australasia) Pty Ltd
PO Box 226
Artarmon, NSW 2064
Australia

Distributed in Philippines/Guam, Middle East, Latin America and Africa by:
Harper & Row International
10 East 53rd Street
New York, NY 10022
USA

Project editors: David Goodfellow
Catherine Moehrle

Design & Illustration: Marie McNestry

Line artists: Marion Tasker
Michael Rabess
Mark Willey

Series design: Michel Laake

Library of Congress Catolog Number: 87-83655

British Library Cataloguing in Publication Data
Marshall, William J.
Illustrated textbook of clinical chemistry.
1. Medicine, Diagnosis, Chemical analysis.
I. Title
616.07'56

ISBN: 0-397-44568-7 (Lippincott/Gower)

Originated in Hong Kong by South Sea International Press Ltd.

Typesetting by IC Dawkins (typesetters) Ltd, London.

Text set in Sabon; captions and figures set in Univers.

Printed in Spain by Imago Publishing Ltd.

Reprinted in Hong Kong in 1989 by Imago Publishing Ltd.

Foreword

Arising from his considerable experience of teaching medical students, Dr William Marshall has produced this textbook — and it is definitely for medical students, although doubtless others will find it useful too.

The author has emphasized both the overriding importance of basic physiological and biochemical facts on which the selection of investigations and their interpretations must be based, and the importance of close liaison between the pathologist and the clinician. In today's climate of cost/benefit assessment, informed use of laboratory services becomes ever more important. Clinical chemistry contributes to the diagnosis and management of disease and to its prevention by early detection. I particularly welcome the numerous brief case histories and their accompanying commentaries which are interspersed throughout the text. The relevance of information to patient care is very important in the education of medical students; without relevance their task can be reduced to the acquirement of a mass of unrelated facts.

The twenty-four chapters cover a wide range of subjects from 'Biochemical Tests in Clinical Medicine' through chapters based on organs, such as 'The Kidney', to such topics as 'Therapeutic Drug Monitoring and Clinical Aspects of Toxicology'. Each chapter has a useful summary at the end and two to four carefully selected and very helpful references for further reading.

The material has been selected carefully with the undergraduate in mind. The importance of such matters as the clinical chemistry of the neonate and old age or the importance of drug monitoring are discussed, but overwhelming detail has been avoided. Other subjects, such as hydrogen ion homoeostasis, which often cause difficulty for the student, are explained with clarity and at some length.

I believe that students will welcome the very high quality of production of this book. The layout is attractive, the tables are clear, the diagrams are excellent and the case histories and commentaries are clearly presented. The author's enthusiasm for his subject and his practical common sense approach are very apparent. I congratulate Dr Marshall on the production of this textbook and suggest it will be only the first of further editions.

B.E.C.

Preface

I have written this book primarily for undergraduate medical students preparing for their qualifying examinations. Clinical chemistry, or chemical pathology, provides a link between medicine and the basic medical sciences, particularly physiology and biochemistry. Each chapter therefore contains a summary of the relevant basic science to assist the reader's understanding of the accounts of pathological derangements which follow. But biochemical data are of little use on their own; tests should be requested, and their results interpreted, in the light of clinical findings and the results of other investigations. I have therefore included more clinical information than is usual in textbooks of pathology. Part of this is in the form of case histories, which illustrate how biochemical data are used in diagnosis and management. These are largely based on my own experience and have been chosen to illustrate the more important conditions or points in the text. Readers may wish to use them as exercises in diagnosis, by reading the clinical summary and results of investigations with the comments covered over. Reading the case histories and the information presented in tables and figures could also help students who are revising their knowledge, and chapter summaries have also been included to this end.

Biochemical tests are only one part of an ever-developing armamentarium of investigative techniques. Inevitably, as techniques develop, so the role of biochemical tests changes; new tests are introduced while others become obsolete or find new roles. My aim throughout has been to discuss biochemical tests in context, that is both in relation to their usefulness in particular clinical circumstances and in relation to other techniques that may be employed.

The inclusion of such material should also make this book attractive to clinical biochemists and medical laboratory scientists preparing for either a Master's degree in clinical biochemistry or professional examinations for medical laboratory scientists and technologists, although it is not intended to cover their requirements fully; analytical methodology is not discussed, except where it has some direct bearing on the interpretation of results. Others who may find it useful will include doctors studying for postgraduate diplomas in medicine, surgery, obstetrics and anaesthetics in which some knowledge of clinical chemistry and metabolic medicine is required.

In writing this book, I have been greatly encouraged by my family and by many friends, students and colleagues. Colleagues who have read individual chapters and provided invaluable advice include Dr Michael Norman, Miss Dianne Baldwin, Miss Joan Butler, Dr Tim Cundy, Mr Doug Hirst, Dr Jerry Jones, Mr Jim Keating, Dr Caje Moniz, Dr Richard O'Donovan and Dr Sarah Rae. Dr Danielle Freedman has read and commented on every chapter and I am indebted to her; many of her suggestions have been incorporated into the text. Any errors or omissions that remain are, however, entirely my own responsibility. Cathy Boreland typed many of the chapters and Robert Willcox never failed to answer calls for help when I was word-processing. At Gower, I have enjoyed working with Catherine Moehrle and David Goodfellow, while Marie McNestry's design has resulted in a book which is as pleasing to the eye as I hope it is to read.

W. J. M.

ACKNOWLEDGEMENTS

Fig.20.7 redrawn from: Dieppe P & Calvert P (1983) *Crystals and Joint Disease*. London: Chapman & Hall.

Fig.22.3 redrawn from: Richens A & Dunlop A (1975) Serum-phenytoin levels in management of epilepsy. *Lancet*, **2**, 247–248.

Contents

1. Biochemical Tests in Clinical Medicine

INTRODUCTION

A central function of the chemical pathology or clinical chemistry laboratory is to provide biochemical information for the management of patients. Such information will be of value only if it is accurate and relevant and if its significance is appreciated by the clinician to enable results to be used appropriately to guide clinical decision-making. This chapter examines how biochemical data is acquired and how it should be used.

USE OF BIOCHEMICAL TESTS

Biochemical tests are used extensively in medicine, both in relation to diseases that have an obvious metabolic basis (diabetes mellitus, hypothyroidism) and those in which biochemical changes are a consequence of the disease (renal failure, malabsorption). Biochemical tests are used in diagnosis, prognosis, monitoring and screening.

Diagnosis

Medical diagnosis is based on the patient's history, if available, the clinical signs found on examination, the results of investigations and, sometimes, retrospectively on the response to treatment. Frequently, a confident diagnosis can be made on the basis of the history combined with the findings on examination. Failing this, it is usually possible to formulate a differential diagnosis, in effect, a short-list of possible diagnoses; biochemical and other investigations may then be used to distinguish between them.

Investigations may be selected to help either confirm or refute a diagnosis and it is important that the clinician appreciates how well an investigation performs in this context. Making a diagnosis, even if incomplete, such as a diagnosis of hypoglycaemia without knowing its cause, allows treatment to be initiated.

Prognosis

Tests used primarily for diagnosis may also provide prognostic information and some are used specifically for this purpose; for example, serial measurements of serum creatinine concentration in progressive renal disease are used to indicate when dialysis may be required. Tests can also predict the risk of developing a particular condition; the measurement of serum cholesterol concentration can, for example, predict the risk of coronary artery disease. However, such risks are calculated from statistical data and cannot give a precise prediction for a particular individual.

Monitoring

In tests which are used to follow the course of an illness and to monitor the effects of treatment, there must be a suitable analyte; for instance, glucose in patients with diabetes mellitus. These tests may also be used to detect complications of treatment, such as hypokalaemia during treatment with diuretics.

Screening

Biochemical tests are widely used to determine whether a condition is present subclinically. The best known example is the mass screening of all newborn babies for phenylketonuria (PKU) which is carried out in many countries, including the United Kingdom and United States. The use of the 'biochemical profile', a battery of biochemical tests usually performed on a multichannel auto-analyzer, is discussed later in this chapter.

SAMPLING

Test request

The sample for analysis must be collected and transported to the laboratory according to a specified procedure if the data are to be of clinical value. This procedure begins with the test request form which should include:

- Patient's name, sex and date of birth
- Ward/clinic/address
- Name of requesting doctor (telephone/page number for urgent requests)
- Clinical diagnosis/problem
- Test(s) requested
- Type of specimen
- Date and time of sampling
- Relevant treatment (for example, drugs)

In practice, vital information is often omitted and this may either cause delay in analysis and reporting or make it impossible to interpret the results.

Relevant clinical information and details of treatment, especially with drugs, are necessary to allow laboratory staff to assess the results in their clinical context. Drugs may interfere with analytical methods *in vitro* or may cause changes *in vivo* that suggest a pathological process; for instance, oestrogens increase thyroid-binding globulin and thus total thyroxine concentration. If there is any doubt as to the appropriate test to request or sample to collect, the laboratory should be contacted for advice.

Patient

Many analytes are little affected by variables such as age and sex, but it may be important to standardize the conditions under which the sample is obtained. Factors of importance in this respect are listed in Fig.1.1 and are discussed further in subsequent chapters.

Sample

The sample provided must be appropriate for the test requested. Many biochemical analyses can be made on either plasma or serum but, in some instances, it is of critical importance which of these is used; for example, serum is necessary for protein electrophoresis and plasma for renin activity. Haemolysis must be avoided when blood is drawn and, if the patient is receiving intravenous therapy, blood must be drawn from a remote site (the opposite arm) to avoid contamination.

The sample can usually be collected into either a glass or plastic container but, in some circumstances, one of these may be preferred or even essential. A

Factor	Example of variable affected
age	alkaline phosphatase
sex	gonadal steroids
pregnancy	thyroxine (total)
posture	proteins
exercise	creatine kinase
stress	prolactin
nutritional state	glucose
time	cortisol

Fig.1.1 Important factors which influence biochemical variables.

preservative may be necessary, such as fluoride to prevent loss of glucose by glycolysis.

All samples must be correctly labelled and transported to the laboratory with the minimum of delay. The serum or plasma is then separated from blood cells and analyzed. When analysis is delayed or when samples are sent to distant laboratories for analysis, degradation of labile analytes must be prevented by refrigeration or freezing of the separated serum or plasma.

Equal care is needed with the collection and transportation of other samples, such as urine and spinal fluid. All samples should be regarded as potentially infectious and special care is required with 'high-risk' samples, for example, from patients infected with hepatitis B virus.

SAMPLE ANALYSIS AND REPORTING OF RESULTS

Analysis

The ideal analytical method is accurate, precise, sensitive and specific. It gives a correct result (accurate) that is the same if repeated (precise; Fig. 1.2). It measures low concentrations of the analyte (sensitive) and is not subject to interference by other substances (specific). In addition, it should preferably be cheap,

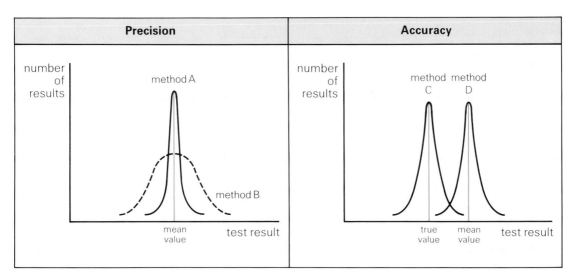

Fig.1.2 Precision and accuracy of biochemical tests. Both graphs show the distribution of results for repeated analysis of the same sample by different methods.

Precision: the mean value is the same in each case but the scatter about the mean is less in method A than in method B. Method A is, therefore, more precise.

Accuracy: both are equally precise but in method D the mean value differs from the true value. The mean for method C is equal to the true value. Both methods are equally precise but method C is more accurate.

simple and quick to perform. In practice, no test is ideal, but the pathologist must ensure that the results are sufficiently reliable to be clinically useful. Laboratory staff make considerable efforts to achieve this and analytical methods are subject to rigorous quality control.

It is important to appreciate that results obtained using different methods may not be strictly comparable. When a comparison between two results is being made, the same analytical method should be used on both occasions.

It is often appropriate to perform a group of related tests on a sample. For example, serum calcium, phosphate and alkaline phosphatase levels all provide information which may be useful in the diagnosis of bone disease; several liver 'function' tests may usefully be grouped together. Such groups are sometimes referred to as 'biochemical profiles'. Labour-saving multichannel auto-analyzers and similar instruments can perform as many as twenty assays simultaneously on a single serum sample. However, although it may be tempting to subject every sample to all tests, this approach generates an enormous amount of information, much of which may be unwanted, ignored or

misinterpreted and, worst of all, may actually prevent the clinician from discerning the important results.

Reporting results

Once analysis has been completed and the necessary quality control checks made and found to be satisfactory, a report can be issued. Computers are being used increasingly for data processing in laboratory medicine and reports may be generated by computers working on- or off-line to the analyzers. The ability of the computer to store and process data facilitates the production of cumulative reports, allowing trends in the data to be picked out at a glance.

INTERPRETATION OF RESULTS

When the result of a biochemical test is obtained, the following points must be taken into consideration:
- Is it normal?
- Is it significantly different from previous results?
- Is it consistent with the clinical findings?

3

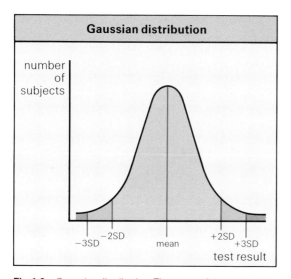

Gaussian distribution

number of subjects

−3SD −2SD mean +2SD +3SD

test result

Fig.1.3 Gaussian distribution. The range of the mean ± 2 standard deviations (SD) encompasses 95.5% of the total number of test results. The range of the mean ± 3 standard deviations encompasses 99.7% of the total number.

Is it normal?

The use of the word 'normal' is fraught with difficulty. Statistically it refers to a distribution of values from repeated measurement of the same quantity and is described by the bell-shaped Gaussian curve (Fig.1.3). Many biological variables show a Gaussian distribution; the majority of individuals within a population will have a value approximating to the mean for the whole and the frequency with which any value occurs decreases with increasing distance from the mean.

Skewed distributions are also often found, for example, that of serum bilirubin concentration, but can often be mathematically transformed to a normal distribution; data distributed with a skew to the right of the mean can often be transformed to a normal distribution if replotted on a semi-logarithmic scale.

If the variable being measured has a normal (Gaussian) distribution in a population, statistical theory predicts that approximately ninety-five percent of the values in the population will lie within the range given by the mean plus or minus two standard deviations (Fig.1.3); of the remaining five percent, half the values will be higher and half will be lower than the limits of this range. Approximately ninety-nine percent of values will be within the range of the mean plus or minus three standard deviations.

When establishing the range of values for a particular variable in healthy people, it is conventional to first examine a representative sample of sufficient size to determine whether or not the values fall into a Gaussian distribution. The range (mean ± two standard deviations) may then be calculated; this is, in statistical terms, the 'normal range'. Several important points arise from this:

- In the population, five percent are, by definition, excluded from the normal range and yet are healthy. This suggests that, if the measurement were to be made in a group of comparable individuals, one in twenty would have a value outside this range.
- The specialized statistical use of the word 'normal' does not equate with what is generally meant by the word, that is, 'habitual' or 'usually encountered'.
- The statistical 'normal' may not be related to another common use of the word, which is to imply 'freedom from risk'. For example, epidemiological evidence indicates that there is an association between increased risk of coronary heart disease and serum cholesterol concentrations, even within the normal range as derived from measurements on apparently healthy men.

Thus, the normal range for an analyte, defined and calculated as described, has severe limitations. It only identifies the range of values that can be expected to occur most often in individuals who are comparable to those in the population for whom the range was derived. It is not necessarily normal in terms of being 'ideal' nor is it associated with no risk of having or developing disease. Further, by definition, it will exclude some healthy individuals.

In all cases, like must be compared with like. When physiological factors affect the concentration of an analyte (see Fig.1.1), an individual's result must be assessed by comparing it with the value expected for comparable healthy people. It may, therefore, be necessary to establish normal ranges for subsets of the population, such as various age groups, or males or females only.

To alleviate the problems associated with the use of the word 'normal', the term 'reference range' has been widely adopted by laboratory staff using numerical values generally based on the mean plus or minus two standard deviations. Results can be compared with the 'reference range' without assumptions being made about the meaning of normal. In practice, the term 'normal range' is still in general use outside laboratories and is used synonymously with 'reference range' in this book. Reference ranges for some of the common analytes, as used in the author's laboratory, are given in *Appendix 1*.

In using normal ranges to assess the significance of a particular result, the individual is being compared with a population. Some analytes show considerable biological variation, but the combined analytical and biological variations will usually be less for an individual than for a population. For example, although the normal range for serum total thyroxine concentration is 60–150mmol/l, the day-to-day variation in an individual is less than this. Thus, it is possible for the test result to be abnormal for an individual, yet still be within the accepted 'normal range'.

An abnormal result does not always indicate a pathological process nor a normal result an absence of pathology. However, the more abnormal a result, that is, the greater its difference from the limits of the reference range, the greater is the probability that it is related to a pathological process.

In practice, there is rarely an absolute demarcation between normal values and those seen in disease; equivocal results must be substantiated by further investigation. If an important decision in the management of a patient is to be based upon a single result, it is vital that the cut-off point, or 'decision level', is chosen to ensure that the test functions efficiently. In screening for PKU, for example, the blood level of phenylalanine selected to indicate a positive result must not exclude infants with the condition; in other words, there must be no false negatives. This inevitably means that some normal children will be test-positive (false positives) and will be subjected to further investigation. Generally, it is unusual to have to determine a patient's management on the basis of one result alone.

It has been explained that five percent of healthy people will, by definition, have a value for a given variable that is outside the reference range. If a second and independent variable is measured, the probability that this result will be 'abnormal' is also 0.05 (five percent). However, the abnormal results may not arise in the same individuals and the overall probability of an abnormal result from at least one test will be greater than five percent. It follows that the more tests that are performed on an individual, the greater the probability that the result of one of them will be abnormal;

for ten independent variables the probability is 0.4, or in other words, at least one abnormal result would be expected in forty percent of healthy people, and for twenty variables, 0.64 or sixty-four percent.

Although biochemical parameters frequently are to some extent interdependent, (e.g. albumin and total protein), the use of multichannel auto-analyzers to produce biochemical profiles inevitably generates a number of spurious 'abnormal results'. Before any decision can be made on the basis of such results, some information is required about their predictive value, that is, the probability that they are related to a pathological process. This topic is discussed on page 7.

Is it different?

If the result of a previous test is available, the clinician will be able to compare the results and decide whether any differences between them are significant. This will depend upon the precision of the assay itself (a measure of its reproducibility) and the natural biological variation. Some examples of variation in common analytes are given in Fig.1.4.

Whatever the biological variation, it can be shown statistically that if two measurements of a particular analyte differ by more than 2.8 times the standard deviation of the method, the probability that this is due to chance alone is only 0.05 or five percent. For example, suppose that the calcium concentration of a serum sample is measured twenty times and the mean of the measurement is found to be 2.60mmol/l with a standard deviation of 0.05mmol/l. Ninety-five percent of the measured values therefore fall into the range of 2.50–2.70mmol/l. Assuming that the standard deviation for the method is the same over the range of concentrations, then the probability of a second result being significantly different from the first is more than 0.95 (ninety-five percent) only if it lies outside the range of 2.46–2.74mmol/l (2.60mmol/l ±2.8SD or ±0.14mmol/l). Whether or not this significant analytical difference is also biologically significant will depend on the extent of normal biological variation.

Is it consistent with clinical findings?

If the result is consistent with clinical findings, it is evidence in favour of the clinical diagnosis. If it is not consistent, the explanation must be sought. There may have been a mistake in the collection, labelling or

Analyte	Analytical variation	Biological variation
sodium	1.1mmol/l	2.0mmol/l
potassium	0.1mmol/l	0.19mmol/l
bicarbonate	0.5mmol/l	1.3mmol/l
urea	0.4mmol/l	0.85mmol/l
creatinine	5.0μmol/l	4.1μmol/l
calcium	0.04mmol/l	0.04mmol/l
phosphate	0.04mmol/l	0.11mmol/l
total protein	1.0g/l	1.66g/l
albumin	1.0g/l	1.44g/l
aspartate transaminase	6.0iu/l	8.0iu/l
alkaline phosphatase	4.0iu/l	15.0iu/l

Fig.1.4 Analytical and biological variation.

Analytical variation: typical standard deviations for repeated measurements made using a multichannel auto-analyzer on a single quality control serum with concentrations in the normal range.

Biological variation: means of standard deviations for repeated measurements made at weekly intervals in a group of healthy subjects over a period of 10 weeks.

analysis of the sample, or in the reporting of the result. In practice, it may be simplest to request a further sample and to repeat the test. If the result is confirmed, the sensitivity and specificity of the test in the clinical context should be considered and the clinical diagnosis itself may have to be reviewed.

SPECIFICITY, SENSITIVITY AND PREDICTIVE VALUE OF TESTS

In using the result of a test, it is important to know how reliable the test is and how suitable it is for its intended purpose. Thus, the laboratory personnel must ensure, as far as is practical, that the data are accurate and precise and the clinician should appreciate how specific and sensitive the test is in the context in which it is used.

Specificity and sensitivity

Specificity is a measure of the incidence of negative results in persons known to be free of a disease, that is

'true negative' (TN). Sensitivity is a measure of the incidence of positive results in patients known to have a condition, that is, 'true positive' (TP). A specificity of ninety percent implies that ten percent of disease-free people would be classified as having the disease on the basis of the test result; ten percent would have a 'false positive' (FP) result. A sensitivity of ninety percent implies that only ninety percent of people known to have the disease will be diagnosed as having it on the basis of that test alone; ten percent will be false negatives (FN).

An ideal test would be one hundred percent sensitive, giving positive results in all diseased subjects, and would be one hundred percent specific, giving negative results in all subjects free of disease. In reality no tests achieve such high standards; all generate false positives and false negatives. Specificity, sensitivity, predictive value and efficiency of tests may be calculated as follows:

1. $\text{Specificity} = \dfrac{\text{TN}}{\text{all without disease (FP + TN)}} \times 100$

2. $\text{Sensitivity} = \dfrac{\text{TP}}{\text{all with disease (TP + FN)}} \times 100$

3. Predictive value of a positive test $= \dfrac{TP}{TP + FP} \times 100$

4. Efficiency $= \dfrac{TP + TN}{\text{total number of tests}} \times 100$

Factors which increase the specificity of a test tend to decrease the sensitivity and vice versa, as there is almost always an overlap between test results seen in health and in disease. To take an extreme example, if it was decided to diagnose thyrotoxicosis only if the serum total thyroxine concentration was 200nmol/l (the upper limit of the reference range is 150nmol/l in the author's laboratory), the test would have one hundred percent specificity; positive results (greater than 200nmol/l) would only be seen in thyrotoxicosis. On the other hand, the test would have a low sensitivity in that many patients with mild thyrotoxicosis would be misdiagnosed. If a concentration of 100nmol/l was used, the test would be very sensitive (all those with thyrotoxicosis would be correctly assigned) but very non-specific, because many normal people would also be diagnosed as having thyrotoxicosis.

Whether it is desirable to maximize specificity or sensitivity depends on the nature of the condition that the test is used to diagnose and the consequences of making an incorrect diagnosis. For example, sensitivity is paramount in a screening test for a harmful condition, but the inevitable false positive results will have to be investigated further. However, in selecting patients for a trial of a new treatment, a highly specific test is more appropriate to ensure that the treatment is being given only to patients with a particular condition.

Efficiency

The efficiency of a test is given by the number of correct results, the sum of true positives plus true negatives, in relation to the number of tests carried out. When sensitivity and specificity are equally important, the test with the greatest efficiency should be used.

Predictive values

A highly specific and sensitive test does not necessarily perform well in a clinical context. This is because the predictive value of a positive test result, which is equal to the percentage of all positive results that are true positives, is dependent upon the prevalence of the disease. If a condition has a low prevalence and the test is less than one hundred percent specific, many false positives will result.

A high predictive value for a positive test is important if the appropriate management of a true positive would be potentially dangerous if applied to a false positive. However, when a test is used for screening, the appropriate management is to perform further confirmatory tests and, although these may cause inconvenience for subjects with false positive results, they are unlikely to be dangerous.

In order not to miss cases, a screening test should have a very high predictive value for a negative result (number of true negatives as a percentage of all negatives). This conclusion follows directly from the fact that the test must be highly sensitive.

For clarity, this discussion has centred on the use of single tests for diagnostic purposes but, in practice, the clinician will combine clinical information and, often, the results of several investigations to make the diagnosis. If the tests are used rationally, the predictive value of positive results will be higher since each test will be used in a patient who has other features suggesting a particular diagnosis (the prevalence of the disease in question will be much higher than in the general population). For example, although Cushing's disease is rare, making the predictive value of a positive test in the general population low, in practice, one would only investigate patients suspected on clinical grounds of having the condition and in whom the prevalence will therefore be higher. This may be self-evident, but doctors frequently order tests on flimsy clinical grounds and fail to appreciate how unhelpful, or even misleading, the results my be.

Although most clinicians and pathologists interpret data intuitively, the most efficient approach can be defined mathematically, especially when there are numerous data to consider. With the increasing availability of computers, such mathematical data analysis should become more widely adopted.

Another approach, useful when several tests are performed, is to combine the results mathematically, usually after each result has been weighted by a multiplier, to produce one or more figures called 'discriminant functions' or 'indices'. These can then be compared with the range of values calculated for a group of patients shown, by an independent definitive technique, to have a particular condition. If the discriminant function for an individual falls within this range, there is a high probability that the individual has the condition in question. This approach has been applied, for example, to the differential diagnosis of hypercalcaemia and obstructive jaundice, but has yet to gain wide acceptance in clinical chemistry.

SCREENING

Screening tests are used to detect disease in groups of apparently healthy individuals. Such tests may be applied to whole populations (the detection of PKU in the newborn); to groups known to be at risk (the detection of hypercholesterolaemia in the relatives of people with premature coronary heart disease); and to groups of people selected for other reasons (biochemical profiling of hospital inpatients and health screening for business executives).

As previously discussed, high sensitivity is essential for screening tests and, to avoid unnecessary further tests of normal people, high specificity is desirable. Screening tests for PKU are designed to maximize sensitivity but are also highly specific. However, PKU has a low incidence so that even with a sensitivity of one hundred percent and specificity of 99.9%, the predictive value of a positive test is only ten percent; that is, nine out of ten positive tests will be shown on further investigation to be false positives. These calculations are made as follows:

1. Incidence of PKU = 1 in 10,000 live births

2. Sensitivity = 100% or $\dfrac{1\ TP}{1\ \text{case of PKU}}$

3. Specificity = 99.9% or $\dfrac{9990\ TN}{9999\ \text{without PKU}}$

4. Number of positive tests per 10,000 infants

   tested $= \dfrac{(100 - 99.9)}{100} \times 10,000 = 10$

5. Number of true positive and false positive results: TP = 1
   FP = 9

6. Predictive value of a positive test
   $= \dfrac{1}{10} \times 100 = 10\%$

In this instance, the predictive value of a negative test will be one hundred percent, confirming that no cases will be missed using the screening test.

Biochemical profiling is based on the use of much less specific or sensitive tests and therefore has a low efficiency for detecting disease. It is also not particularly efficient at detecting minor abnormalities as the more tests that are performed, the greater the probability that a result will be abnormal.

When multichannel auto-analyzers are used to generate biochemical data and an unexpected abnormality is found, a decision must be made as to what action to take. The abnormality may be considered insignificant in some clinical circumstances but, if it is not, further investigations must be made. Although these may be of ultimate benefit to the patient, their cost and economic consequences may be considerable. At the very least, the test should be repeated to ensure that the abnormality was not due to analytical error.

The ready availability of an investigation often leads to it being used unnecessarily or inappropriately. Doctors should be encouraged to be selective in test requests. They should also join with laboratory staff in critically examining all current tests and investigative techniques to ensure that they are using these tests to their best advantage in medical practice.

SUMMARY

Biochemical tests are used in diagnosis, monitoring patients' progress, screening for disease and for prognosis. The results of some biochemical tests provide specific diagnostic information but, in many instances, biochemical changes reflect pathological processes which are common to a number of diseases.

In selecting a test or tests, it is essential to consider what type of information is required and whether the test is capable of providing it. Samples for analysis must be collected under appropriate conditions and the analytical methods must be reliable. In assessing the significance of a test result, the clinical circumstances and the possible contribution of any analytical or biological variation must be considered. When a test result is assessed by comparison with a reference range of values expected for healthy people, this reference range should be based on analyses of subjects who are comparable in, for example, age and sex.

Since there is rarely a clear distinction between test results which can occur in health and in disease, it is essential to appreciate the statistical basis on which such comparisons can be made. This is also necessary when a result is compared with one obtained previously.

FURTHER READING

Fraser C G (1986) *Interpretation of Clinical Chemistry Laboratory Data*. Oxford: Blackwell Scientific Publications.

Galen R S & Gambino S R (1975) *Beyond Normality: The Predictive Value and Efficiency of Medical Diagnosis*. New York: John Wiley.

2. Water and Sodium

WATER AND SODIUM DISTRIBUTION

Water distribution

Water accounts for approximately sixty percent of body weight in men and fifty-five percent in women, reflecting a greater body fat content in women. Approximately sixty-six percent of this water is in the intracellular fluid (ICF) and thirty-three percent in the extracellular fluid (ECF); only eight percent of total body water is in the plasma (Fig.2.1). Water is not actively transported in the body. It is, in general, freely permeable through the ICF and ECF and its distribution is determined by the osmotic contents of these compartments. Except in the kidney, the osmotic concentrations, or osmolalities, of these compartments are always equal; they are isotonic. Any change in the solute content of a compartment engenders a shift in water which restores isotonicity.

The major contributors to the osmolality of the ECF are sodium and its associated anions, mainly chloride and bicarbonate; in the ICF, the predominant cation is potassium. Other determinants of ECF osmolality include glucose and urea. Protein makes a numerically small contribution of approximately 0.5%. However, since the capillary endothelium is relatively impermeable to protein and since the protein concentration of interstitial fluid is much less than that of serum, the osmotic effect of the protein is an important factor in determining water distribution between these two compartments. The contribution of protein to the osmotic pressure of serum is known as the colloid osmotic pressure or oncotic pressure (see *Chapter 15*).

Under normal circumstances, the amounts of water taken into the body and lost from it are equal over a period of time. Water is obtained from the diet and oxidative metabolism and is lost through the kidneys, skin, lungs and gut (Fig.2.2). The minimum volume of urine necessary for normal excretion of waste products is about 500ml/day but, as a result of obligatory losses by other routes, the minimum daily water intake necessary for the maintenance of water balance is 1100ml. This increases if the losses are abnormally large, for example, with excessive sweating and diarrhoea. Water intake is usually considerably greater than this minimum requirement but the excess is easily removed by the kidneys.

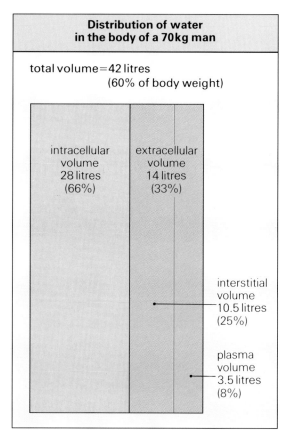

Fig.2.1 Distribution of water in the body of a 70kg man. The distribution is similar in women although the amount of water as a percentage of body weight is less. In children and infants, total body water is 75–80% of body weight, with a higher ECF:ICF volume ratio than in adults, but the proportion of the total body water contained in the plasma is the same.

Obligatory losses		Sources	
skin	500ml		
lungs	400ml		
gut	100ml	water from oxidative metabolism	400ml
kidneys	500ml	minimum in diet	1100ml
total	1500ml	total	1500ml

Fig.2.2 Daily water balance in an adult. The minimum intake necessary to maintain balance is approximately 1100ml.

Sodium distribution

The body of an adult man contains approximately 3000mmol of sodium, seventy percent of which is freely exchangeable with the remainder complexed in bone. The majority of the exchangeable sodium is extracellular; the normal ECF sodium concentration is 135–145mmol/l while that of the ICF is only 4–10mmol/l. Most cell membranes are permeable to sodium and the gradient is maintained by active pumping of sodium from the ICF to the ECF.

As with water, sodium input and output normally are balanced. The normal daily intake of sodium in the Western world is 100–200mmol/day but the obligatory sodium loss, via the kidneys, skin and gut, is less than 10mmol/day. Thus, the sodium intake necessary to maintain sodium balance is much less than the normal intake and excess sodium is excreted in the urine. Evidence that excessive sodium intake is harmful, especially as an aetiological factor in hypertension, is at present conflicting.

It is important to appreciate that there is a massive internal turnover of sodium. Sodium is secreted into the gut at the rate of approximately 1000mmol/day and filtered by the kidneys at a rate of 25,000mmol/day, the vast majority being regained by reabsorption in the gut and renal tubules. If there is even a partial failure of this reabsorption, sodium homoeostasis will be compromised.

WATER AND SODIUM HOMOEOSTASIS

Water and ECF osmolality

Changes in body water content independent of the amount of solute will alter the osmolality (Fig.2.3). The osmolality of the ECF is normally maintained in the range 280–295mmol/kg of water. Any loss of water from the ECF, such as occurs with water deprivation, will increase its osmolality and result in movement of water from the ICF to the ECF. However, a slight increase in ECF osmolality will still occur, stimulating the hypothalamic thirst centre, which promotes a desire to drink, and the hypothalamic osmoreceptors, which cause the release of vasopressin (antidiuretic hormone or ADH).

Vasopressin renders the renal collecting ducts permeable to water, permitting water reabsorption and concentration of the urine; the maximum urine concentration that can be achieved in humans is about 1200mmol/l. The osmoreceptors are highly sensitive to osmolality, responding to a change of as little as one percent. Vasopressin is undetectable in the plasma at a plasma osmolality of 280mmol/kg, but its concentration rises sharply if the plasma osmolality increases above this level (Fig.2.4).

If the ECF osmolality falls, there is no sensation of thirst and vasopressin secretion is inhibited. A dilute urine is produced, allowing water loss and restoration

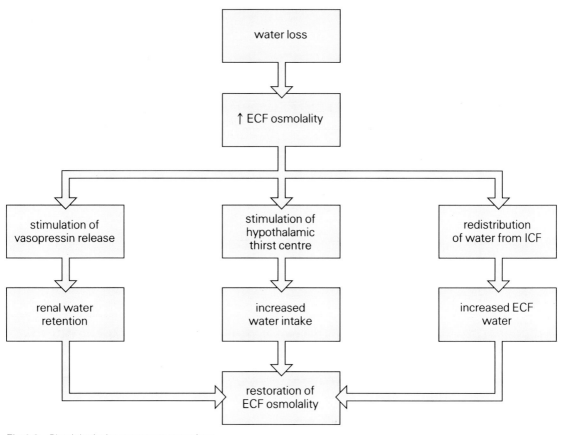

Fig.2.3 Physiological responses to water loss.

of the ECF osmolality to normal. If an increase in ECF osmolality occurs as a result of a solute, such as urea, which diffuses readily across cell membranes, the ICF osmolality is also increased and osmoreceptors are not stimulated.

Other stimuli affecting vasopressin secretion (Fig.2.5) include angiotensin and arterial and venous baroreceptors or volume receptors. If there is a decrease in serum volume of more than ten percent, hypovolaemia becomes a powerful stimulus to vasopressin release (see Fig.2.4) and osmolar controls are overridden. In other words, ECF volume is defended at the expense of a decrease in osmolality.

Sodium and ECF volume

The volume of the ECF is directly dependent upon the total body sodium since water intake and loss are regulated to hold the concentration of sodium in the ECF constant and because sodium is virtually confined to the ECF.

Sodium balance is maintained by regulation of its renal excretion. Sodium excretion is dependent upon glomerular filtration, but the glomerular filtration rate (GFR) appears to become an important determinant of sodium excretion only at extremely low rates of filtration (sodium retention is a late feature of chronic renal failure). Normally, approximately seventy percent of filtered sodium is actively reabsorbed in the proximal convoluted tubule, with

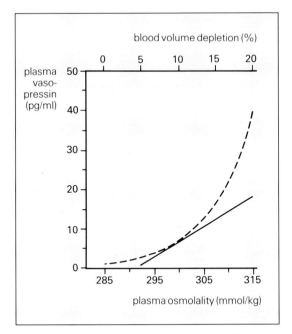

Fig.2.4 Vasopressin concentration in relation to blood volume and osmolality. Vasopressin secretion increases linearly (solid line) with increasing plasma osmolality if blood volume remains constant. A small decrease in blood volume under iso-osmotic conditions has little effect on the secretion of vasopressin (broken line) whereas a fall of more than 8% results in a massive increase in vasopressin secretion.

Control of vasopressin secretion	
Stimulating factors	**Inhibiting factors**
increased ECF osmolality	decreased ECF osmolality
severe hypovolaemia (via angiotensin II and arterial and venous volume receptors)	hypervolaemia
	alcohol
stress, including pain	
exercise	
drugs: narcotic analgesics, nicotine, some sulphonylureas, carbamazepine, clofibrate and vincristine	

Fig.2.5 Factors affecting vasopressin secretion.

further reabsorption in the loop of Henle. Less than five percent of filtered sodium reaches the distal convoluted tubule. Aldosterone, released from the adrenal cortex in response to activation of the renin-angiotensin system, stimulates sodium reabsorption in the distal convoluted tubules and collecting ducts.

Other factors must also be involved in the control of sodium reabsorption since patients with adrenal insufficiency, on a fixed replacement dose of mineralo-corticoids, maintain sodium balance even though their serum mineralocorticoid levels are not controlled by sodium status. In such subjects, chronic loading with mineralocorticoids causes sodium retention only for a short period; thereafter, sodium balance is regained, albeit with an increased ECF volume.

The existence of a natriuretic peptide hormone in man is now well established. The hormone is secreted by the cardiac atria, apparently in response to ECF volume expansion, and it may act through an intermediate hormone, kallikrein. It increases renal sodium excretion by inhibiting sodium reabsorption in the distal nephron and, possibly, by an effect on the GFR. The human atrial natriuretic hormone can be measured by radioimmunoassay but, as yet, its precise role in sodium homoeostasis has not been elucidated.

There is also evidence for a natriuretic factor with a low molecular weight of less than 500 daltons, possibly having a structure resembling a cardiac glycoside and acting through the inhibition of sodium/potassium ATPase. However, this substance has yet to be purified and characterized.

Water depletion	
Causes	**Clinical features**

Increased loss
 	
from kidneys:
    renal tubular disorders
    diabetes insipidus
    increased osmotic load due to diabetes mellitus,
        osmotic diuretics or high protein intake
from skin:
    sweating
from lungs:
    hyperventilation
from gut:
    diarrhoea (in infants)
 	
Decreased intake
 	
infancy   dysphagia
old age   restriction of
unconsciousness   oral intake
	Symptoms
	thirst
	dryness of mouth
	difficulty in swallowing
	weakness
	confusion
	
	Signs
	weight loss
	dryness of mucous membranes
	decreased saliva secretion
	loss of skin turgor
	decreased urine volume (early)

Fig.2.6 Causes and clinical features of predominant water depletion. In infantile gastroenteritis and in acclimatization to high temperatures, some sodium is lost from the gut and skin but the effects of water loss may predominate.

In general, the control mechanisms for ECF volume respond less rapidly and are less precise than the control mechanisms for ECF osmolality. Except at extremes of hypovolaemia, maintenance of osmolality takes precedence.

WATER AND SODIUM DEPLETION

Water depletion or combined water and sodium depletion will occur if losses are greater than intake. Pure water depletion is seen much less frequently than depletion of both water and sodium. As sodium cannot be excreted from the body without water, sodium loss never occurs alone but is always accompanied by water loss. The fluid loss may be isotonic or hypotonic with respect to the serum.

The clinical and biochemical features of pure water depletion and of isotonic sodium and water loss are quite different, as are the physiological responses. In clinical practice, however, states of fluid depletion encompass the whole spectrum between these two extremes and the clinical and biochemical features will reflect this. Further, it should be appreciated that they may have been modified by previous treatment.

Water depletion

Water depletion will occur if water intake is inadequate or if losses are excessive (Fig.2.6). Excessive loss of water without any sodium loss is unusual, except in diabetes insipidus, but providing the sodium loss is small the clinical consequences will be related primarily to the water depletion (Fig.2.6).

Loss of water from the ECF causes a rise in the concentrations of sodium, urea and protein in the serum and a rise in the haematocrit. A water loss of three litres in an adult is sufficient to cause detectable hypernatraemia. The increase in ECF osmolality stimulates thirst and secretion of vasopressin. Signs of a reduced ECF volume are noticeably absent, except in severe water depletion, because the loss is borne by the body water in total and not just the ECF (Fig.2.7).

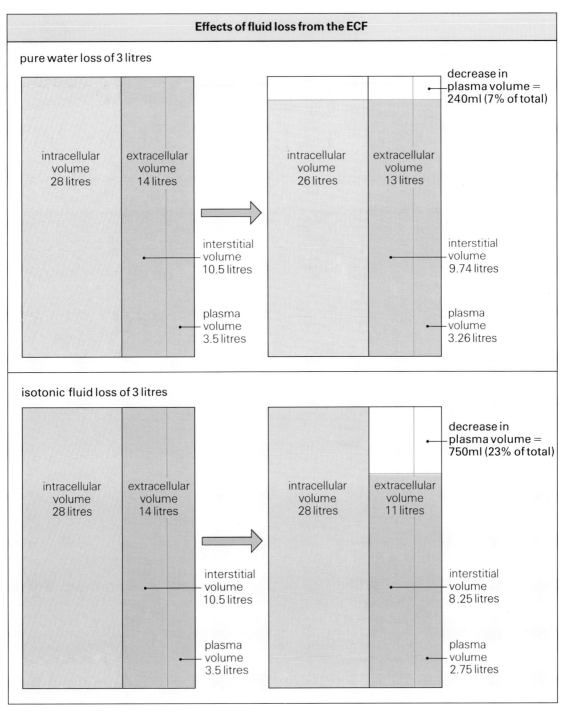

Effects of fluid loss from the ECF

pure water loss of 3 litres

intracellular volume 28 litres

extracellular volume 14 litres

interstitial volume 10.5 litres

plasma volume 3.5 litres

intracellular volume 26 litres

extracellular volume 13 litres

decrease in plasma volume = 240ml (7% of total)

interstitial volume 9.74 litres

plasma volume 3.26 litres

isotonic fluid loss of 3 litres

intracellular volume 28 litres

extracellular volume 14 litres

interstitial volume 10.5 litres

plasma volume 3.5 litres

intracellular volume 28 litres

extracellular volume 11 litres

decrease in plasma volume = 750ml (23% of total)

interstitial volume 8.25 litres

plasma volume 2.75 litres

Fig.2.7 Comparison of the effects of pure water loss and isotonic fluid loss from the extracellular compartment. When 'pure' water is lost from the ECF, the increase in osmolality causes water to move from the ICF to minimize the decrease in plasma volume. When isotonic fluid is lost from the ECF, no osmotic imbalance is produced, there is no movement of water from the ICF and the effect on plasma volume is, therefore, much greater.

15

Sodium depletion	
Causes	**Clinical features**

Excessive loss
 	
from kidneys:
    diuretic phase of 'acute tubular necrosis'
    diuretic therapy
    mineralocorticoid deficiency
    other salt-losing states
from skin:
    massively increased sweating
    cystic fibrosis
    widespread dermatitis
    burns
from gut:
    vomiting, diarrhoea
    fistulae
    ileus
    intestinal obstruction
 	
Inadequate intake
 	
sodium depletion will occur whenever
    intake is inadequate to balance excessive
    losses; inadequate intake alone is rarely
    a cause of depletion
	Symptoms
	weakness
	apathy
	postural dizziness
	syncope
	
	Signs
	weight loss
	related to decreased plasma volume:
	    tachycardia
	    hypotension
	    peripheral circulatory failure
	    oliguria
	related to decreased interstitial fluid:
	    decreased intraocular pressure
	    decreased skin turgor

Fig.2.8 Causes and clinical features of predominant sodium depletion. There is absence of thirst and presence of the early signs of hypovolaemia and oliguria of gradual onset. This is because it is primarily a result of decreased GFR and not of vasopressin secretion.

Furthermore, the increasing colloid osmotic pressure of the serum tends to hold extracellular water in the vascular compartment. Circulatory failure may be a very late feature of water depletion; it is much more likely to occur if sodium depletion is also present.

Severe water depletion induces cerebral dehydration which may cause cerebral haemorrhage through tearing of blood vessels. Such damage can also occur if rehydration is too rapid. If dehydration persists, the brain cells synthesize osmotically active compounds and cerebral oedema may then follow rapid fluid replacement.

The management of water depletion involves treatment of the underlying cause and replacement of the fluid deficit. Water should preferably be given either orally or via a nasogastric tube. If this is not possible, either five percent dextrose or hypotonic saline, particularly if there is sodium loss, should be given intravenously. The aim is to correct approximately two-thirds of the deficit in the first twenty-four hours and the remainder in the next twenty-four hours.

WATER AND SODIUM DEPLETION

Mechanism of sodium depletion	Plasma sodium concentration
sodium and water loss, water loss predominating; e.g. excessive sweating	increased
isotonic sodium and water loss; e.g. burns, haemorrhage	normal
sodium loss with water retention; e.g. treatment of isotonic sodium depletion with low sodium fluids	decreased

Fig.2.9 Status of plasma sodium concentration following various causes of sodium depletion. The plasma sodium concentration alone is not a guide to ECF sodium status.

Sodium depletion

Sodium depletion is seldom due to inadequate oral intake alone, but sometimes inadequate parenteral input is responsible. Most often, however, sodium depletion is a consequence of excessive sodium loss (Fig.2.8). Sodium is lost from the body in either isotonic or hypotonic fluid. In either case, there will be a decrease in ECF volume (see Fig.2.7), but this will be less with hypotonic loss since some of the water loss will then be shared with the ICF. The clinical features of sodium depletion (Fig.2.8) are primarily a result of the decrease in ECF volume.

The normal responses to hypovolaemia are an increase in aldosterone secretion, stimulating renal sodium retention, and a fall in urine volume as a consequence of a decreased GFR. Increased vasopressin secretion, which stimulates the production of a highly concentrated urine, only occurs with severe ECF volume depletion (see Fig.2.4).

As the fluid loss is largely confined to the ECF, peripheral circulatory failure occurs much earlier in sodium depletion than in water depletion. A prerenal uraemia (see *Case History 5.1*) is frequently observed but, in contrast to water depletion, there is no haemoconcentration.

The serum sodium concentration can give an indication of the relative amounts of water and sodium that have been lost; the serum sodium will be normal if fluid is lost isotonically and increased if it is lost hypotonically. With severe sodium depletion, increased vasopressin secretion, secondary to the resulting hypovolaemia, may cause water retention; plasma volume is then maintained at the expense of osmolality and hyponatraemia develops. Thus, the plasma sodium concentration in a sodium depleted patient may be low, normal or high (Fig.2.9).

Although sweat has a low sodium concentration, with massive losses, for example, people working in hot environments and endurance athletes, sodium depletion will become significant and both sodium and water losses will have to be replaced.

Management of sodium depletion involves treatment of the underlying cause but, if severe, it is essential to restore the intravascular volume by giving isotonic fluid ('normal' saline, plasma or blood) by intravenous infusion. This can usually be achieved rapidly, but any associated free water deficit requires more cautious correction.

Excess body water	
Causes	**Clinical features**
Increased intake	behavioural disturbance
compulsive water drinking	confusion
excessive parenteral fluid administration	headache
water absorption during bladder irrigation	convulsions
	coma
Decreased excretion	muscle twitching
renal failure (severe)	extensor plantar responses
cortisol deficiency	
inappropriate or ectopic secretion of vasopressin	
drugs:	
stimulating vasopressin release	
potentiating the action of vasopressin,	
e.g. chlorpropamide	
agonists of vasopressin,	
e.g. oxytocin	
interfering with renal diluting capacity,	
e.g. diuretics	

Fig.2.10   Causes and clinical features of excess body water.

WATER AND SODIUM EXCESS

Excess of water and sodium may result from a failure of normal excretion or from excessive intake. The latter is often iatrogenic. As with the syndromes of depletion, pure water excess and sodium excess with isotonic retention of water can be considered as separate conditions although, in practice, there is often a degree of overlap.

Water excess

This is usually related to an impairment of water excretion (Fig.2.10). However, the limit to the ability of the healthy kidney to excrete water is about 20ml/min and, occasionally, excessive intake is sufficient to cause water intoxication. Hyponatraemia is invariably present. The excessive water load is shared by the ICF and ECF.

The clinical features of water intoxication (Fig.2.10) are related to cerebral over-hydration, the incidence and severity depending upon the extent of the water excess and its time course. A patient with a serum sodium concentration of 120mmol/l, in whom water retention has occurred gradually over several days, may be asymptomatic while another, in whom this is an acute phenomenon, may show signs of severe water intoxication.

Sodium excess

Sodium excess usually occurs secondarily to impaired renal sodium excretion although increased input may also play a part (Fig.2.11). The clinical features are related to an expansion of ECF volume, with an increase in volume of the plasma and of the interstitial fluid, generating oedema (Fig.2.11). The inappropriate use of hypertonic saline may induce a rapid shift of

Sodium excess	
Causes	**Clinical features**
Increased intake	Related to increased interstitial fluid
excessive parenteral administration	oedema
absorption from saline emetics	
	Related to increased plasma volume
Decreased excretion	dyspnoea
glomerular filtration:	pulmonary oedema ⎤
acute or chronic renal failure	venous congestion ⎬ features of congestive cardiac failure
tubular reabsorption:	hypertension ⎥
primary mineralocorticoid excess:	effusions ⎦
Cushing's syndrome	
Conn's syndrome	
secondary mineralocorticoid excess:	
congestive cardiac failure	
nephrotic syndrome	
hepatic cirrhosis with ascites	Related to the overall increase in ECF volume
renal artery stenosis	weight gain

Fig.2.11   Causes and clinical features of predominant sodium excess.

fluid from the intracellular compartment, causing cerebral dehydration.

Sodium overload is most frequently seen in patients who are unable to excrete sodium normally but who do not have an intrinsic renal lesion (Fig.2.11). The precise mechanism of this sodium retention is unclear. Plasma aldosterone levels are increased (secondary hyperaldosteronism, see page 134), indicating that although the ECF volume is overloaded it is actually perceived as being low. It has been suggested that the 'effective' plasma volume is reduced, perhaps due to venous pooling of blood.

Patients with a sodium overload are usually hyponatraemic, implying a defect in free water excretion, possibly because the reduced effective plasma volume also stimulates vasopressin secretion. The decrease in GFR and increase in proximal tubular sodium reabsorption lead to a decrease in the delivery of sodium and chloride to the loop of Henle and distal tubule. This reduces the kidney's diluting ability and therefore compromises free water excretion. Loop diuretics, which decrease chloride absorption in the loop of Henle, and thiazides, which decrease sodium reabsorption in the distal convoluted tubule, also impair urinary dilution. The 'sick cell syndrome' probably contributes to the hyponatraemia, particularly in congestive cardiac failure and liver disease. Although patients with secondary hyperaldosteronism tend to be hyponatraemic, other patients with sodium overload can have either normal or high serum sodium concentrations, depending upon the underlying cause.

Management should be directed towards the cause, where possible. In addition, diuretics may be used to promote sodium excretion and sodium intake must be controlled. Dialysis may be necessary if renal function is poor and is occasionally necessary in acute sodium overload associated with the use of hypertonic fluids.

LABORATORY ASSESSMENT OF WATER AND SODIUM STATUS

The serum sodium concentration is dependent upon the amount of sodium and water in the serum. In isolation, therefore, the serum sodium concentration provides no information about the sodium content of the ECF. It may be either raised, normal or low, in states of sodium excess or depletion, according to the amount of water in the ECF.

The serum sodium concentration is one of the most frequent measurements made in clinical chemistry laboratories but the genuine indications for its measurement are few and results are often misinterpreted. Serum sodium concentration should be measured in the following:

- Patients with dehydration or excessive fluid loss as a guide to appropriate replacement.
- Patients on parenteral fluid replacement who are unable to indicate or respond to thirst (e.g. the comatose, infants and elderly).
- Patients with unexplained confusion, abnormal behaviour or signs of CNS irritability.

In the assessment of a patient's water and sodium status, clinical observations, such as the measurement of central venous pressure, fluid balance charts and body weight, may all provide vital information. An increase in the concentration of serum proteins or of the haematocrit suggests haemoconcentration. Other abnormal results may suggest specific conditions; for example, hyperkalaemia in a hyponatraemic patient with clinical evidence of sodium depletion suggests adrenal failure.

Analysis of urine can provide valuable information but results may be misleading. It should be established whether the urine volume and composition are physiologically appropriate for the patient's water and sodium status. If they are not, the reason should be sought. Thus, a low urinary sodium excretion is an appropriate response in a patient with hyponatraemia who is sodium depleted. However, a natriuresis in such a patient would imply either a failure of aldosterone secretion or a failure of the kidney to respond to the hormone (see Case History 2.4).

Sodium measurement

Sodium concentration has traditionally been measured by flame photometry, which determines the number of sodium atoms in a defined volume of solution. Sodium is now increasingly being measured by ion-selective electrodes, which determine the activity of sodium; that is, the number of atoms which act as true ions in a defined volume of water.

Under most circumstances the two techniques give results that are, for practical and clinical purposes, the same. However, as activity is a measure of sodium in the water fraction of plasma (normally ninety-three percent by volume), significant discrepancies between activity and concentration may arise if the fractional plasma water content is decreased, such as in severe hyperlipidaemia and hyperproteinaemia. The sodium concentration, determined by flame photometry in millimoles per litre of plasma, will be less than the concentration inferred from the activity. This is because although the concentration of sodium in plasma water is unchanged, there is less water and thus less sodium in a given volume of plasma.

Analyzers employing electrodes for which the plasma is diluted before measurement also give a spuriously low result. This spurious result, or pseudohyponatraemia, is only seen with massive hyperlipidaemia when the plasma is turbid to the naked eye (see *Case History 16.2*) and with large increases in total protein due to paraproteinaemia. If it is suspected, the plasma osmolality should be measured; it is osmolality that is regulated by the hypothalamus through the release of vasopressin. Plasma osmolality should be normal in a patient with pseudohyponatraemia.

Osmolality and osmolarity measurement

Given that it is osmolality, rather than sodium concentration, that is controlled by the hypothalamus, it might appear logical to measure plasma osmolality rather than sodium concentration. The measurement of osmolality is, however, less precise than that of sodium and is not easily automated. It is nevertheless useful under certain circumstances.

Measurement of osmolality may help in the interpretation of a low serum sodium concentration and is necessary in water deprivation tests. If it is suspected that an abnormal osmotically active moiety is present in the plasma, as in a case of poisoning, for instance with ethylene glycol, it may be revealed by a discrepancy between measured osmolality and calculated osmolarity. Osmolality is a measure of concentration per kilogram of water and osmolarity a measure of concentration per litre of solution. Plasma osmolarity can be calculated from the formula:

$$2 \times [Na^+] + [urea] + [glucose]$$

where concentrations are measured in millimoles per litre (mmo/l). It is normally numerically very similar to measured osmolality. Discrepancies occur with hyperlipidaemia and hyperproteinaemia, however, and if osmotically active species, such as ethanol or mannitol, are present in the plasma.

Anion measurement

A change in plasma sodium concentration must be matched by a change in anion concentration. The major anions of the ECF are chloride and bicarbonate. Bicarbonate (strictly total carbon dioxide) is frequently measured since it reflects the extracellular buffering capacity, but the measurement of serum chloride rarely adds to the information that can be derived from knowledge of the sodium concentration alone and few laboratories in the United Kingdom now measure plasma chloride concentration for this reason. However, it may occasionally be helpful in the diagnosis of patients with non-respiratory acidosis and those with rare chloride-losing states.

HYPONATRAEMIA

A slightly low serum sodium concentration is a frequent finding. The mean serum sodium concentration of hospital inpatients is 5mmol/l lower than in healthy controls. Mild hyponatraemia is seen with a wide variety of illnesses and is most probably a result of the sick cell syndrome. It is essentially a secondary phenomenon which merely reflects the presence of disease; treatment should be directed at the underlying cause and not at the hyponatraemia. Severe hyponatraemia does sometimes warrant primary treatment, but usually only when it is associated with clinical features of water intoxication (see Fig.2.10).

Causes

Hyponatraemia frequently has more than one cause, but is more often related to an excess of water than to sodium loss; indeed, it may occur in states where total body sodium is increased. Several pathological mechanisms may lead to hyponatraemia:

1. A decreased fractional water content of plasma due to hyperproteinaemia and hyperlipidaemia (pseudohyponatraemia). This has been discussed previously in this chapter.

2. A decrease in the total negative charge on plasma proteins, which contributes to the anion gap, can displace sodium from the plasma. This is unusual, but it may contribute to hyponatraemia in severe hypoalbuminaemia and in paraproteinaemias if the paraprotein is positively charged.

3. Addition of a solute to the plasma which is confined to the ECF will tend to increase ECF osmolality. Acutely, this will cause a shift of water from the ICF to the ECF, lowering the ECF sodium concentration. The resulting increase in ECF volume inhibits aldosterone secretion, leading to natriuresis.

CASE HISTORY 2.1

An insulin-dependent diabetic patient woke feeling hypoglycaemic and drank two glasses of a sugar-rich drink which abolished the symptoms. She had a hospital appointment that morning and, worried that she might become hypoglycaemic while driving, decided to omit her usual injection of insulin. She felt quite well on arrival at the hospital. Blood was taken for biochemical tests.

Investigations

blood glucose	28mmol/l
serum: sodium	126mmol/l
osmolality	290mmol/kg

serum urea, potassium and bicarbonate were normal

Comment

The hyponatraemia is dilutional. It is the result of a movement of water from the ICF to the ECF to maintain isotonicity as the serum glucose concentration rose. In that short time, there was no significant osmotic diuresis and thus no dehydration.

Hyponatraemia may occur for the same reason when glucose is administered intravenously at a rate greater than it can be metabolized, for example, during parenteral nutrition. It can also occur following mannitol infusion. Mannitol may be given in cerebral oedema, to reduce intracellular water content, and is also used as an osmotic diuretic.

Movement of water from the ICF to the ECF does not occur in uraemia. In renal failure, the rate of increase in serum urea concentration is slow, allowing time for urea to equilibrate between the ECF and the ICF, and thus preventing an osmotic imbalance.

4. An iso-osmotic shift of water between the ICF and the ECF is probably one of the mechanisms of hyponatraemia in the sick cell syndrome. This is a term used to describe the defect in cell membrane permeability, possibly due to a decreased energy supply, that occurs in many patients with generalized illness. If, as a result of increased cell membrane permeability, normally non-diffusible solutes leak into the extracellular compartment, a concomitant shift of water from the ICF will keep the ECF osmolality normal.

It is often said that the hyponatraemia of the sick cell syndrome is due to intracellular movement of sodium. While this certainly occurs, it would not on its own cause hyponatraemia since any shift of sodium must be accompanied by a shift of water.

5. Loss of intracellular solute from the body (effectively potassium depletion) may also contribute to the hyponatraemia of the sick cell syndrome. If intracellular solute content is chronically decreased and this solute is not retained in the body, ICF osmolality will fall and that of the ECF must fall also. Under such circumstances, the control mechanism for osmolality, the 'osmostat', may become reset so that osmolality is still controlled but at a lower value than normal.

In practice, however, the mechanism for the hyponatraemia of the sick cell syndrome is unimportant. The hyponatraemia reflects the presence of underlying disease and it is this that must be treated, not the hyponatraemia.

6. Water excess gives rise to a dilutional hyponatraemia with reduced plasma osmolality. Since osmolality is normally precisely controlled, the persistence of a dilutional hyponatraemia implies either continued production of vasopressin or impairment of the ability of the kidneys to dilute the urine.

Normal kidneys are capable of excreting one litre of water per hour, and therefore, water intoxication and hyponatraemia due to excessive intake alone are only seen when very large quantities of fluid are ingested rapidly, as in some psychotics and heavy beer drinkers. Reduction of water intake in these circumstances allows the excess to be excreted and the plasma sodium concentration to return to normal.

CASE HISTORY 2.2

An elderly man was admitted to hospital in an acute confusional state. No history was available but the nicotine stains on his fingers indicated that he was a heavy smoker. Physical examinations revealed he had digital clubbing and signs of a right-sided pleural effusion but no other obvious abnormality was detected. He was neither dehydrated nor oedematous.

Investigations

serum:	sodium	114 mmol/l
	potassium	3.6 mmol/l
	bicarbonate	22 mmol/l
	urea	2.5 mmol/l
	glucose	4.0 mmol/l
	total protein	48 g/l
	osmolality	236 mmol/kg
urine:	osmolality	350 mmol/kg
	sodium	50 mmol/l

A chest radiograph confirmed the presence of the effusion and showed a mass in the right lower zone with an appearance typical of a carcinoma.

Comment

There is severe hyponatraemia. The patient is not clinically dehydrated and the biochemical results with low serum protein and urea suggest that the hyponatraemia is dilutional. The serum osmolality is equal to the calculated osmolarity, militating against the presence of additional solute in the serum. The normal response to a low serum osmolality is for vasopressin secretion to be inhibited, resulting in the production of a dilute urine. However, in this case, the urine is inappropriately concentrated in relation to the serum, indicating continuing vasopressin secretion. The chest radiograph indicates the likely source; ectopic secretion of vasopressin by a bronchial carcinoma, an example of the syndrome of inappropriate antidiuretic hormone secretion (SIADH).

The diagnostic features of SIADH are:
- Hyponatraemia
- Decreased plasma osmolality
- Inappropriately concentrated urine
- Continued natriuresis [>20mmol/l]
- No oedema
- Normal renal function
- Normal adrenal function
- Clinical and biochemical response to fluid restriction

In SIADH, there is a continued natriuresis despite the low serum sodium concentration because serum volume is maintained by water retention and there is therefore no hypovolaemic stimulus to aldosterone secretion. Hyponatraemia with natriuresis can also occur in adrenal failure and in renal disorders and these must be excluded before a diagnosis of SIADH can be made.

Water intoxication should always be considered as a possible cause of a confusional state, especially in the elderly, and this is one of the few situations in which emergency measurements of serum sodium concentration is genuinely indicated.

The diagnosis of SIADH is frequently made on insufficient evidence without regard to other possible causes of hyponatraemia. It is difficult to measure vasopressin in the serum and the diagnosis must usually be made on clinical and other laboratory data. It is essential to measure urine and serum osmolalities; the urine may not be more concentrated than the serum but must be less than maximally dilute (osmolality > 50mmol/l). Oedema is not a feature of SIADH; the excess of water is shared by the ICF and the ECF and the effect on ECF volume is insufficient to cause oedema.

There is undoubtedly more than one type of SIADH. Tumours may produce the hormone (ectopic production) but patients with many other conditions (Fig.2.12) also fulfill the criteria for SIADH. In some of these there may be an inappropriate stimulus to vasopressin release, such as stimulation of volume receptors during artificial ventilation, and in others the 'osmostat' appears to be reset so that osmolality is still controlled but at a low level. Certain drugs stimulate vasopressin release (see Fig.2.5) or have a vasopressin-like action on the kidney.

Conditions associated with SIADH
Ectopic secretion
bronchial carcinomas
other tumours, e.g. thymus and prostate
Inappropriate secretion
pulmonary diseases: pneumonia tuberculosis positive pressure mechanical ventilation
cerebral diseases: head injury encephalitis tumours aneurysms
miscellaneous: pain, e.g. postoperative intermittent acute porphyria Guillain–Barré syndrome hypothyroidism drugs, e.g. narcotics, chlorpropamide, carbamazepine and vinca alkaloids

Fig.2.12 Conditions associated with the syndrome of inappropriate antidiuretic hormone secretion (SIADH).

The logical treatment of this condition is to reduce the patient's water intake to less than that required to maintain normal water balance, to 400ml/day, for example. This form of treatment is unpleasant and is impractical in chronic cases. An alternative is to administer the drug, demeclocycline, which antagonizes the action of vasopressin on the renal collecting ducts. Occasionally, in severe water intoxication, it may be necessary to infuse hypertonic saline with a diuretic. This is potentially dangerous, since the ECF can become overloaded, and must be done cautiously with careful monitoring.

CASE HISTORY 2.3

Blood was taken for biochemical tests from a man who had undergone major abdominal surgery thirty-six hours before.

Investigations

serum: sodium   127mmol/l
       urea   4.0mmol/l

Serum potassium and bicarbonate were normal. The patient was alert and appeared neither under- nor overhydrated.

Comment

Hyponatraemia is a very common finding in postoperative patients on intravenous drips. It is usually a reflection of excessive administration of hypotonic fluids (five percent dextrose or 'dextrose- saline') at a time when the ability of the body to excrete water is depressed as part of the normal metabolic response to trauma. It may also be due, in part, to the sick cell syndrome. If, as is usually the case, there are no clinical features of water intoxication, the only necessary action is an adjustment in the fluid input. This patient had been given a total of 3.5 litres of dextrose-saline since his operation and a check on the fluid balance chart showed that he had a positive balance of two litres.

7. Sodium depletion is a relatively unusual cause of

hyponatraemia. Isotonic sodium loss should not alter plasma osmolality. Hyponatraemia will only be seen either when the sodium loss causes such severe hyponatraemia that this, appropriately, stimulates vasopressin secretion or when sodium-containing fluids, such as sweat and diarrhoea, have been lost from the body and replaced by fluid containing insufficient sodium. A case of adrenal failure with hyponatraemia as a result of sodium depletion is presented in *Case history 9.1*.

The management of sodium depletion is discussed on page 17. The hyponatraemia will be corrected as the ECF volume is restored to normal. The hypovolaemic stimulus to vasopressin release will be cut off and sensitivity to inhibition by the low plasma osmolality will be restored.

CASE HISTORY 2.4

A fifty-year-old woman with a long history of rheumatoid disease complained of fainting episodes following an attack of gastroenteritis and, on examination, was found to have postural hypotension.

Investigations

serum: sodium   118mmol/l
      potassium   3.9mmol/l
      urea   9.1mmol/l

short Synacthen test: normal cortisol response to ACTH

plasma aldosterone (recumbent)   720pmol/l
24h urine sodium excretion   118mmol/l

Comment

Postural hypotension may be due to hypovolaemia, autonomic neuropathy or hypotensive drugs. This patient was not taking such medication and there was no other evidence of neuropathy. The hyponatraemia with a slightly raised urea is consistent with sodium depletion producing hypovolaemia. The Synacthen test is normal, excluding adrenal failure, and indeed the aldosterone is appropriately raised. The patient's urinary sodium excretion is excessive; although the input was not assessed, normal kidneys should retain sodium in a sodium depleted patient with hypovolaemia.

It was concluded that the patient had a renal salt-losing state such that the kidneys could not respond to the normal physiological stimuli to retain sodium. She only became symptomatic when diarrhoea and vomiting caused further fluid loss. This was later confirmed by sodium balance studies and the patient was found to have renal papillary necrosis, an occasional complication of the use of certain analgesic drugs, which principally affects renal tubular function.

Investigation of hyponatraemia

It should be apparent from the previous section that, in many instances, the cause of hyponatraemia can often be recognized clinically and that additional investigation adds nothing to the management of the patient. Even in apparently obscure cases, careful clinical evaluation and the study of fluid balance charts, if properly kept, will often indicate the underlying mechanism or mechanisms, and thus point the way to a diagnosis. The investigations which may help in elucidating the cause of hyponatraemia are shown in Fig.2.13. It must be emphasized that an appreciation of the underlying principals is vital for the correct interpretation of their results.

Management of hyponatraemia

Hyponatraemia is essentially a sign of a disturbance or a disorder involving water or sodium or both. Measures to treat the causative condition may have to be supplemented by direct measures to correct the imbalance of sodium and water. It is essential to understand the pathogenesis of the hyponatraemia if the measures are to be appropriate. Hyponatraemia itself usually only requires treatment if features of water intoxification are present.

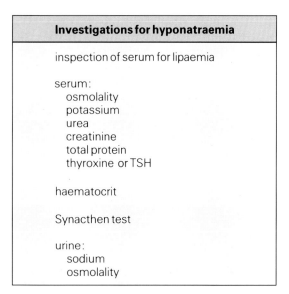

Investigations for hyponatraemia

inspection of serum for lipaemia

serum:
    osmolality
    potassium
    urea
    creatinine
    total protein
    thyroxine or TSH

haematocrit

Synacthen test

urine:
    sodium
    osmolality

Fig.2.13 Laboratory investigations of value in the investigation of hyponatraemia.

HYPERNATRAEMIA

Hypernatraemia is much less common than hyponatraemia, but is much more frequently of clinical significance. The causes include pure water depletion combined sodium and water depletion, with water loss predominating, or sodium excess; of these, excess body sodium is the least common.

CASE HISTORY 2.5

Following surgery for major abdominal injuries sustained in a knife-fight, a young man was fed parenterally and artificially ventilated. On the fifth day after his operation, serum biochemical results, which had been normal the day before, were as follows:

Investigations

serum:	sodium	150mmol/l
	potassium	4.2mmol/l
	urea	10.2mmol/l
	glucose	25mmol/l

During the previous twenty-four hours he had become pyrexial and positive blood cultures were subsequently obtained. His fluid intake had been 3000ml, urine output had been steady at 90–100ml/h and 300ml of fluid had been aspirated via a nasogastric tube. The sodium intake had been 70mmol.

Comment

Sodium input is not excessive; water depletion is the more likely cause of the hypernatraemia. His measured net fluid intake is only 400ml. This is insufficient to balance insensible losses, which will have been increased by the pyrexia and possibly by ventilation. The urine output has not decreased and there has therefore also been excessive renal water loss. This is due to an osmotic diuresis as a result of glycosuria and a high urea output.

Glucose intolerance may be a problem in patients receiving parenteral nutrition and can be exacerbated by sepsis, which causes insulin resistance. Parenteral administration of excessive nitrogen will result in increased formation of urea which will also contribute to an osmotic diuresis; this patient was receiving amino acids equivalent to over 100g of protein per day, more than his probable requirements. Inadequate humidification of inspired air may also be a causative factor in water depletion in such circumstances.

In most cases of hypernatraemia, the cause is obvious from the history and clinical observations. Diabetes insipidus is an important cause and the investigation of patients suspected of having this condition is considered in *Chapter 8*.

Regardless of its cause, hypernatraemia should be treated by administration of hypotonic fluids such as water (orally) or five percent dextrose (parenterally). In patients with sodium overload, measures to remove excess sodium may have to be considered. As already emphasized, it is important not to correct hypernatraemia due to water depletion too rapidly.

SUMMARY

Sodium and water homoeostasis are closely linked. Sodium is the principal extracellular cation and the amount of sodium in the body is the major determinant of ECF volume. Sodium is transported actively in the body but water moves passively in response to changes in the solute concentration of the body's fluid compartments. Sodium excretion is controlled by aldosterone. This hormone is released in response to a decrease in ECF volume and causes renal sodium retention. Water excretion is controlled by vasopressin (antidiuretic hormone). This is secreted in response to an increase in ECF osmolality and a decrease in ECF volume and promotes water retention.

Primary disturbances of either water or sodium homoeostasis produce characteristic clinical and biochemical features. Such disturbances are less common than those in which both water and sodium are affected. The features may then be less clear cut and will depend on which abnormality predominates.

Changes in plasma sodium concentration require careful interpretation. In general, they are more frequently due to changes in body water content than sodium content. Thus, hyponatraemia is more often related to an excess of water than a deficit of sodium. In patients who are asymptomatic, hyponatraemia may not require specific treatment; it is often either a non-specific consequence of underlying disease or a result of normal physiological adaptation. However, the cause of the imbalance may require treatment. Hypernatraemia is less common than hyponatraemia and is usually related to predominant loss of water. This should be replaced cautiously and appropriate measures taken to treat the underlying cause.

FURTHER READING

Beck LH (1981) Body fluid and electrolyte disorders. *The Medical Clinics of North America*, **65**, 247–451.

Gill GV & Flear CTG (1985) Hyponatraemia. *Recent Advances in Clinical Biochemistry*, **3**, 149–159.

Morgan DB (ed.) (1984) Electrolyte disorders. *Clinics in Endocrinology and Metabolism*, **13**, 231–434.

3. Potassium

INTRODUCTION

Potassium is the predominant intracellular cation. Ninety percent of the total body potassium is free and therefore exchangeable, whilst the remainder is bound in red blood cells, bone and brain tissue. However, only approximately two percent (50–60mmol) of the total is located in the extracellular compartment (Fig.3.1) where it is readily accessible for measurement. As a consequence, the concentration of potassium in the plasma is not an accurate index of total body potassium status. The potassium concentration of serum is 0.2–0.3mmol/l higher than that of plasma, due to the release of potassium from platelets during clot formation, but this difference is not of practical significance.

HOMOEOSTASIS

Extracellular potassium balance is controlled primarily by the kidney and, to a lesser extent, by the gastrointestinal tract. In the kidney, filtered potassium is almost completely reabsorbed in the proximal tubule.

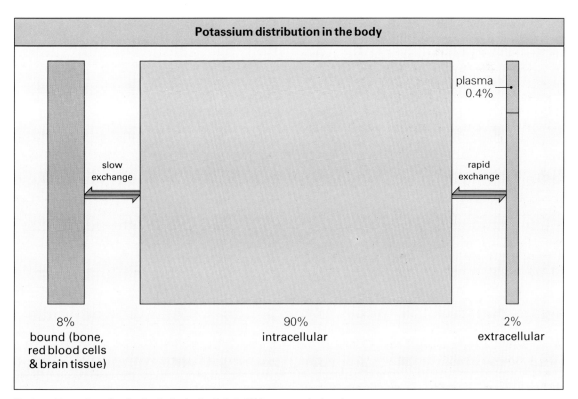

Potassium distribution in the body

plasma
0.4%

slow
exchange

rapid
exchange

8%
bound (bone,
red blood cells
& brain tissue)

90%
intracellular

2%
extracellular

Fig.3.1   Potassium distribution in the body. Only 0.4% is present in the plasma.

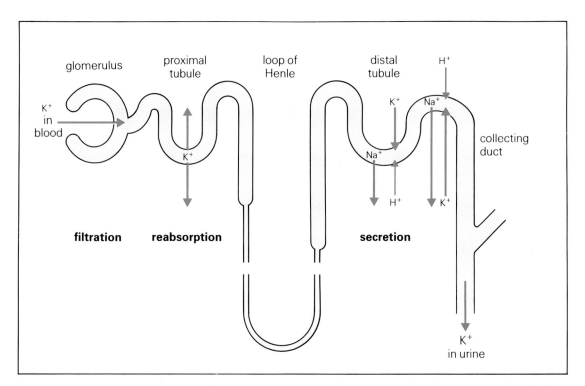

Fig.3.2 Potassium handling in the kidney. The bulk of filtered potassium is reabsorbed in the proximal convoluted tubule. In the distal tubule and collecting duct, potassium and hydrogen ions are exchanged for sodium ions. Sodium reabsorption is stimulated by aldosterone.

Some active potassium secretion takes place in the most distal part of the distal convoluted tubule but potassium excretion is primarily a passive process. The active reabsorption of sodium generates a membrane potential which is neutralized by the movement of potassium and hydrogen ions from tubular cells into the lumen of the tubule (Fig.3.2). Thus, urinary potassium excretion depends upon several factors:

- The amount of sodium available for reabsorption in the distal convoluted tubule and the collecting duct.
- The relative availability of hydrogen and potassium ions in the cells of the distal convoluted tubule and collecting duct.
- The ability of these cells to secrete hydrogen ions.
- The circulating concentration of aldosterone.

Aldosterone stimulates potassium excretion both indirectly, by increasing the active reabsorption of sodium in the distal convoluted tubule and the collecting duct, and directly, by increasing active potassium secretion in the distal part of the distal convoluted tubule. Aldosterone secretion from the adrenal cortex is stimulated indirectly, by activation of the renin-angiotensin system in response to hypovolaemia (see pages 123 and 124), and directly, by hyperkalaemia.

Since both hydrogen and potassium ions may neutralize the membrane potential generated by active sodium reabsorption, there is a close relationship between potassium and hydrogen ion homoeostasis. In a state of acidosis, hydrogen ions will tend to be secreted in preference to potassium; in alkalosis, fewer hydrogen ions will be available for excretion and there will be an increase in potassium excretion. Thus, there is a tendency to hyperkalaemia in acidosis and to hypokalaemia in alkalosis. An exception to this tendency is renal tubular acidosis caused by defective renal hydrogen ion excretion (see page 71). In this condition, because of the decrease in hydrogen ion excretion, potassium secretion must increase to balance sodium reabsorption. The result is the unusual combination of hypokalaemia with acidosis.

The relationship between the excretion of hydrogen

Causes of hypokalaemia	
Decreased K$^+$ intake	**Increased K$^+$ loss**
oral (rare)	renal:
parenteral	diuretics
	diuretic phase of acute renal failure
	mineralocorticoid excess:
	primary hyperaldosteronism
	secondary hyperaldosteronism
Transcellular K$^+$	Cushing's syndrome
movement	carbenoxolone and liquorice
alkalosis	extrarenal:
insulin administration	diarrhoea
rapid cellular proliferation	purgative abuse
	villous adenoma of the rectum
	vomiting
	enterocutaneous fistulae
	excessive sweating

Fig.3.3 Causes of hypokalaemia. Carbenoxolone and liquorice are aldosterone agonists, carbenoxolone has been used in the management of peptic ulceration.

and potassium ions also explains why potassium depletion tends to produce alkalosis. If there is insufficient potassium available for excretion as sodium is reabsorbed, then the excretion of hydrogen ions will be increased.

The healthy kidney is less efficient at conserving potassium than sodium; even on a potassium-free intake, urinary excretion remains at 10–20mmol/24h. Since there is also an obligatory loss from the skin and gut of approximately 15–20mmol/24h the kidneys cannot compensate if intake falls much below 40mmol/24h. The average diet contains more potassium than this. However, potassium depletion can occur, even on a normal diet, if there are increased losses from the body.

Potassium is secreted in gastric juice and much of this, along with dietary potassium, is reabsorbed in the small intestine. In the colon and rectum, potassium is secreted in exchange for sodium, partly under the control of aldosterone. Stools normally contain some potassium, but considerable amounts may be lost in patients with fistulae or chronic diarrhoea, or in patients who are losing gastric secretions through persistent vomiting or nasogastric aspiration.

The movement of potassium between the intracellular and extracellular compartments can have a profound effect on the plasma potassium concentration. The cellular uptake of potassium is stimulated by insulin. Potassium ions move passively into cells from the extracellular fluid (ECF) in exchange for sodium which is actively excluded by a membrane-bound, energy-dependent sodium pump. Hyperkalaemia may result either if the activity of this sodium pump is impaired or if there is damage to cell membranes.

In renal tubular cells (and in all other cells in the body) there is, in general, a reciprocal relationship between hydrogen and potassium ions. In a systemic acidosis, intracellular buffering of hydrogen ions results in the displacement of potassium into the ECF. In alkalosis, there is a shift of hydrogen ions from the intracellular fluid (ICF) to the ECF and a net movement of potassium ions in the opposite direction which tends to produce hypokalaemia.

HYPOKALAEMIA

Hypokalaemia may develop due to one or more of three basic causes: a decrease in potassium intake, increased loss from the kidneys or gut and redistribution (Fig.3.3). Decreased potassium intake is rarely the sole cause of hypokalaemia, except in patients who are fasting, since potassium is available in many foods and the normal dietary intake is 60–200mmol/24h. Hypokalaemia is often related to drug therapy.

CASE HISTORY 3.1

A sixty-seven-year-old woman presented with severe muscle weakness. She had been in the habit of taking large amounts of purgatives and recently had been prescribed a thiazide diuretic for mild heart failure.

Investigations

serum: potassium	2.4mmol/l
bicarbonate	36mmol/l

Comment

The patient is severely hypokalaemic and the high total bicarbonate concentration reflects the associated extracellular alkalosis.

Purgative abuse can cause considerable potassium loss from the gut. Thiazides act by decreasing chloride reabsorption, and thus sodium reabsorption, in the distal part of the ascending limb of the loop of Henle and in the first part of the distal convoluted tubule. As a result, there is an increase in the amount of sodium delivered to and available for reabsorption from the distal tubule; this will tend to increase potassium excretion from the kidneys. Loop diuretics similarly increase renal potassium excretion, though to a lesser extent. With either type of diuretic, however, serum potassium levels tend to stabilize unless, as in this case, other causes of hypokalaemia are present.

CASE HISTORY 3.2

A sixty-year-old man underwent total gastrectomy for a carcinoma. He was malnourished prior to surgery and it was decided to provide parenteral nutrition postoperatively. On the fifth day, his serum potassium concentration was 3.0mmol/l despite the provision of 60mmol/24h in the intravenous feed.

Comment

The patient is hypokalaemic in spite of the provision of sufficient potassium to cover normal obligatory losses.

Potassium excretion increases during the metabolic response to trauma and yet, once a patient becomes anabolic, the body's requirements increase as potassium is taken up into cells. Furthermore, during total parenteral nutrition, glucose is often the predominant energy source and thus provides a considerable stimulus to insulin release. Potassium requirements may, therefore, be much greater than normal because insulin stimulates its uptake into cells.

This patient had recently undergone abdominal surgery and an ileus is usual in these circumstances. This will result in decreased reabsorption of any potassium secreted into the gut and may also contribute to the loss of potassium from the ECF.

Potassium supplements are often prescribed at the same time as diuretics; combined preparations are widely used but they are generally expensive and typically provide less than 10mmol of potassium per tablet. Hypokalaemia potentiates digoxin toxicity and this is an important practical consideration since diuretics and digoxin are very often prescribed together. However, in general, the routine use of potassium supplements is to be deprecated. They are probably unnecessary unless the serum level is below 3.0mmol/l and they are potentially dangerous in patients with renal impairment since hyperkalaemia may result.

Clinical features

Severe hypokalaemia may be asymptomatic. When symptoms are present, they are related primarily to disturbances of neuromuscular function (Fig.3.4); muscular weakness, constipation and paralytic ileus are common problems.

Management

Although the serum potassium concentration is a poor guide to total body potassium, a serum concentration of 3.0mmol/l generally implies a deficit of the order of

300mmol. However, since this deficit is almost entirely from the ICF and since administered potassium first enters the ECF, replacement must be undertaken with care, particularly when the intravenous route is used.

As a guide, the following potassium dosages should not be exceeded without good reason: a rate of 20mmol/h, a concentration of 40mmol/l in the intravenous fluid or a total of 140mmol/24h. Thorough mixing with the bulk of the fluid to be infused is vital. Serum concentrations should be monitored during treatment. If unusually large amounts of potassium are necessary and particularly if there is impaired renal function, electrocardiographic (ECG) monitoring is useful since characteristic changes in the waveform occur with changing serum potassium concentrations (Fig.3.5).

Clinical features of hypokalaemia	
Disorder	**Feature**
neuromuscular	weakness ileus hypotonia depression confusion

cardiac	arrhythmias
	potentiation of digoxin toxicity
	ECG changes (ST depression,
	  T depression/inversion and
	  prominent U wave)

renal	impaired concentrating ability
	  leading to polyuria and
	  polydipsia

metabolic	alkalosis

Fig.3.4 Clinical features of hypokalaemia. The resting potential of excitable membranes is reduced in hypokalaemia, thereby decreasing excitability.

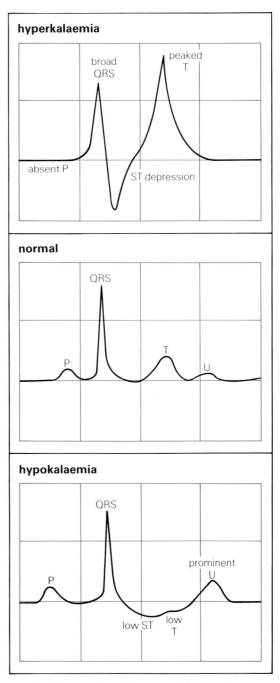

Fig.3.5 Characteristic ECG changes in hyper- and hypokalaemia. Each sinus discharge produces atrial depolarization (P wave) followed by ventricular depolarization (QRS complex) and ventricular repolarization (T wave). The U wave, of unknown cause, is present in most normal ECG's.

HYPERKALAEMIA

The mechanisms which can be responsible for hyperkalaemia are the opposite of those which can give rise to hypokalaemia and include excessive input of potassium (virtually always iatrogenic and parenteral), movement from the ICF to the ECF and decreased output (Fig.3.6). As in hypokalaemia, more than one cause is often present. Spurious hyperkalaemia, due to the leakage of potassium from blood cells, often occurs. If hyperkalaemia is found unexpectedly, the possibility that it is spurious should be eliminated by repeating the measurements on a fresh sample. Spurious hyperkalaemia may be present in the absence of frank haemolysis.

CASE HISTORY 3.3

A young man was admitted to hospital after sustaining a fractured femur and ruptured spleen in a motorcycle accident. He underwent splenectomy and was put in traction. Twenty-four hours after admission, he had passed only 300ml of urine.

Investigations

serum: urea   21.5mmol/l
       potassium   6.5mmol/l

Comment

Since the patient is oliguric with a high serum urea he is by definition in renal failure; this might be reversible, that is prerenal (see page 59 and 60). The hyperkalaemia is due to a combination of decreased renal excretion, as a result of decreased renal perfusion (hypovolaemic shock), and the release of potassium either from cells damaged directly by trauma or from cells whose membrane integrity was impaired by hypoxaemia.

Similar results may be seen in patients who have sustained a gastrointestinal haemorrhage. This may itself cause hypovolaemic shock, affecting renal function. In addition, there will be absorption of potassium from red blood cells undergoing lysis in the gut and increased synthesis of urea from the amino acids released.

Causes of hyperkalaemia

Excessive K^+ intake

oral (rare except with K^+-sparing diuretics taken simultaneously)

parenteral

stored blood

Transcellular K^+ movement

tissue damage

catabolic states

systemic acidosis

Decreased K^+ loss

acute renal failure

chronic renal failure (late)

K^+-sparing diuretics

mineralocorticoid deficiency:
    Addison's disease
    adrenalectomy

Spurious

haemolysis

delayed separation of serum

contamination

Fig.3.6   Causes of hyperkalaemia.

CASE HISTORY 3.4

Blood from an outpatient on diuretics was received in the laboratory for urea and electrolyte determination. The serum potassium concentration was 6.7mmol/l. There was no visible haemolysis and the blood was freshly drawn.

Comments

The patient was recalled and asked to bring all her tablets with her. It transpired that she had initially been prescribed a loop diuretic and potassium supplements for congestive cardiac failure. However, at an outpatient attendance she had been prescribed spironolactone, a potassium-sparing diuretic which is an antagonist of aldosterone, instead of the potassium supplements. She had misunderstood the instructions given to her and continued to take both the supplements and the diuretic.

She surrendered the potassium supplements and her plasma potassium concentration was normal when checked one week later.

Clinical features

Cardiac arrest with ventricular fibrillation may be the first sign of hyperkalaemia. It is therefore necessary to be alert for this disorder in appropriate circumstances, for instance, in acute renal failure, to ensure that effective early management is instituted. Characteristic ECG changes precede the onset of ventricular fibrillation (see Fig.3.5). Peaking of T waves occurs first, followed by loss of P waves and, finally, the development of abnormal QRS complexes.

Management

Intravenous calcium gluconate (10ml of a ten percent solution) affords some degree of immediate protection to the myocardium by antagonizing the effect of hyperkalaemia on myocardial excitability. Intra-venous glucose and insulin, for example, 500ml of twenty percent dextrose with twenty units of soluble insulin given over thirty minutes, promote intracellular potassium uptake. In the acidotic patient, hyperkalaemia can be controlled temporarily by bicarbonate infusion.

In acute renal failure and in other circumstances where the hyperkalaemia is uncontrollable, dialysis will be required. In chronic renal failure, the restriction of potassium intake and the administration of oral ion-exchange resins are often successful in preventing dangerous hyperkalaemia until such time as dialysis becomes necessary for other reasons.

ECG monitoring can be valuable in patients with hyperkalaemia. Changes in the plasma potassium concentration are reflected by changes in the ECG waveform more rapidly than could be determined by biochemical measurement.

SUMMARY

Potassium is the major intracellular cation, with less than two percent of the total body potassium in the extracellular fluid. The plasma potassium concentration is therefore a poor guide to overall potassium status. The plasma potassium concentration is controlled mainly by the kidney. Its excretion by the kidney is regulated by aldosterone, but also depends upon the glomerular filtration rate, extracellular hydrogen ion concentration and sodium and water excretion.

Changes in plasma potassium concentration can be due to changes in both potassium intake and output, and to redistribution between the intracellular and extracellular fluids. Hypokalaemia is most frequently due to excessive gastrointestinal or renal loss and may be exacerbated by a poor intake. It results in skeletal muscle weakness and impairment of myocardial contractility and renal concentrating ability. It also potentiates digoxin toxicity.

Hyperkalaemia is most frequently due to decreased renal excretion or to loss of potassium from cells; excessive intake should be avoidable, since it is usually iatrogenic. Spurious hyperkalaemia, due to the release of potassium from cells *in vitro*, is common. The danger of hyperkalaemia *in vivo* is that it can cause cardiac arrest; this may occur in the absence of any warning clinical signs.

FURTHER READING

Allison SP (1984) Potassium. *British Journal of Hospital Medicine*, **32**, 19–22.

Beck LH (1981) Body fluid and electrolyte disorders. *The Medical Clinics of North America*, **65**, 247–451.

Morgan DB (ed.) (1984) Electrolyte disorders. *Clinics in Endocrinology and Metabolism*, **13**, 231–434.

4. Hydrogen Ion Homoeostasis and Blood Gases

INTRODUCTION

The normal processes of metabolism result in the net formation of 40–80mmol of hydrogen ion per day, principally from the oxidation of sulphur-containing amino acids, and approximately 15,000mmol of carbon dioxide per day, generated from the oxidation of carbon in energy-yielding reactions. Although carbon dioxide is not itself an acid, in the presence of water it undergoes hydration to form a weak acid, carbonic acid (Equation 4.1).

$$(4.1) \quad CO_2 + H_2O \rightleftharpoons H_2CO_3$$

Carbon dioxide is removed from the body in expired air. Hydrogen ions can only be excreted in the urine.

The homoeostatic mechanisms for hydrogen ion and carbon dioxide are very efficient and, in health, the rates of excretion and production are equal. Temporary imbalance is absorbed by buffering and, as a result, the hydrogen ion concentration of the body is maintained within narrow limits; for example, 36–43 nmol/l (pH 7.35–7.46) in extracellular fluid (ECF). The intracellular hydrogen ion concentration is slightly higher but is also rigorously controlled. In disease, an imbalance between the rates of production and excretion may occur, the hydrogen ion concentration then becomes abnormal and a state of acidosis or alkalosis results.

Buffering of hydrogen ions

As hydrogen ions are generated they are buffered, thus limiting the rise in hydrogen ion concentration which will otherwise occur. A buffer system consists of a weak acid, that is one which is incompletely dissociated, and its conjugate base. If hydrogen ions are added to a buffer, some will combine with the conjugate base and convert it to the undissociated acid. Thus, the addition of hydrogen ions to the bicarbonate–carbonic acid system (Equation 4.2) drives the reaction to the right, increasing the amount of carbonic acid and using up the bicarbonate ions.

$$(4.2) \quad H^+ + HCO_3^- \rightleftharpoons H_2CO_3$$

Conversely, if the hydrogen ion concentration falls, carbonic acid dissociates, thereby generating hydrogen ions.

The efficacy of any buffer is limited by its concentration and by the position of the equilibrium. A buffer operates most efficiently at hydrogen ion concentrations which result in approximately equal concentrations of undissociated acid and conjugate base. The bicarbonate buffer system is the most important in the ECF, yet at normal ECF hydrogen ion concentrations the concentration of carbonic acid is about 1.2mmol/l while that of bicarbonate is twenty times greater. However, the capacity of the bicarbonate system in the body is greatly enhanced by the fact that carbonic acid can readily be formed from carbon dioxide or disposed of by conversion into carbon dioxide and water (Equation 4.1).

For every hydrogen ion buffered by bicarbonate, a bicarbonate ion is consumed (Equation 4.2). To maintain the capacity of the buffer system, the bicarbonate must be regenerated. Yet, when bicarbonate is formed from carbonic acid (indirectly from carbon dioxide and water), equimolar amounts of hydrogen ion are formed simultaneously (Equation 4.2). Bicarbonate formation can only continue if these hydrogen ions are removed. This process occurs in the cells of the renal tubules, where hydrogen ions are secreted into the urine while bicarbonate is generated and retained in the body.

Proteins, including intracellular proteins, are also involved in buffering. The proteinaceous matrix of bone is an important buffer in chronic acidosis. Phosphate is a minor buffer in the ECF but is of fundamental importance in the urine. The special role of haemoglobin is considered on page 38.

Bicarbonate reabsorption and hydrogen ion excretion

The glomerular filtrate contains the same concentration of bicarbonate ions as the plasma. If this bicarbonate is not reabsorbed, copious amounts will be excreted in the urine, depleting the body's buffering capacity so that an acidosis may develop. In health, at normal serum bicarbonate concentrations, virtually all the filtered bicarbonate is reabsorbed.

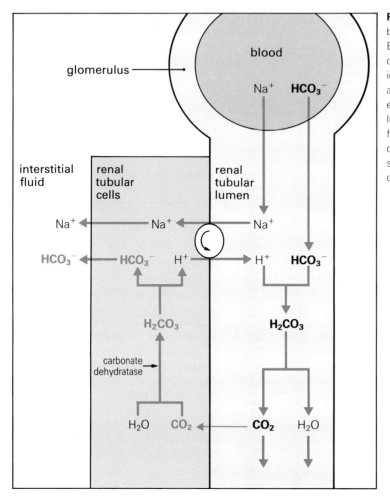

Fig.4.1 Reabsorption of filtered bicarbonate by renal tubular cells. Bicarbonate cannot be reabsorbed directly. Hydrogen and bicarbonate ions are generated in renal tubular cells and the hydrogen ions are secreted in exchange for sodium into the tubular lumen where they combine with filtered bicarbonate to form carbon dioxide and water. Bicarbonate ions are secreted with sodium from the tubular cells into the extracellular space.

The luminal surface of renal tubular cells is impermeable to bicarbonate and therefore direct reabsorption cannot occur. Within the renal tubular cells, carbonic acid is formed from carbon dioxide and water (Fig.4.1). This reaction (Equation 4.1) is catalyzed in the kidney by the enzyme, carbonate dehydratase (carbonic anhydrase). The carbonic acid thus formed dissociates to give hydrogen and bicarbonate ions. The bicarbonate ions pass across the basal border of the cells into the interstitial fluid. The hydrogen ions are secreted across the luminal membrane in exchange for sodium ions, which accompany bicarbonate into the interstitial fluid (Fig.4.1). The formation of bicarbonate and hydrogen ions is promoted by their continuous removal and by the presence of carbonate dehydratase.

In the tubular fluid, hydrogen ions combine with bicarbonate to form carbonic acid, most of which dissociates into carbon dioxide and water. Some of the carbon dioxide diffuses back into the renal tubular cells while the remainder is excreted in the urine. This whole process effectively results in the reabsorption of filtered bicarbonate.

Although hydrogen ions are secreted into the tubular fluid, there is no net hydrogen ion excretion, as the formation of hydrogen ions provides a device for the reabsorption of bicarbonate. Hydrogen ion excretion depends upon the same reactions occurring in the renal tubular cells but, in addition, requires the presence of a suitable buffer system in the urine. The principal urinary buffer is phosphate. This is present in the glomerular filtrate, approximately eighty percent

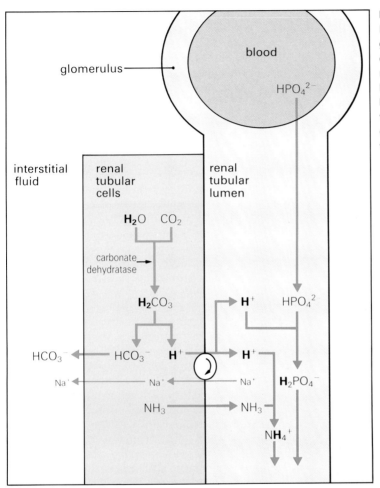

Fig.4.2 Renal hydrogen ion excretion. Hydrogen and bicarbonate ions are generated in renal tubular cells from carbon dioxide and water by the reversal of the buffering reaction. The hydrogen ions are excreted in the urine, buffered by phosphate and ammonia, while the bicarbonate enters the extracellular fluid replacing that which was consumed in buffering.

being in the form of the divalent anion, HPO_4^{2-}. This combines with hydrogen ions and is converted to $H_2PO_4^-$ (Equation 4.3).

(4.3) $HPO_4^{2-} + H^+ \leftrightharpoons H_2PO_4^-$

At the minimum urinary pH (4.6), virtually all the phosphate is in the $H_2PO_4^-$ form. About 30–40mmol of hydrogen ions are normally excreted in this way every twenty-four hours.

Ammonia, produced by the deamination of glutamine in renal tubular cells, is also an important urinary buffer. The enzyme which catalyzes this reaction, glutaminase, is induced in states of chronic acidosis, allowing increased ammonia production and, hence, increased hydrogen ion excretion via ammonium ions. Ammonia can readily diffuse across cell membranes but ammonium ions, formed when ammonia buffers hydrogen ions (Equation 4.4), cannot. Passive reabsorption of ammonium ions is therefore prevented.

(4.4) $NH_3 + H^+ \leftrightharpoons NH_4^+$

At normal intracellular hydrogen ion concentrations, most ammonia is present as the ammonium ion. Diffusion of ammonia out of the cell disturbs the equilibrium causing more ammonia to be formed. The simultaneous production of hydrogen ions would seem to negate the process. However, these ions can be used up in gluconeogenesis when they combine with glutamate formed by the deamination of glutamine. Urinary hydrogen ion excretion is summarized in Fig.4.2.

It will be apparent that hydrogen and bicarbonate ions are generated in equimolar amounts in renal tubular cells. This is essential for the reabsorption of filtered bicarbonate but also means that when a hydrogen ion is excreted in the urine, a bicarbonate ion is produced and retained. This process effectively regenerates the bicarbonate ions consumed when hydrogen ions are buffered.

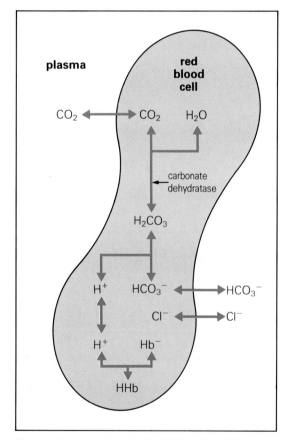

Fig.4.3 Transport of carbon dioxide in the blood. In capillary beds, carbon dioxide diffuses into red blood cells and combines with water to form carbonic acid; the reaction is catalyzed by carbonate dehydratase. The carbonic acid dissociates to form hydrogen ions, which are buffered by haemoglobin, and bicarbonate, which diffuses out of the cell; chloride diffuses in to maintain electrochemical neutrality. In the alveoli, the process reverses; carbon dioxide is produced from bicarbonate and is excreted in the expired air.

Transport of carbon dioxide

Carbon dioxide, produced by aerobic metabolism, diffuses out of cells and dissolves in the ECF. A small amount combines with water to form carbonic acid, thereby increasing the hydrogen ion concentration of the ECF.

In red blood cells, metabolism is anaerobic and little carbon dioxide is produced. Carbon dioxide thus diffuses into red cells down a concentration gradient and carbonic acid is formed, facilitated by carbonate dehydratase (Fig.4.3). Haemoglobin buffers the hydrogen ions formed when the carbonic acid dissociates. Haemoglobin is a more powerful buffer when in the deoxygenated state and the proportion in this state increases during the passage of blood through capillary beds as oxygen is lost to the tissues.

The overall effect of this process is that carbon dioxide is converted to bicarbonate in red blood cells. This bicarbonate diffuses out of the red cells, because a concentration gradient develops, and electrochemical neutrality is maintained by inward diffusion of chloride ions (the chloride shift). In the lungs, the reverse process occurs because of the low partial pressure of carbon dioxide in the alveolar capillaries. Carbon dioxide is produced from bicarbonate and diffuses into the alveoli to be excreted in the expired air.

Most of the carbon dioxide in the blood is present in the form of bicarbonate. Dissolved carbon dioxide, carbonic acid and carbamino compounds (compounds of carbon dioxide and protein) account for less than 2.0mmol/l in a total carbon dioxide concentration of approximately 26mmol/l. The terms 'bicarbonate' and 'total carbon dioxide' are frequently used synonymously. They are not strictly the same but may be considered to be for most practical clinical purposes. It is technically difficult to measure bicarbonate concentration alone; most analytical techniques for bicarbonate actually measure total carbon dioxide.

CLINICAL AND LABORATORY ASSESSMENT OF HYDROGEN ION STATUS

As will be seen, many conditions are associated with abnormalities of blood hydrogen ion concentration and partial pressure of carbon dioxide ($P\text{CO}_2$). The clinical features associated with these abnormalities and those with an altered partial pressure of oxygen ($P\text{O}_2$) are shown in Fig.4.4.

It is usual to measure hydrogen ion concentration and $P\text{CO}_2$ in arterial blood, anticoagulated with

	Increase	**Decrease**
P_{CO_2}	peripheral vasodilation headache bounding pulse papilloedema flapping tremor ⎤ drowsiness, coma ⎦ — late signs	paraesthesiae dizziness muscle cramps headache tetany

P_{O_2}	pulmonary and retinal
	  fibrosis (only with
	  prolonged use of high
	  inspiratory P_{O_2},
	  particularly in infants)
		breathlessness
		cyanosis
		drowsiness, confusion and coma
		pulmonary hypertension (in
		  chronic hypoxaemia)

$[H^+]$	hyperventilation
	increased catecholamine release
	hyperkalaemia
	decreased myocardial ⎤ severe
	  contractility and ⎬ acidosis
	  CNS depression ⎦ only
		hypoventilation
		paraesthesiae
		muscle cramps
		dizziness
		headache
		tetany
		drowsiness, confusion and coma

Fig.4.4   Effects of increased or decreased values of P_{CO_2}, P_{O_2} and hydrogen ion concentration in the blood. Paraesthesiae, dizziness, muscle cramps and tetany are related to a decrease in ionized calcium.

heparin. The arteriovenous gradient for hydrogen ion concentration is small (<2nmol/l), but the gradient is significantly different for P_{CO_2} (approximately 1.1kPa or 8mmHg higher in venous blood) and P_{O_2} (approximately 7.5kPa or 56mmHg lower in venous blood).

It is vital that air is excluded from the syringe, both before and after drawing blood, and that analysis is, if possible, performed immediately. If the blood sample has to be transported, the syringe, capped with a blind hub and enclosed in a plastic bag, should be chilled in ice-water. Analytical instruments measure hydrogen ion concentration (strictly, activity), P_{CO_2} and P_{O_2} using specific electrodes; these measurements are together known as 'blood gases'.

By the law of mass action it follows, from the equations describing the dissociation of carbonic acid (Equations 4.1 and 4.2), that hydrogen ion concentration is directly proportional to P_{CO_2} and inversely proportional to bicarbonate concentration; that is, it is determined by the ratio of P_{CO_2} to bicarbonate (Equation 4.5).

$$(4.5) \quad [H^+] = K \frac{P_{CO_2}}{[HCO_3{}^-]}$$

The constant, K, embraces the dissociation constants for Equations 4.1 and 4.2 and the solubility coefficient of carbon dioxide, which governs the concentration of the gas in solution at a given partial pressure. When hydrogen ion concentration is measured in nmol/l, bicarbonate in mmol/l and P_{CO_2} in kilopascals (kPa), the value of K is approximately 180 at 37°C; if P_{CO_2} is measured in mmHg, the value of K is 24.

It follows that it is possible to calculate the bicarbonate concentration from the hydrogen ion concentration and P_{CO_2} alone. In blood gas analyzers, the bicarbonate concentration is derived by calculation in this way and is not measured. It is not the same as the bicarbonate (strictly, total carbon dioxide) measured by autoanalyzers. There has been considerable argument over whether it is valid to derive a bicarbonate concentration in this way, given that the values of the constants involved are based upon observations in supposedly ideal solutions, which biological fluids are

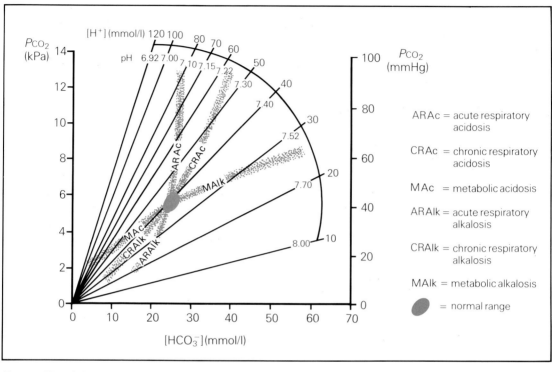

Fig.4.5 The relationship between P_{CO_2}, hydrogen ion concentration and bicarbonate concentration. The shaded areas represent the ranges of values found in simple disturbances of acid–base homoeostasis. Data falling outside these areas indicate mixed disturbances.

not. However, for most practical purposes the derivation is an acceptable one.

An appreciation of the relationship between hydrogen ion concentration, bicarbonate concentration and P_{CO_2} is of fundamental importance to an understanding of pathophysiology of hydrogen ion homoeostasis. It will be apparent from Equation 4.5 that the relationships between hydrogen ion concentration and P_{CO_2} and between bicarbonate concentration and P_{CO_2} are linear. These relationships have been quantified by measurements made *in vivo* and it is therefore possible to predict the effect of a change in one variable on another; for example, the effect of an acute rise in P_{CO_2} on hydrogen ion concentration. This information is an important aid in the interpretation of acid–base data.

The relationships between hydrogen ion concentration, P_{CO_2} and bicarbonate concentration are plotted in Fig.4.5. This may be useful as an *aide-mémoire* to the interpretation of acid–base data but should not be used as a substitute for a full understanding of the underlying principles.

Many instruments for blood gas analysis generate other data such as standard bicarbonate and base excess. The meanings, uses and misuses of these terms are described later.

DISORDERS OF HYDROGEN ION HOMOEOSTASIS

There are four stages in the pathophysiology of hydrogen ion disorders (acid–base disorders):
- Generation
- Buffering
- Compensation
- Correction

It is helpful to consider the various stages separately although in reality they occur concurrently, albeit with different time courses.

Acid–base disorders are classified as either respiratory or non-respiratory (metabolic) according to whether or not there is a primary (causative) change in P_{CO_2}. The term 'acidosis' signifies a tendency for the hydrogen ion concentration to be above normal and 'alkalosis' signifies below normal.

Mixed primary acid–base disorders, that is disorders of combined respiratory and non-respiratory origin, are common. However, the secondary, or compensatory, responses to a primary disorder of hydrogen ion homoeostasis, may produce changes indistinguishable from those seen in disorders of mixed origin.

Non-respiratory (metabolic) acidosis

The primary abnormality in non-respiratory acidosis is either an increased production or a decreased excretion of hydrogen ions. In some cases, both of these may contribute. Loss of bicarbonate from the body can also, indirectly, cause an acidosis. Common causes of non-respiratory acidosis are given in Fig.4.6.

Excess hydrogen ions are buffered by bicarbonate (Equation 4.2) and other buffers. The carbonic acid thus formed dissociates (Equation 4.1) and the carbon dioxide is lost in the expired air. This buffering limits the potential rise in hydrogen ion concentration at the expense of a reduction in bicarbonate concentration.

Compensation is effected by hyperventilation which increases the removal of carbon dioxide and lowers the P_{CO_2}. The P_{CO_2} to bicarbonate ratio is reduced, thus reducing the hydrogen ion concentration (Equation 4.5). Hyperventilation is a direct result of the increased hydrogen ion concentration stimulating the respiratory centre. Respiratory compensation cannot completely normalize the hydrogen ion concentration since it is the high concentration itself that stimulates the compensatory hyperventilation. Furthermore, the increased work of the respiratory muscles produces carbon dioxide, thereby limiting the extent to which the P_{CO_2} can be lowered.

If the cause of the acidosis is not corrected, a new steady state may be attained, with a raised hydrogen ion concentration, low bicarbonate and low P_{CO_2}. In the steady state, the decrease in P_{CO_2} attributable to respiratory compensation is approximately 0.17kPa (1.3mmHg) for each 1mmol/l decrement in bicarbonate concentration. The extent to which compensation can take place will be limited if respiratory function is

---

Causes of non-respiratory acidosis

Increased H⁺ formation

ketoacidosis (usually diabetic, also alcoholic)
lactic acidosis
poisoning: e.g. ethanol, methanol, ethylene glycol and salicylate
inherited organic acidoses

Acid ingestion

acid poisoning
excessive parenteral administration of amino acids: e.g. arginine, lysine and histidine

Decreased H⁺ excretion

renal tubular acidosis
generalized renal failure
carbonate dehydratase inhibitors

Loss of bicarbonate

diarrhoea
pancreatic, intestinal and biliary drainage

Fig.4.6 Principal causes of non-respiratory (metabolic) acidosis.

---

compromised. Even with normal respiratory function, it is exceptional for a P_{CO_2} of less than 1.5kPa (11.3mmHg) to be recorded, however severe the non-respiratory acidosis.

In a healthy person, hyperventilation would produce a respiratory alkalosis. In general, the compensatory mechanism for any acid–base disturbance involves the generation of a second, opposing disturbance. In the case of a metabolic acidosis, compensation is through the generation of a respiratory alkalosis; in a respiratory acidosis, it is through the generation of a metabolic alkalosis (see following).

If renal function is normal in a patient with non-respiratory acidosis, excess hydrogen ions can be excreted by the kidneys. However, in many cases there is impairment of renal function, although this is not necessarily the primary cause of the acidosis.

The complete correction of a non-respiratory acidosis requires reversal of the underlying cause, for example, rehydration and insulin for diabetic ketoacidosis (see *Case history 12.2*) and removal of salicylate in salicylate overdosage. It is important to maintain adequate renal perfusion to maximize renal hydrogen ion excretion. The use of exogenous bicarbonate to buffer hydrogen ions is discussed below and on page 176.

Increased production of hydrogen ions

This is the cause of the acidosis in ketoacidosis (diabetic, alcoholic), lactic acidosis and acidosis seen in poisoning, for example, with salicylates and ethylene glycol.

CASE HISTORY 4.1

A sixty-year-old man was admitted to hospital with severe abdominal pain which had begun two and a half hours earlier. He was not taking any drugs. On examination, he was shocked and had a distended rigid abdomen; neither femoral pulse was palpable.

Investigations

arterial blood:

hydrogen ion	90nmol/l (pH 7.05)
P_{CO_2}	3.5kPa (26.3mmHg)
P_{O_2}	12kPa (90mmHg)
bicarbonate (derived)	7mmol/l

Comment

The patient is acidotic (raised hydrogen ion concentration) and this must be non-respiratory in origin since the P_{CO_2} is not raised. Indeed, P_{CO_2} is decreased, reflecting compensatory hyperventilation. The hyperventilation may be clinically obvious (Kussmaul's respiration, see *Case history 12.2*). An even lower P_{CO_2} might have been expected as a result of respiratory compensation but, in this case, splinting of the abdominal muscles (the abdomen is rigid) has restricted respiratory movements. The low bicarbonate concentration reflects the primary abnormality; bicarbonate is consumed as hydrogen ions are buffered. If there is no respiratory component to the acidosis, the serum bicarbonate concentration is a good guide to the severity of a metabolic acidosis.

The clinical diagnosis (confirmed at laparotomy) is a ruptured abdominal aortic aneurysm. The patient is severely shocked following extravasation of blood from the aneurysm. Impaired tissue perfusion has led to inadequate oxygenation, despite the normal arterial P_{O_2}, with consequently increased anaerobic metabolism of glucose to lactic acid, rather than oxidative metabolism.

Lactic acid is a normal metabolite of muscle and is converted back to glucose in the liver (the Cori cycle). However, with greatly increased production and possible impairment of hepatic metabolism due to poor perfusion, lactic acid accumulates. If renal function is compromised, for instance, by hypoperfusion, the ability of the kidneys to excrete excess hydrogen ions is also impaired.

Other causes of lactic acidosis are given in Fig.4.7.

Decreased excretion of hydrogen ions

Acidosis occurs in renal glomerular failure, when the decreased glomerular filtration causes a reduction in the amount of sodium that is filtered and therefore available for exchange with hydrogen ions. The amount of phosphate filtered and available for buffering also decreases (see *Case history 5.2*). The acidosis of renal tubular failure is discussed in *Chapter 5*.

Loss of bicarbonate

Loss of bicarbonate and retention of hydrogen ions may result in acidosis in patients losing alkaline secretions from the small intestine, through fistulae, for example. In the stomach, bicarbonate generated from carbon dioxide and water is retained and hydrogen ions are secreted into the lumen (Fig.4.8). In the pancreas and small intestine, the movements of bicarbonate and hydrogen ions occur in the opposite directions (Fig.4.8). Therefore, hydrogen ions which are secreted into the stomach lumen are neutralized by bicarbonate in the small intestine.

Since most of the fluid and ions secreted into the gut are reabsorbed, the gut is effectively a closed system with regard to acid–base balance, under normal circumstances. If, however, alkaline secretions are lost, the patient is at risk of becoming acidotic. Increased renal hydrogen ion excretion may prevent this, but excessive fluid loss from the gut may deplete the ECF to such an extent that the glomerular filtration rate falls and the kidneys will no longer be able to compensate.

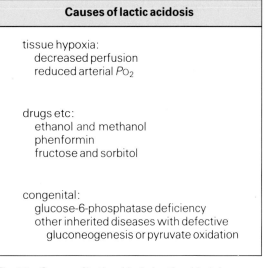

Causes of lactic acidosis

tissue hypoxia:
    decreased perfusion
    reduced arterial P_{O_2}

drugs etc:
    ethanol and methanol
    phenformin
    fructose and sorbitol

congenital:
    glucose-6-phosphatase deficiency
    other inherited diseases with defective
        gluconeogenesis or pyruvate oxidation

Fig.4.7 Causes of lactic acidosis. Lactic acidosis is sometimes classified as type A (tissue hypoxia) and type B (all other causes).

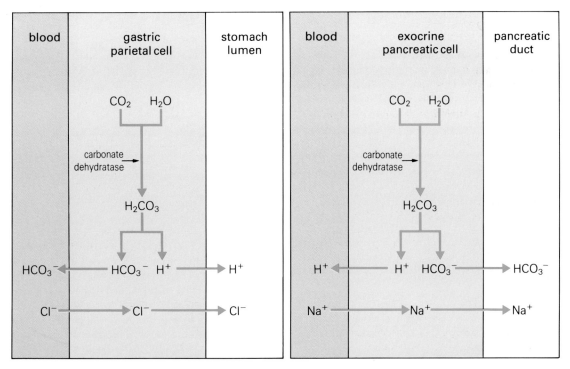

Fig.4.8 Generation of acidic gastric and alkaline pancreatic secretions. Hydrogen and bicarbonate ions are generated from carbon dioxide and water, catalyzed by carbonate dehydratase. In the stomach, the hydrogen ions are secreted while bicarbonate is retained. The reverse process occurs in the pancreas.

43

The anion gap

When the bicarbonate concentration falls in a non-respiratory acidosis, electrochemical neutrality must be maintained by other anions. In many cases, anions are produced simultaneously and equally with hydrogen ions, for example, acetoacetate and β-hydroxybutyrate in diabetic ketoacidosis, and lactate in lactic acidosis. When this does not occur, the deficit is met by chloride ions.

The difference between the sums of the concentrations of the principal cations (sodium and potassium) and of the principal anions (chloride and bicarbonate) is known as the 'anion gap'.

$$\text{Anion gap} = ([Na^+] + [K^+]) - ([Cl^-] + [HCO_3^-])$$

In health, the anion gap has a value of 14–18mmol/l and mainly represents the unmeasured net negative charge on plasma proteins.

In an acidosis in which anions other than chloride are increased, the anion gap is increased. In contrast, in an acidosis due to loss of bicarbonate, for example, renal tubular acidosis, the plasma chloride concentration is increased and the anion gap is normal. It has therefore been suggested that calculation of the anion gap is of value in the diagnosis of acidosis. In the majority of cases of acidosis, however, the cause is obvious clinically and can be confirmed by the results of simple tests. The anion gap may be useful in the analysis of complex acid–base disorders, as shown by *Case history 4.7*, but some laboratories do not measure chloride and the anion gap cannot then be calculated.

The characteristic biochemical changes seen in non-respiratory acidosis can be summarized as follows:

Non-respiratory acidosis

$[H^+]$	↑
pH	↓
P_{CO_2}	↓
$[HCO_3^-]$	↓

The decrease in P_{CO_2} is a compensatory change; the decrement in P_{CO_2} is approximately 0.17kPa (1.3mmHg) per millimole decrease in the concentration of bicarbonate.

Changes due to the underlying condition will also be present. Hyperkalaemia is common in acidotic patients, except in bicarbonate-wasting conditions, for reasons discussed in *Chapter 3*.

Causes of respiratory acidosis

Airway obstruction

chronic obstructive airway disease (bronchitis, emphysema)
bronchospasm, e.g. in asthma
aspiration

Depression of respiratory centre

anaesthetics
sedatives
cerebral trauma
tumours

Neuromuscular disease

poliomyelitis
Guillain-Barré syndrome
motor neuron disease
tetanus, botulism
neurotoxins, curare

Pulmonary disease

pulmonary fibrosis
severe pneumonia
respiratory distress syndrome

Extrapulmonary thoracic disease

flail chest
severe kyphoscoliosis

Fig.4.9 Major causes of respiratory acidosis.

Management

The management of non-respiratory acidosis must be directed at reversal of the underlying cause. Where this is not immediately possible, bicarbonate may be given to buffer hydrogen ions although there is no general agreement as to when bicarbonate should be used. Many would consider it prudent to give bicarbonate when the arterial hydrogen ion concentration is greater than 100nmol/l (pH <7) and there is no immediate prospect of lowering it by other means, particularly in a patient whose clinical condition is generally poor. However, when bicarbonate is used it should be given in small aliquots and the effect on the arterial hydrogen ion concentration measured. Large amounts of bicarbonate, perhaps calculated from a formula, given rapidly in an attempt to correct an acidosis, may be dangerous.

Respiratory acidosis

Some of the many conditions associated with the development of respiratory acidosis are shown in Fig.4.9. They are all characterized by an increase in P_{CO_2}. The hydrogen ions produced are mainly buffered by intracellular buffers, especially haemoglobin. For every hydrogen ion that is produced a bicarbonate is generated. With an acute rise in P_{CO_2}, every 1kPa (7.5mmHg) increase is associated with a concomitant increase in bicarbonate concentration of just under 1mmol/l and hydrogen ion concentration of approximately 5.5nmol/l. In chronic carbon dioxide retention, when renal compensation is maximal, the hydrogen ion concentration is increased by only 2.5nmol/l for each 1kPa (7.5mmHg) rise in P_{CO_2}.

A respiratory acidosis can only be corrected by means which restore the P_{CO_2} to normal but, if a high P_{CO_2} persists, compensation occurs through increased renal hydrogen ion excretion.

CASE HISTORY 4.2

A young man sustained injury to the chest in a road traffic accident. Effective ventilation was compromised by a large flail segment.

Investigations

arterial blood: P_{O_2} 8kPa (60mmHg)
           P_{CO_2} 8kPa (60mmHg)
           hydrogen ion   58nmol/l (pH 7.24)
           bicarbonate
             (derived)   25mmol/l

Comment

There is a severe acidosis and the raised P_{CO_2} indicates that it is respiratory in origin. The

magnitude of the increase in hydrogen ion concentration suggests that no renal compensation has occurred. Such compensation can take several days to become fully effective, in contrast to the rapid respiratory compensation in non-respiratory disorders.

In acute respiratory acidosis, the bicarbonate concentration is slightly raised but is usually still within the normal range. If the bicarbonate concentration is more than slightly raised in a respiratory acidosis, either a more chronic course with renal compensation (see *Case History 4.3*) or a coexisting non-respiratory alkalosis is suggested. A low bicarbonate would suggest a coexisting non-respiratory acidosis.

CASE HISTORY 4.3

A seventy-year-old man, known to suffer from chronic obstructive airways disease, was admitted to hospital with an acute exacerbation of his illness. Arterial blood analysis was carried out on admission (results A). In spite of vigorous physiotherapy and medical treatment his condition deteriorated (results B) and it was decided to start artificial ventilation. Analysis was repeated after six hours (results C). After twelve hours he had a generalized fit (results D).

Investigations

arterial blood:	A	B	C	D
P_{CO_2} (kPa)	9.5	11.0	7.8	5.7
(mmHg)	71.3	82.5	58.5	42.8
hydrogen ion (nmol/l)	50	58	40	29
pH	7.30	7.24	7.40	7.54
bicarbonate (mmol/l) (derived)	35	35	34	35

Comment

The results on admission (A) indicate an acidosis. This is of respiratory origin since the P_{CO_2} is raised. However, the hydrogen ion concentration is only slightly elevated, indicating that renal compensation is occurring as would be expected in a case of chronic carbon dioxide retention. The raised bicarbonate, which suggests a non-respiratory alkalosis is the result of the compensatory increase in renal hydrogen ion excretion. Indeed, the commonest causes of raised serum bicarbonate concentration in the elderly are chronic carbon dioxide retention and diuretic-induced potassium depletion (see page 30).

A more severe acidosis (B) subsequently develops, commensurate with the rise in P_{CO_2}. This is a result of further carbon dioxide retention with no corresponding increase in renal hydrogen ion excretion.

Artificial ventilation lowers the P_{CO_2} rapidly (C); the hydrogen ion concentration is now normal although the P_{CO_2} is still elevated. This represents this patient's normal steady state, in which there is an almost complete renal compensation of the acidosis.

Continued ventilation reduces the P_{CO_2} (D) to within the normal range for a healthy subject but to below this patient's normal. He has become alkalotic and suffers a fit as a consequence. The alkalosis is due to the continued high rate of renal hydrogen ion excretion in response to the chronically raised P_{CO_2}. Adaptation of renal hydrogen ion excretion in response to a change in P_{CO_2} takes several days. Thus, rapid reduction of the P_{CO_2} exposes the compensatory, secondary response, which then appears to be the sole acid–base abnormality. The compensatory mechanism in respiratory acidosis involves the generation of a metabolic alkalosis.

Management

The aim when treating respiratory acidosis is to improve alveolar ventilation and lower the P_{CO_2}. In acute alveolar hypoventilation, however, it is usually hypoxaemia rather than hypercapnia that poses the main threat to life, unless the P_{O_2} is being maintained by the supply of additional oxygen. If ventilation ceases abruptly, death from hypoxaemia will occur in approximately four minutes; the P_{CO_2} by comparison rises at such a rate that it would take more than ten minutes for it to reach a lethal level.

In chronic respiratory acidosis, it is seldom possible to remove the underlying cause and treatment is directed at maximizing alveolar ventilation by, for example, utilizing physiotherapy, bronchodilators and antibiotics. If artificial ventilation becomes necessary, it is vital to monitor the patient's arterial blood gases and hydrogen ion in order to avoid over-correction of the respiratory acidosis.

Oxygen may safely be used at high concentrations in patients with acute respiratory failure. In many patients with chronic carbon dioxide retention, however, the respiratory centre becomes insensitive to carbon dioxide and hypoxaemia provides the main stimulus to respiration. Oxygen administration in such patients must be carefully controlled to prevent abolition of this stimulus.

It is important to appreciate that, on the basis of the data alone, it would not be possible to tell whether results 'C' in *Case history 4.3* represented a state of either compensated chronic carbon dioxide retention or acute carbon dioxide retention developing in a patient with a pre-existing metabolic alkalosis. The management of these two states would not be the same.

The characteristic biochemical changes in acute and chronic acidosis can be summarized as follows:

Respiratory acidosis

	acute	chronic
$[H^+]$	↑	normal or slight ↑
pH	↓	normal or slight ↓
P_{CO_2}	↑	↑
$[HCO_3^-]$	slight ↑	↑

Non-respiratory (metabolic) alkalosis

Non-respiratory alkalosis is characterized by a primary increase in the ECF bicarbonate concentration, with a consequent reduction in the hydrogen ion concentration (see *Equation 4.5*). In a normal subject, an increase in plasma bicarbonate concentration leads to incomplete renal tubular bicarbonate reabsorption and excretion of bicarbonate in the urine. Massive quantities of bicarbonate must be ingested to produce a sustained alkalosis.

It is necessary both to consider the mechanisms which cause non-respiratory alkalosis and those which can perpetuate it. In contrast to non-respiratory acidosis and to respiratory disorders of acid–base balance, a non-respiratory alkalosis often persists even after the primary cause has been corrected.

Causes of non-respiratory alkalosis

Loss of unbuffered hydrogen ion

gastrointestinal:
   gastric aspiration
   vomiting with pyloric stenosis
   congenital chloride-losing diarrhoea

renal:
   mineralocorticoid excess:
      Cushing's syndrome
      Conn's syndrome
   drugs with mineralocorticoid activity,
     e.g. carbenoxolone
   diuretic therapy (not K^+-sparing)
   rapid correction of chronically raised P_{CO_2}
   potassium depletion

Administration of alkali

inappropriate treatment of acidotic states
chronic alkali ingestion

Fig.4.10 Causes of non-respiratory alkalosis.

Causes of a non-respiratory alkalosis are shown in Fig.4.10. Alkali loading causes only a transient alkalosis unless there are additional factors operating to sustain it. Disproportionate loss of chloride, for example, during diuretic-induced mobilization of oedema fluid when little bicarbonate is excreted in the urine, can cause a non-respiratory alkalosis but, if this is the sole mechanism, the disturbance is always mild.

The maintenance of a non-respiratory alkalosis requires inappropriately high (as far as hydrogen ion homoeostasis is concerned) renal bicarbonate reabsorption. Factors which may be responsible for this include a decrease in ECF volume, mineralocorticoid excess and potassium depletion.

In hypovolaemia, there is an increased stimulus to renal sodium reabsorption (see page 17). Sodium reabsorption is dependent upon the availability of adequate anions. If there is a relative deficit of chloride as, for example, with loss of gastric juice and treatment with some diuretics, electrochemical neutrality during sodium reabsorption is maintained by increased bicarbonate reabsorption and by hydrogen

and potassium ion excretion. In states of mineralocorticoid excess, alkalosis is perpetuated by the increased hydrogen ion excretion in the nephron which occurs secondarily to the increased sodium reabsorption.

The correction of a non-respiratory alkalosis requires reversal both of the primary cause and of the mechanism responsible for its perpetuation. Compensation is effected by hypoventilation (the respiratory centre is inhibited by a low hydrogen ion concentration), which causes carbon dioxide retention and increases the ratio of P_{CO_2} to bicarbonate concentration (see *Equation 4.5*, page 39).

CASE HISTORY 4.4

A forty-five-year-old man was admitted to hospital with a history of persistent vomiting. He had a long history of dyspepsia but had never sought advice for this, preferring to treat himself with proprietary remedies. On examination he was obviously dehydrated and his respiration was shallow.

Investigations

arterial blood:		
	hydrogen ion	28nmol/l (pH 7.56)
	P_{CO_2}	7.2kPa (54mmHg)
	bicarbonate	
	(derived)	45mmol/l
serum:	sodium	146mmol/l
	potassium	2.8mmol/l
	urea	34.2mmol/l

A barium meal, performed after this metabolic imbalance had been corrected, showed pyloric stenosis, thought to be due to scarring caused by peptic ulceration.

Comment

The patient is alkalotic and since the P_{CO_2} is high, this must be non-respiratory in origin. The increase in P_{CO_2} is a result of compensatory hypoventilation leading to carbon dioxide retention. Although hypercapnia and hypoxaemia, the latter of which can also result from hypoventilation, are both respiratory stimulants, they do not always dampen the respiratory compensation. In chronic non-respiratory alkalosis, as in this case, each increment of 1 mmol/l in bicarbonate concentration typically gives rise to an increase in P_{CO_2} of approximately 0.1kPa (0.8mmHg).

The alkalosis is a result of loss of unbuffered hydrogen ions in gastric juice with a concomitant retention of bicarbonate. The raised urea is consistent with the clinical signs of dehydration resulting from fluid loss. Fluid loss stimulates renal sodium reabsorption, but sodium can only be reabsorbed either with chloride or in exchange for hydrogen and potassium ions. Gastric juice has a high concentration of chloride and patients losing gastric secretions become hypochloraemic. It appears that the defence of ECF volume takes precedence over acid–base homoeostasis and therefore, rather than being limited by the lack of chloride, sodium ions are reabsorbed in exchange for hydrogen ions (perpetuating the alkalosis) and potassium ions (causing potassium depletion). This explains the apparently paradoxical finding of an acidic urine in patients with severe non-respiratory alkalosis. Also, since the gastric juice contains potassium, such patients are frequently hypokalaemic and yet are losing potassium in the urine.

A non-respiratory alkalosis due to loss of gastric acid may also occur in patients undergoing nasogastric aspiration. It is not usually a feature of vomiting if the pylorus is patent, since the additional loss of alkaline secretions from the upper small intestine counteracts the effect of the retention of bicarbonate ions generated by gastric mucosal cells. Although vomiting with pyloric stenosis is an unusual cause of non-respiratory alkalosis, other causes rarely result in such a severe disturbance.

Management

The management of a non-respiratory alkalosis depends upon the severity of the condition and upon the cause. When hypovolaemia and hypochloraemia are present, they can be simultaneously corrected by an infusion of sodium chloride solution ('normal saline')

which will also improve renal perfusion and allow excretion of the bicarbonate load. In such cases, it is common practice to provide potassium supplements although often this is not strictly necessary. It is very rarely necessary to attempt rapid correction of non-respiratory alkalosis, for example, by administration of ammonium chloride.

The mild alkalosis commonly associated with potassium depletion which may, for example, be diuretic induced, rarely requires treatment *per se* although the hypokalaemia may require correction. The management of a state of mineralocorticoid excess is considered in *Chapter 9*.

The biochemical features of non-respiratory alkalosis can be summarized as follows:

Non-respiratory alkalosis

$[H^+]$	↓
pH	↑
P_{CO_2}	↑
$[HCO_3^-]$	↑

Respiratory alkalosis

The main causes of respiratory alkalosis are shown in Fig.4.11. The common feature and cause of the alkalosis is a fall in P_{CO_2}, which reduces the ratio of P_{CO_2} to bicarbonate concentration (see *Equation 4.5*, page 39). In acute respiratory alkalosis, the hydrogen ion concentration falls by approximately 5.5 mmol/l for each 1.0kPa (7.5mmHg) fall in P_{CO_2}.

The fall in P_{CO_2} causes a small decrease in bicarbonate concentration. Compensation occurs through a reduction in renal hydrogen ion excretion, which further decreases the serum bicarbonate concentration. Renal compensation in a respiratory alkalosis develops slowly, as it does in a respiratory acidosis. If a steady P_{CO_2} is maintained, maximal compensation with a new steady state develops within 36–72 hours.

CASE HISTORY 4.5

As part of a class experiment in physiology, a medical student volunteered to have a sample of arterial blood taken. The demonstrator took some time to explain the procedure to the class, during which time the student became increasingly anxious. As the blood was being drawn she complained of tingling in her fingers and toes.

Investigations

arterial blood:

hydrogen ion	30nmol/l (pH 7.52)
P_{CO_2}	3.5kPa (26.3mmHg)
bicarbonate (derived)	22mmol/l

Comment

The student is alkalotic with a low P_{CO_2}, thus the disturbance is respiratory in origin. The extent of the decrease in hydrogen ion concentration indicates that there is neither compensation nor additional acid–base disturbance. The low P_{CO_2} is a result of anxiety-induced hyperventilation and no compensation would be expected to have occurred in this short time. The symptoms are a result of a decrease in the concentration of ionized calcium, an effect of alkalosis.

Causes of respiratory alkalosis

Hypoxia

high altitude
severe anaemia
pulmonary disease

Increased respiratory drive

respiratory stimulants, e.g. salicylates
cerebral disturbances, e.g. trauma,
    infection and tumours
hepatic failure
Gram-negative septicaemia
primary hyperventilation syndrome
voluntary hyperventilation

Pulmonary disease

pulmonary oedema
pulmonary embolism

Mechanical overventilation

Fig.4.11 Major causes of respiratory alkalosis.

CASE HISTORY 4.6

A young woman, admitted to hospital, was unconscious following a head injury. A skull fracture was demonstrated on radiography and a computerized tomography (CT) scan revealed extensive cerebral contusions. The respiratory rate was 38/min. Three days after admission, the patient's condition was unchanged and arterial blood was analyzed.

Investigations

arterial blood:

hydrogen ion	36nmol/l (pH 7.44)
P_{CO_2}	3.9kPa (29.3mmHg)
bicarbonate (derived)	19mmol/l

Comment

This is a compensated respiratory alkalosis. The P_{CO_2} is reduced as a result of hyperventilation. The hydrogen ion concentration is at the lower limit of normal. Abnormalities of respiration (hypo- and hyperventilation) are common in patients with head injuries. Hyperventilation can occur with injuries involving the brain stem and as a result of raised intracranial pressure.

Even though a low P_{CO_2} is also characteristic of the respiratory compensation in non-respiratory acidosis, the history and normal hydrogen ion concentration preclude this diagnosis. Also, a much lower bicarbonate concentration would be expected in a non-respiratory acidosis.

Management

As with other disturbances of acid–base homoeostasis, the management of patients with respiratory alkalosis should be directed towards the underlying cause although this is frequently not possible. Fortunately, a chronic compensated respiratory alkalosis is not, in itself, dangerous. Increasing the inspired P_{CO_2} by making the patient re-breathe into a paper bag may abort the clinical features of acute hypocapnia in acute hyperventilation, but this is only a temporary measure.

The biochemical features of respiratory alkalosis can be summarized as follows:

Respiratory alkalosis

	acute	chronic
$[H^+]$	↓	slight ↓ or normal
pH	↑	slight ↑ or normal
P_{CO_2}	↓	↓
$[HCO_3^-]$	slight ↓	↓

INTERPRETATION OF ACID–BASE DATA

A thorough understanding of the pathophysiology of acid–base homoeostasis is essential for the correct interpretation of laboratory data, but these data should always be considered in the clinical context.

The starting point in any evaluaton should be the hydrogen ion concentration, or pH. This will indicate whether the predominant disturbance is an acidosis or an alkalosis. However, a normal value does not exclude an acid–base disorder. There may be either a fully compensated disturbance or two primary disturbances, where effects on hydrogen ion concentration cancel each other out.

If the P_{CO_2} is abnormal, there must be a respiratory component to the disturbance; if the P_{CO_2} is raised in an acidosis, the acidosis is respiratory and comparison of the hydrogen ion concentration with that predicted for an acute change in P_{CO_2} will indicate whether there is an additional metabolic component; this may be compensatory. If the P_{CO_2} is low in an acidosis, the acidosis is non-respiratory and there is an additional, often compensatory, respiratory component. A similar rationale applies to alkalotic states.

Since the derived bicarbonate is calculated from the P_{CO_2} and hydrogen ion concentration, it does not provide any more information than these two measurements. However, knowledge of the bicarbonate level may simplify the interpretation of acid–base data. Its concentration is always decreased in non-respiratory acidosis and increased in non-respiratory alkalosis, regardless of whether or not there is compensation.

CASE HISTORY 4.7

A young woman was admitted to hospital, eight hours after she had taken an overdose of aspirin.

Investigations

arterial blood:

hydrogen ion	30nmol/l (pH 7.53)
P_{CO_2}	2.0kPa (15mmHg)

Comment

The patient is alkalotic and the low P_{CO_2} indicates a respiratory origin. However, the hydrogen ion concentration is not as low as would have been expected as a result of an acute fall in P_{CO_2}. The data would be appropriate for a chronic compensated respiratory alkalosis, but such a low P_{CO_2} would be exceptional and this interpretation is not compatible with the history. The alternative is that there is an acute respiratory alkalosis with a coexistent non-respiratory acidosis. This combination is characteristic of salicylate poisoning where initial respiratory stimulation causes a respiratory alkalosis but later the metabolic effects of salicylate tend to predominate, producing an acidosis.

This case history illustrates the importance of considering the clinical setting when analyzing acid–base data. Calculation of the anion gap might have been helpful here. It would have been increased by the presence of organic anions, indicating a coexisting non-respiratory acidosis, but normal in a compensated respiratory alkalosis.

	Acidosis			Alkalosis		
	Non-respiratory	Respiratory		Non-respiratory	Respiratory	
		acute	chronic		acute	chronic
[H$^+$] (pH)	↑ (↓)	↑ (↓)	normal or slight ↑ (normal or slight ↓)	↓ (↑)	↓ (↑)	normal or slight ↓ (normal or slight ↑)
P_{CO_2}	↓	↑	↑	↑	↓	↓
[HCO$_3^-$]	↓	slight ↑	↑	↑	slight ↓	↓

Fig.4.12 Biochemical changes characteristic of disturbances of acid–base homoeostasis.

CASE HISTORY 4.8

An elderly man was admitted to hospital in a confused state. He was dyspnoeic and had a cough productive of sputum. He was unable to give a coherent history but one of the casualty officers knew him to be an insulin-dependent diabetic patient with a long history of chronic bronchitis.

Investigations

arterial blood:

hydrogen ion	66nmol/l (pH 7.18)
P_{CO_2}	7.4kPa (55.5mmHg)

Comment

The patient is acidotic and the raised P_{CO_2} indicates a respiratory component. However, the hydrogen ion concentration is higher than would be expected in an acute respiratory acidosis with a P_{CO_2} at this level. Therefore, there must be a non-respiratory component to the acidosis.

On the basis of these data alone, it is not possible to determine whether the respiratory disturbance is acute or chronic. These results could, for example, represent the results of the concurrent development of an acute respiratory and a non-respiratory acidosis. On the other hand, they are also compatible with the presence of severe non-respiratory acidosis in a patient with chronic carbon dioxide retention. Given that the patient is known to suffer from chronic bronchitis, the second interpretation is more likely.

Mixed acid–base disturbances are common and appear complex. Correct diagnosis requires a logical approach and a clear understanding both of the relevant pathophysiology and of the quantitative relationships between hydrogen ion concentration and P_{CO_2}. The biochemical changes which are characteristic of the various acid–base disturbances are shown in Fig.4.12. With this physiological approach, calculated parameters such as 'standard bicarbonate' and 'base excess' are obsolete.

The standard bicarbonate is a calculated estimate of the bicarbonate concentration that would be present if the P_{CO_2} were normal and thus reflects only non-respiratory influences on bicarbonate. The base excess is a calculated estimate of the non-respiratory influences on total buffering capacity. These parameters were introduced with a view to distinguishing between the respiratory and non-respiratory components in acid–base disorders, but they take no account of normal physiological responses. An abnormal standard bicarbonate or base excess indicates the presence of a non-respiratory acidosis or alkalosis. It does not, however, indicate whether this is either part of a mixed disturbance of acid–base homoeostasis or related to normal physiological compensation.

Hypoxaemia	
Cause	**Mechanism**
Alveolar hypoventilation depression of respiratory centre neuromuscular disease	low alveolar P_{O_2}
Venous-to-arterial shunt cyanotic congenital heart disease	dilution of arterial blood (high P_{O_2}) with venous blood (low P_{O_2})
Impaired diffusion pulmonary fibrosis pulmonary oedema	inadequate arterial oxygenation despite normal alveolar P_{O_2}
Ventilation/perfusion imbalance chronic obstructive airways disease	blood perfuses non-aerated areas of lung and is not oxygenated

Fig.4.13 Causes and mechanisms of hypoxaemia.

OXYGEN TRANSPORT AND ITS DISORDERS

In patients with respiratory disorders, a disturbance of the partial pressure of oxygen (P_{O_2}) may be of greater clinical significance than either an abnormal P_{CO_2} or abnormal hydrogen ion concentration. Although both oxygen and carbon dioxide are transported between the alveoli and the bloodstream, albeit in opposite directions, their respective partial pressures do not necessarily change in a reciprocal fashion. There are two reasons for this: first, carbon dioxide is generally more diffusible than oxygen, with the result that, in pulmonary oedema and interstitial lung disease, hypoxaemia develops but the P_{CO_2} may not increase; and secondly, very little oxygen is carried in physical solution in the blood while haemoglobin is normally nearly fully saturated with oxygen. Hyperventilation cannot, therefore, increase the arterial P_{O_2} significantly, but can reduce the P_{CO_2}.

A raised P_{O_2} is only seen in patients given supplementary oxygen which results in an increased inspired P_{O_2}. There are many causes of hypoxaemia (Fig.4.13). The reasons for the hypoxaemia associated with hypoventilation, venous-to-arterial shunting and impaired diffusion are self-evident. However, in many respiratory diseases, such as lung collapse and pneumonia, there is an imbalance between ventilation and perfusion of the alveoli. Blood leaving poorly ventilated, well perfused alveoli will have a low P_{O_2} and a raised P_{CO_2}. The effect on P_{CO_2} can be compensated in normally perfused and ventilated alveoli by hyperventilation. This removes additional carbon dioxide, but cannot compensate for the low P_{O_2} since the haemoglobin will be fully saturated and thus the amount of oxygen carried cannot be increased. The poorly perfused alveoli are effectively dead space. With moderate degrees of ventilation/perfusion imbalance, the P_{O_2} is reduced and the P_{CO_2} is either normal or even reduced. With severe imbalance, hyperventilation cannot compensate through increased removal of carbon dioxide from normally ventilated and perfused alveoli and the P_{CO_2} becomes elevated.

SUMMARY

The body is a net producer of hydrogen ions and also of carbon dioxide which can be hydrated to form carbonic acid. Hydrogen ion homoeostasis depends upon buffering in the tissues and bloodstream, excretion of hydrogen ions in urine and removal of carbon dioxide through the lungs in the expired air. The hydrogen ion concentration of the blood is directly proportional to the partial pressure of carbon dioxide (PCO_2) and inversely proportional to the concentration of bicarbonate, the principal extracellular buffer.

In an acidosis, the hydrogen ion concentration is greater than normal; in an alkalosis, it is below normal. Both acidosis and alkalosis can be either primarily respiratory or non-respiratory in origin. Thus respiratory acidosis is the result of carbon dioxide retention and respiratory alkalosis of excessive carbon dioxide loss. A non-respiratory, or metabolic, acidosis is the result of either increased hydrogen ion production or decreased excretion or both, while a non-respiratory or metabolic alkalosis is usually a result of excessive hydrogen ion excretion from the body.

Compensatory mechanisms may counteract a tendency to acidosis or alkalosis. Compensation in effect results from the physiological generation of an opposing disturbance. Therefore, in a respiratory acidosis, compensation is through increased renal hydrogen ion excretion which, in a normal subject, would produce a non-respiratory alkalosis. Ultimate correction of a disturbance of hydrogen ion homoeostasis usually requires correction of the underlying cause.

Mixed disturbances, with respiratory and non-respiratory components, occur frequently. Even in these cases, a diagnosis can be made based on clinical assessment and logical consideration of the arterial hydrogen ion concentration and partial pressure of carbon dioxide.

FURTHER READING

Cohen JJ & Kassiven JP (1982) *Acid-Base*. Boston: Little Brown & Company.

Kurtzman NA & Batalle DC (eds) (1983) Acid-base disorders. *The Medical Clinics of North America*, **67**, 751–932.

5. The Kidney

INTRODUCTION

The kidneys have three major functions: (i) excretion of waste, (ii) maintenance of extracellular fluid (ECF) volume and composition and (iii) hormone synthesis. Each kidney consists of approximately one million functional units, the nephrons (Fig.5.1).

The kidneys have a rich blood supply and normally receive about twenty-five percent of the cardiac out-put. Most of this is distributed initially to the capillary tufts of the glomeruli which act as high pressure filters. Blood is separated from the lumen of the nephron by three layers: the capillary endothelial cells, the basement membrane and the epithelial cells of the nephron. The endothelial and epithelial cells are in intimate contact with the basement membrane; the

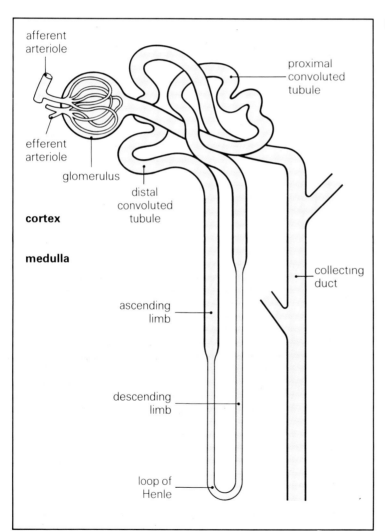

Fig.5.1 Structure of a nephron.

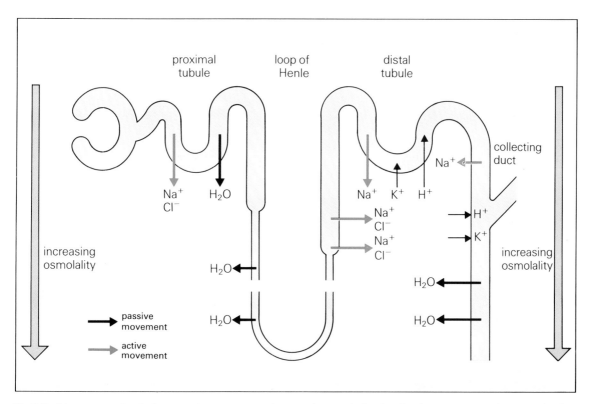

Fig.5.2 Movements of major ions, passive movement of water and changes in osmolality in the nephron. In the ascending loop of Henle, chloride ions are actively transported and sodium ions accompany them to maintain electrochemical neutrality.

cells are fenestrated so that the membrane is exposed to blood on one side and to the lumen of the nephron on the other.

The glomerular filtrate is an ultrafiltrate, which means that it has a similar composition to plasma except that it is almost free of proteins. This is because the endothelium provides a barrier to red and white blood cells and the basement membrane, though permeable to water and low molecular weight substances, is largely impermeable to macromolecules. This impermeability is related to both molecular size and electrical charge. Proteins with molecular weights lower than albumin (68,000 daltons) are filterable; negatively charged molecules are less easily filtered than those bearing a positive charge. Almost all the protein entering the glomerular filtrate is reabsorbed in the proximal convoluted tubule with the result that normal urinary protein excretion is less than 150mg/24h.

Filtration is a passive process. The total filtration rate of the kidneys is mainly determined by the difference between the blood pressure in the glomerular capillaries and the hydrostatic pressure in the lumen of the nephron, by the nature of the glomerular basement membrane and by the number of glomeruli. The normal glomerular filtration rate (GFR) is approximately 120ml/min, equivalent to a volume of about 170l/day. However, urine production is only 1–2l/day, depending on fluid intake; the bulk of the filtrate is reabsorbed further along the nephron.

Much of the glomerular filtrate is reabsorbed in the proximal convoluted tubule. Under normal circumstances, all the glucose, amino acids, potassium and bicarbonate, and about seventy-five percent of the sodium, is reabsorbed isotonically here by energy-dependent mechanisms.

Medullary hyperosmolality, which is vital for the further reabsorption of water, is generated by the counter-current system, summarized in Fig.5.2. Chloride ions, accompanied by sodium, are pumped out of the ascending limb of the loop of Henle into the surrounding interstitial fluid, and thence diffuse into the descending limb. Since the ascending limb of the loop of Henle is impermeable to water, the net effect is an exchange of sodium and chloride ions between the ascending and descending limbs. This alters the osmolality of both the fluid within the nephron and the surrounding interstitial fluid. A gradient of osmolality is set up between the isotonic corticomedullary junction and the extremely hypertonic (approximately 1200mmol/l) deep medulla.

The tubular fluid becomes increasingly dilute as it passes up the ascending limb of the loop of Henle, as a result of the continued removal of chloride and sodium ions. Fluid entering the distal convoluted tubule is therefore hypotonic (approximately 150mmol/l) with respect to the glomerular filtrate.

Approximately ninety percent of the filtered sodium and eighty percent of the filtered water has been reabsorbed from the glomerular filtrate by the time it reaches the beginning of the distal convoluted tubule. In the distal tubule, further sodium reabsorption takes place, controlled by aldosterone, while potassium and hydrogen ions are secreted. Ammonia is also secreted in the distal tubule and buffers hydrogen ions, being excreted as ammonium ions (see page 37).

Whereas the proximal tubule is responsible for bulk reabsorption of the glomerular filtrate, the distal tubule exerts a fine control over the composition of the tubular fluid, depending upon the requirements of the body.

Tubular fluid then passes into the collecting ducts which extend through the hypertonic renal medulla and discharge urine into the renal pelvis. The cells lining the collecting ducts are normally impermeable to water. Vasopressin (antidiuretic hormone or ADH) renders them permeable allowing water to be reabsorbed passively due to the osmotic gradient between the duct lumen and the interstitial fluid. Thus, in the absence of vasopressin, a dilute urine is produced; in its presence, the urine is concentrated. Some reabsorption of sodium also occurs in the collecting ducts under the stimulus of aldosterone.

Since the normal GFR is approximately 120ml/min, a volume of fluid equivalent to the entire ECF is filtered every two hours. Disease processes affecting the kidney therefore have an enormous potential for affecting water, salt and hydrogen ion homoeostasis and the excretion of waste products.

The kidneys are also important endocrine organs, producing renin, erythropoietin and calcitriol. The secretion of these hormones may be altered in renal disease. In addition, several other hormones are either inactivated or excreted by the kidneys and hence their concentrations in the blood can also be affected by renal disease.

BIOCHEMICAL TESTS OF RENAL FUNCTION

Formal tests of tubular function are used less frequently than those of glomerular function. In glomerular disease, there can either be a change in GFR (usually a decrease) or glomerular permeability (usually an increase), or both. Tests of glomerular function are, therefore, used either to measure the rate of filtration or to assess permeability. The GFR declines with age, to a greater extent in males than females, and this must be taken into account when interpreting results.

Measurement of glomerular filtration rate (GFR)

Clearance

Because urine is derived from the glomerular filtrate, measurement of the GFR is extremely important in the assessment of renal function. An estimate of the GFR can be made by measuring the urinary excretion of a substance which is completely filtered from the blood by the glomeruli and which is not secreted, reabsorbed or metabolized by the renal tubules. Experimentally, inulin has been found to meet these requirements. The volume of blood from which inulin is cleared or completely removed in one minute is known as the inulin clearance and is equal to the GFR.

Measurement of inulin clearance requires the infusion of inulin into the blood and is therefore not suitable for routine clinical use. The clearance of creatinine, an endogenous substance normally present in the blood and excreted after glomerular filtration, may be measured instead (Equation 5.1).

$$(\textbf{5.1}) \quad \text{Clearance} = \frac{U \times V}{P} \text{ ml/min}$$

U = urinary creatinine concentration (μmol/l)
V = urine volume [ml/min or (l/24h)/1.44]
P = serum creatinine concentration (μmol/l)

Creatinine is derived largely from the turnover of creatine phosphate in muscle and the daily production is relatively constant, being a function of total muscle mass. A small amount is derived from meat in the diet. Creatinine clearance in adults is normally of the order of 120ml/min, corrected to a standard body surface area of $1.73m^2$.

The accurate measurement of creatinine clearance is difficult, especially in outpatients, since it is necessary to obtain a complete and accurately timed sample of urine. The usual collection time is twenty-four hours, but patients may forget the time or may forget to include some urine in the sample. Incontinent patients may find it impossible to make a urine collection. Patients have been known to add water or some other person's urine to their own collection, hoping to gain the doctor's approval for having been so prolific.

Another problem is that creatinine is actively secreted by the renal tubules and, as a result, the creatinine clearance is higher than the true GFR. This is of considerable significance when the GFR is very low (less than 10ml/min) since tubular secretion can then make a major contribution to creatinine excretion. Creatinine breakdown in the gut also becomes significant when the GFR is very low. Furthermore, in the calculation of creatinine clearance, two measurements of creatinine concentration and one of urine volume are required. Each of these has an inherent imprecision which contributes to the inaccuracy of the overall result.

Although measurements of creatinine clearance are made frequently in clinical chemistry laboratories, they are potentially unreliable and should not be carried out unless there is a definite indication. In fact, accurate measurement of the GFR is required infrequently. Indications for its measurement include: assessment of potential kidney donors; investigation of patients with minor abnormalities of renal function; and calculation of the initial dose of a potentially toxic drug that is eliminated from the body by renal excretion. The majority of patients with established renal disease do not require repeated measurements of creatinine clearance. In most cases, their renal function may be more reliably monitored by serial measurements of the serum creatinine concentration (see following).

In hospitals with facilities for handling radioactive isotopes and measuring radioactivity, the technique of choice for measuring GFR is based on the injection of ^{51}Cr-labelled EDTA (ethylenediaminotetra-acetic acid). This substance is completely filtered by the glomeruli and is neither secreted nor reabsorbed by the tubules. Serial blood samples are taken after injection of the isotope and the GFR can be calculated from the rate of fall of serum radioactivity as the isotope is cleared.

Serum creatinine

The serum creatinine concentration is the most reliable, simple biochemical test of glomerular function. Ingestion of meat can increase the serum creatinine concentration by as much as thirty percent seven hours after a normal meal and ideally blood samples should be collected after an overnight fast. Strenuous exercise also causes a mild transient increase in serum creatinine concentration. Serum creatinine concentration is related to muscle bulk and therefore a concentration of 120μmol/l could be normal for an athletic young man but would suggest renal functional impairment, though not necessarily of clinical significance, in a thin, seventy-year-old woman. Although muscle bulk tends to decline with age, so too does the GFR and hence the serum creatinine levels remain fairly constant.

The normal range for serum creatinine in the adult population is 60–120μmol/l, but the day-to-day variation in an individual is much less than this range. In the clearance formula (see *Equation 5.1*), the GFR is inversely related to the serum creatinine. Consequently, a normal serum creatinine does not necessarily imply normal renal function, although a raised creatinine does usually indicate impaired renal function (Fig.5.3). Further, a change in serum creatinine concentration, provided it is outside the limits of analytical error, usually indicates a change in GFR.

Changes in serum creatinine concentration can occur, independently of renal function, due to changes in muscle mass. Thus a decrease can occur in starvation and wasting diseases, immediately after surgery and in patients treated with corticosteroids; an increase can occur during re-feeding. However, changes in creatinine concentration for these reasons rarely lead to diagnostic confusion.

In pregnancy the GFR increases. This usually more than balances the effect of increased creatinine synthesis during pregnancy and results in a decrease in serum creatinine concentration.

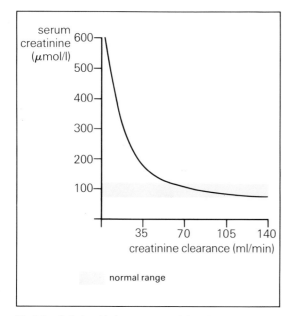

Fig.5.3 Relationship between creatinine clearance and serum creatinine concentration.

Serum urea

Urea is synthesized in the liver, primarily as a by-product of the deamination of amino acids. Its elimination in the urine represents the major route for nitrogen excretion. It is filtered from the blood at the glomerulus but passive tubular reabsorption occurs to a significant extent, especially at low rates of urine flow.

Although serum urea concentration is often used as an index of renal glomerular function, measurement of serum creatinine provides a more accurate assessment. Urea production is increased by a high protein intake, in catabolic states, and by the absorption of amino acids and peptides after gastrointestinal haemorrhage. Conversely, production is decreased in patients with a low protein intake and sometimes in patients with liver disease. As a consequence of tubular reabsorption, serum urea concentration rises during dehydration even when renal function is normal.

Factors affecting the ratio of serum urea to creatinine are summarized in Fig.5.4. Changes in serum urea are a feature of renal impairment but it is important to consider possible extrarenal influences on urea concentrations before ascribing any such changes to an alteration in renal function.

Urea diffuses readily across dialysis membranes and during renal dialysis a fall in serum urea concentration is a poor guide to the efficacy of the process in removing other toxic substances from the blood.

Serum β₂-microglobulin

β_2-microglobulin is a small peptide (molecular weight 11,800 daltons) which forms part of the class I proteins of the major histocompatibility complex. It is present on the surface of most cells and in low concentrations in the plasma. It is completely filtered by the glomerulus and is reabsorbed and catabolized by proximal tubular cells.

The serum concentration of β_2-microglobulin is a good index of GFR in normal people, being unaffected by diet or muscle mass. However, it is increased in certain malignancies and inflammatory diseases. Since it is normally reabsorbed and catabolized in the tubules, measurement of β_2-microglobulin excretion by radioimmunoassay is a sensitive, if expensive, method of assessing tubular integrity.

Assessment of glomerular integrity

Impairment of glomerular integrity results in the filtration of large molecules which are normally retained and is manifest as proteinuria. Proteinuria can, however, occur for other reasons (see page 68).

With severe glomerular damage, red blood cells are detectable in the urine (haematuria). Whilst haematuria can occur as a result of lesions anywhere in the urinary tract, the red cells often have an abnormal morphology in glomerular disease. The presence of red cell casts (cells embedded in a proteinaceous matrix) in urinary sediment is strongly suggestive of glomerular dysfunction.

Tests of renal tubular function

Formal tests of renal tubular function are performed less frequently than tests of glomerular function. The presence of glycosuria in a subject with a normal blood glucose level implies proximal tubular malfunction which may be either isolated (renal glycosuria) or part of a generalized tubular defect (Fanconi syndrome). Aminoaciduria can occur with tubular defects and can be investigated by amino acid chromatography. Tests of proximal tubular bicarbonate

Causes of abnormal serum urea to creatinine ratio	
Increased	**Decreased**
high protein intake	low protein intake
gastrointestinal bleeding	dialysis
hypercatabolic state	muscle wasting
dehydration	amputation
urinary stasis	severe liver disease

Fig.5.4 Causes of an abnormal serum urea to creatinine ratio.

reabsorption are required in the assessment of proximal renal tubular acidosis.

The only tests of distal tubular function in widespread use are the water deprivation test, to assess renal concentrating ability (see page 120) and tests of urinary acidification, to diagnose distal renal tubular acidosis (see page 71).

RENAL DISORDERS

Failure of renal function may occur rapidly, producing the syndrome of acute renal failure. This is potentially reversible since, if the patient survives the acute illness, normal renal function can be regained. However, chronic renal failure, more accurately termed 'end-stage' renal failure, develops insidiously, often over many years, and is irreversible. Patients with this condition will eventually either require renal dialysis for the rest of their lives or a renal transplant in order to survive.

The term 'glomerulonephritis' encompasses a group of renal diseases which are characterized by pathological changes in the glomeruli of an immunological basis, such as immune complex deposition. Glomerulonephritis may present in many ways: for example, as an acute nephritic syndrome with haematuria, hypertension and oedema; as acute or end-stage renal failure; or as proteinuria leading to the nephrotic syndrome (proteinuria, hypoproteinaemia and oedema).

Many disorders primarily affect renal tubular function, but most are rare. Their metabolic and clinical consequences range from the trivial, such as isolated renal glycosuria, to the very severe, for example, cystinuria.

Acute renal failure

Acute renal failure is characterized by a rapid loss of renal function, with retention of urea, creatinine, hydrogen ions and other metabolic products and, usually but not always, oliguria (<400ml urine/24h). Although potentially reversible, the consequences to homoeostatic mechanisms are so profound that this condition continues to be associated with a high mortality. Furthermore, acute renal failure often develops in patients who are already severely ill.

Acute renal failure is conventionally divided into three categories, according to whether renal functional impairment is related to a decrease in renal blood flow (prerenal), to intrinsic damage to the kidneys (intrarenal) or to urinary tract obstruction (postrenal). Should any of these occur in a patient whose renal function is already impaired, the consequences are likely to be more serious.

The term 'uraemia' is often used as a synonym for renal failure but, although it is a characteristic feature of renal failure, it is an unreliable indicator of its severity. 'Azotaemia' is used in a similar context and refers to an increase in the blood concentration of all nitrogenous compounds.

Prerenal acute renal failure

This is caused by circulatory insufficiency, as may occur with severe haemorrhage, burns or fluid loss, and leads to renal hypoperfusion and a decrease in GFR. Renal hypoperfusion induces intense renal vasoconstriction; there is a redistribution of renal blood flow which results in a decrease in the GFR with preservation of tubular function. If left untreated,

	Prerenal failure	Intrinsic failure
Urine sodium concentration	<20mmol/l	>40mmol/l
Urine:serum urea concentration	>10:1	<3:1
Urine:serum osmolality	>1.5:1	<1.1:1

Fig.5.5 Biochemical values in oliguria due to prerenal and intrinsic renal failure.

prerenal uraemia may progress to intrinsic renal failure (acute tubular necrosis). This may be prevented, however, if renal perfusion can be restored before structural damage has occurred.

Prerenal uraemia is essentially the result of a normal physiological response to hypovolaemia or a fall in blood pressure. Stimulation of the renin-angiotensin-aldosterone system and of vasopressin secretion, results in the production of a small volume of highly concentrated urine with a low sodium concentration. Renal tubular function is normal, but the decreased GFR results in the retention of substances normally excreted by filtration, such as urea and creatinine. The decreased delivery of sodium to the distal tubule impairs hydrogen ion and potassium excretion; acidosis and hyperkalaemia are characteristic features of acute renal failure.

CASE HISTORY 5.1

A young man sustained multiple injuries in a motorcycle accident. He received blood transfusions and underwent surgery; twenty-four hours after admission he had only passed 500ml of urine. He was clinically dehydrated and his blood pressure was 90/50mmHg.

Investigations

serum: potassium   5.6mmol/l
        urea   21.0mmol/l
        creatinine   140μmol/l

urine: sodium   5mmol/l
        urea   480mmol/l

Comment

The diagnosis is prerenal uraemia. The urine contains little sodium and the urea has been concentrated by a factor of twenty-two. These are normal physiological responses, implying that intrinsic renal function is intact and that the ability of the kidneys to function normally is constrained only by hypoperfusion. Osmolality was not measured, but in prerenal uraemia the urine to plasma osmolality ratio is characteristically greater than 1.5:1.

The distinguishing features of prerenal as opposed to intrinsic renal failure are listed in Fig.5.5; some overlap may be encountered in practice. All these values are invalidated by the use of diuretics and osmolalities are invalidated by X-ray contrast media. A concentrated, sodium-poor urine is a more reliable indicator of prerenal uraemia than a dilute sodium-containing urine is of intrinsic renal failure, since the latter is appropriate for a well-hydrated healthy person. However, oliguria, although usually present, is not a constant feature of renal failure.

The increase in serum urea concentration in this patient is greater than the increase in creatinine. This is due in part both to passive reabsorption of urea and to increased synthesis from amino acids released as a result of tissue damage. The patient was given extra fluid intravenously and this resulted in a diuresis. His serum urea and creatinine were normal forty-eight hours later.

Intrinsic acute renal failure	
Causes	**Examples**
specific renal diseases and systemic disease affecting kidneys	rapidly progressive glomerulonephritis systemic lupus erythematosus
nephrotoxins	aminoglycosides cis-platinum
renal hypoperfusion	hypotension haemorrhage septicaemia low cardiac output
obstruction	Bence–Jones protein prostatic hypertrophy

Fig.5.6   Causes of acute renal failure.

Intrinsic acute renal failure

A wide variety of conditions can cause intrinsic acute renal failure (Fig.5.6). Most cases are due to either nephrotoxins, including several drugs such as aminoglycosides and some cephalosporins, or renal ischaemia, both of which lead to renal tubular necrosis. Less common causes of acute renal failure include specific renal diseases and systemic conditions in which renal involvement is a prominent feature. The pathogenesis of this condition is incompletely understood and in any individual case several factors may be important.

Although glomerular damage is uncommon in acute tubular necrosis, glomerular function is affected because of decreased perfusion and increased hydrostatic pressure in the tubules, for example, due to oedema and obstruction of tubule lumina with debris.

There are typically three phases to the course of acute tubular necrosis: (i) initial oliguric phase; (ii) diuretic phase, in which urine output increases as the GFR increases but functional abnormalities persist; and (iii) recovery phase during which normal function returns.

In acute renal failure there is often a history of severe trauma, bleeding in relation to surgery, sepsis or the use of potentially nephrotoxic drugs. Although in prerenal uraemia the urine is concentrated, after the development of tubular damage the ability of the tubules to concentrate urine is lost and the ionic composition of the urine tends to resemble that of plasma. Proteinuria is always present and the urine is often dark due to the presence of haem pigments from the blood.

The characteristic biochemical changes in the serum in acute renal failure are summarized in Fig.5.7.

Biochemical changes in serum in acute oliguric renal failure	
Increased	**Decreased**
potassium urea creatinine phosphate magnesium hydrogen ion urate	sodium bicarbonate calcium

Fig.5.7   Biochemical changes in serum in acute oliguric renal failure.

Hyponatraemia is common; contributory factors include increased water formation from oxidative metabolism, continued intake of water or injudicious fluid administration, decreased excretion and, possibly, loss of intracellular solute. Hyperkalaemia occurs as a result of a decreased excretion of potassium. This is combined with both a loss of intracellular potassium to the ECF (due to tissue breakdown) and intracellular buffering of retained hydrogen ions. Decreased hydrogen ion excretion causes a non-respiratory acidosis.

Retention of phosphate and leakage of intracellular phosphate into the interstitial fluid leads to hyperphosphataemia which inhibits the 1α-hydroxylation of 25-hydroxycholecalciferol to calcitriol (see page 193). The resulting decreased plasma concentration of calcitriol leads to impaired calcium absorption from the gut and thus to hypocalcaemia. Although parathyroid hormone (PTH) secretion is increased as a result of the hypocalcaemia, skeletal resistance to the action of this hormone prevents correction of the hypocalcaemia. Hypermagnesaemia is also often present as a result of decreased magnesium excretion.

If patients survive the acute illness, the urine output usually begins to increase after a period of time, which may be only a few days but can be several weeks (mean 10–11 days), and indicates the onset of the diuretic phase. This is due to an increase in GFR and initially there is little improvement in tubular function. Therefore, the composition of the urine is similar to that of protein-free plasma. During this phase, the urine volume may exceed five litres per day and, because of its high ionic concentration, there is a considerable risk of dehydration and sodium and potassium deficiency.

Although the onset of the diuretic phase often heralds clinical improvement, serum concentrations of urea and creatinine do not fall immediately since the GFR is still low and, therefore, insufficient to allow excretion of the surplus. The persisting high urea concentration in the blood, and hence in the glomerular filtrate, contributes to the diuresis by an osmotic effect. The acidosis also persists until tubular function is restored. The serum calcium concentration may rise during this phase, particularly after crush injuries, due to the release of calcium from damaged muscle. Calcium absorption from the gut may increase as calcitriol synthesis resumes, stimulated by the temporarily raised serum level of PTH.

Gradually, in the recovery phase, as the tubular cells regenerate and tubular function is restored, the diuresis subsides and the various abnormalities of renal function resolve. Patients who survive the acute illness usually recover completely. Some residual impairment of renal function is often demonstrable but it is not usually of clinical significance and may not be apparent from simple tests.

In very severe cases of acute renal failure, renal cortical necrosis may occur and there is no recovery of renal function.

CASE HISTORY 5.2

A young man was admitted to hospital with severe abdominal injuries after being knocked down by a car. On examination, he was severely shocked with a swollen tender abdomen. He was given intravenous fluids and blood, and was taken to the operating theatre. At laparotomy, his spleen was found to be ruptured; splenectomy was performed. There was also mesenteric damage and a tear in the duodenum; the damaged gut was resected.

Three days later he became hypotensive and pyrexial and was taken back to theatre. Free fluid was present in the peritoneal cavity and a leak was found in a segment of gangrenous small intestine. Appropriate surgical procedures were performed. Following this, the patient became oliguric despite adequate hydration.

Investigations

serum:	sodium	128mmol/l
	potassium	5.9mmol/l
	bicarbonate	16mmol/l
	urea	22.0mmol/l
	creatinine	225μmol/l
	calcium	1.72mmol/l
	phosphate	2.96mmol/l
	albumin	28g/l
urine:	urea	50mmol/l
	sodium	80mmol/l

Comment

These findings are typical of acute renal failure in a septic, catabolic patient (see Fig.5.7). He was treated with regular haemodialysis and

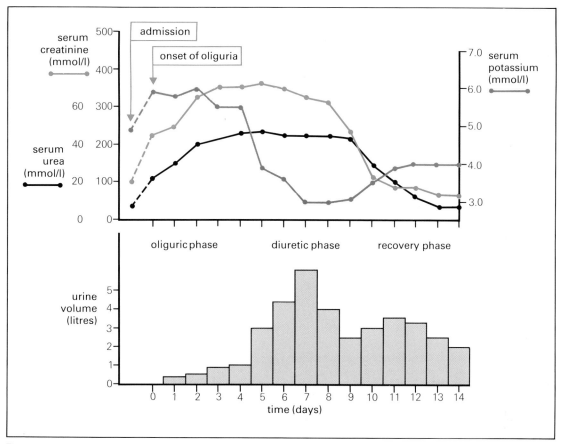

Fig.5.8 Serum urea, creatinine and potassium in a case of acute renal failure (see *Case History 5.2*).

parenteral nutrition; antibiotics were continued and his pyrexia settled. Eight days after the accident, the patient's urine output began to increase as shown in Fig.5.8. The biochemical changes that occurred before and after the diuretic phase, until recovery of normal renal function, are also shown in Fig.5.8.

Postrenal renal failure

Obstruction to the flow of urine leads to an increase in hydrostatic pressure which acts in opposition to glomerular filtration and, if prolonged, leads to secondary renal tubular damage. Obstruction which occurs above the level of the vesicourethral junction must be bilateral to have a major effect on urine flow. Com-

plete anuria is rare even in acute tubular necrosis and therefore it is strongly indicative of the presence of obstruction. More often, however, obstruction is either intermittent or incomplete and urine production may even be normal in obstruction with overflow. The degree of reversibility of renal damage in obstructive renal failure depends to some extent on how long-standing it has been. It is more likely to be reversible if the obstruction is acute.

Management of acute renal failure

Obstruction should always be excluded in a patient with renal failure, for example, by ultrasound examination. If present, obstruction should either be relieved or, if this is not immediately possible, bypassed by an appropriate procedure.

End-stage renal disease		
Metabolic features	**Biochemical changes in serum**	
impairment of urinary concentration and dilution impairment of electrolyte and hydrogen ion homoeostasis retention of waste products of metabolism impaired vitamin D metabolism decreased erythropoietin synthesis	Increased potassium urea creatinine hydrogen ion phosphate magnesium	Decreased sodium bicarbonate calcium

Fig.5.9 Metabolic and biochemical consequences of end-stage renal disease.

Many cases of intrinsic renal failure are preventable and, if a patient is judged to be in the prerenal phase, it is important to attempt to halt the progression to acute tubular necrosis by measures to expand the ECF volume and improve renal perfusion. Volume repletion should preferably be monitored by measurements of central venous pressure. Additional measures that can be employed include: the judicious use of mannitol, an osmotic diuretic; frusemide, a loop diuretic (both of these are also renal vasodilators); and dopamine, at a dose appropriate for a renal effect, to promote renal vasodilation.

If oliguria persists and acute tubular necrosis is diagnosed, it becomes necessary to minimize the severe adverse consequences of renal failure. The general principles of treatment include: strict control of sodium and water intake, to prevent overload; nutritional support, with some limitation of protein but adequate carbohydrate to minimize endogenous protein breakdown; prevention of metabolic complications, such as hyperkalaemia and acidosis; and prevention of infection.

When renal failure is short-lived, conservative measures alone may suffice. However, the majority of patients will require either peritoneal dialysis or extracorporeal haemodialysis. Specific indications for dialysis include: a dangerously high plasma potassium concentration, fluid overload, severe acidosis, a rapidly rising serum urea concentration or a general deterioration in the patient's condition.

Dialysis may have to be continued into the early part of the diuretic phase when the GFR has recovered sufficiently for the serum concentration of creatinine

to start falling. The main problem during the diuretic phase is to supply sufficient water and electrolytes to compensate for the excessive losses. Fluid replacement should not automatically be isovolaemic since the diuresis is partly due to mobilization and excretion of excess ECF. From the onset of acute renal failure until its resolution, it is essential to monitor the patient's serum creatinine, sodium, potassium, bicarbonate, calcium and phosphate, and to monitor urinary volume and sodium and potassium excretion.

The general principles of management are similar whatever the cause of acute renal failure. In addition, specific measures may be indicated for certain diseases, for example, the control of infection or hypertension and the use of immunosuppressive drugs in immunologically mediated renal disease.

End-stage (chronic) renal failure

Many disease processes can lead to progressive impairment of renal function, but they all effectively result in a decrease in the number of functioning nephrons. Thus, the major pathological and clinical features tend to be similar in all patients with end-stage renal failure.

Consequences

The important metabolic features of chronic renal failure are summarized in Fig.5.9. Although there is impairment of urinary concentration, polyuria is never

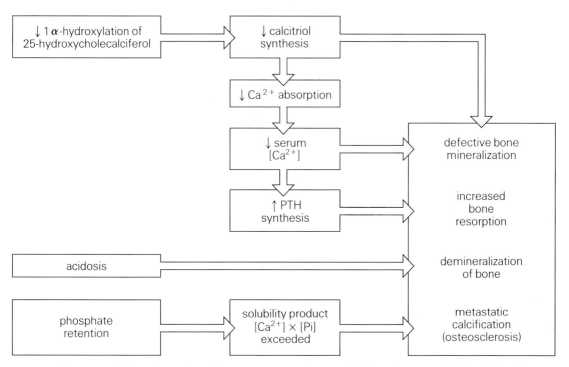

Fig.5.10 Pathogenesis of renal osteodystrophy. Aluminium toxicity may also be an important factor in patients on dialysis or those treated with oral aluminium salts.

gross, that is, not more than four litres per day, because the GFR is too low. The urine tends to be of a fixed specific gravity. The lack of urinary concentration is particularly noticed by the patient at night and nocturia is a common complaint. The ability to dilute the urine may be lost late in the course of renal failure and patients become very sensitive to the effects of either fluid loss or overload.

Sodium balance is usually maintained until the GFR falls below 20ml/min. The majority of patients tend to retain sodium but severe renal sodium wasting is occasionally seen. This syndrome of 'salt-losing nephritis' occurs most often in patients whose renal disease particularly affects the tubules; for example, analgesic nephropathy, polycystic disease and chronic pyelonephritis.

Hyperkalaemia is a late feature of chronic renal failure; it may be precipitated by a sudden deterioration in renal function and by the injudicious use of potassium-sparing diuretics.

Patients with chronic renal failure tend to be acidotic. The urinary buffering capacity is impaired as a result of decreased phosphate excretion and ammonia synthesis. The ability of individual nephrons to reabsorb filtered bicarbonate is often also impaired, probably, in part as an effect of the raised blood level of parathyroid hormone. However, although the plasma hydrogen ion concentration is increased and the bicarbonate is decreased, these changes progress only slowly, due to buffering of excess hydrogen ions in bone.

Most patients with chronic renal failure become hypocalcaemic and, in time, many develop renal osteodystrophy. The pathogenesis of this bone disease is complex (Fig.5.10). Renal production of calcitriol is decreased, as a result of the decrease in renal mass, and this leads to hypocalcaemia and secondary hyperparathyroidism. Buffering of hydrogen ions by bone leads to bone demineralization.

Aluminium may cause osteomalacia; it is present in dialysis fluid prepared from tap water in some areas and aluminium may be absorbed from the gut when given orally to bind phosphate and prevent hyperphosphataemia.

The increase in parathyroid hormone secretion causes a decrease in phosphate reabsorption from each nephron. However, as the GFR falls it becomes the limiting factor in phosphate excretion and hyperphosphataemia develops; this will inhibit calcitriol synthesis. If the concentration of phosphate is so high that the solubility product of calcium and phosphate ($[Ca^{2+}] \times [Pi]$) is exceeded, metastatic calcification may occur. This is seen particularly in blood vessels and also in bone where it causes sclerotic deposits.

In addition to the effect on calcitriol synthesis, other endocrine consequences of chronic renal failure include: decreased testosterone and oestrogen synthesis; abnormalities of thyroid function tests (seldom, however, associated with clinical thyroid disease); and abnormal glucose tolerance with hyperinsulinaemia due to insulin resistance. However, insulin-dependent diabetic patients who develop renal disease, often have decreased insulin requirements since insulin is metabolized in the kidney.

A normochromic normocytic anaemia is usual in end-stage renal failure, due to depression of bone marrow function by retained toxins and a decrease in the renal production of erythropoietin. A bleeding tendency may also be present and bleeding may exacerbate the anaemia.

CASE HISTORY 5.3

A fifty-six-year-old man presented to his family practitioner with weight loss, generalized weakness and lethargy of six months duration. During this time, he had been passing more urine than usual, particularly at night. He had become impotent. On examination, the patient was slightly anaemic and had a blood pressure of 180/110mmHg. His urine contained protein but no glucose. A blood sample was taken for analysis.

Investigations

serum:		
	sodium	130mmol/l
	potassium	5.2mmol/l
	bicarbonate	16mmol/l
	urea	43.0mmol/l
	creatinine	640μmol/l
	glucose (random)	6.4mmol/l
	calcium	1.92mmol/l
	phosphate	2.42mmol/l
	alkaline phosphatase	205iu/l
haemoglobin		9.1g/dl

Comment

The practitioner's first thought was that this patient had diabetes mellitus, but the absence of glycosuria militated against this and the results are typical of chronic renal failure. The history suggests slowly progressive, rather than acute, renal failure. The presence of anaemia and the raised alkaline phosphatase (due to renal osteodystrophy) are consistent with this diagnosis, although they are not specific findings.

The presence of hypertension, and the demonstration of small kidneys on abdominal radiography or ultrasonography, would also suggest that a patient had chronic renal disease.

Many other clinical features may be present in patients with end-stage renal disease (Fig.5.11). The causes of many of these features are unknown, but they are presumably related to the retention of toxins which cannot be excreted.

Management of chronic renal failure

Identification and subsequent treatment of the cause of chronic renal failure may prevent, or at least delay, further deterioration. In most cases, however, the cause is either unknown or untreatable and patients eventually require either maintenance dialysis or renal transplantation. Before dialysis or transplantation becomes necessary, considerable amelioration of symptoms and biochemical abnormalities can be obtained by conservative measures.

Since the kidneys become unable to control water and sodium balance, it is essential that intake is matched to the obligatory losses. Diuretics are often used to promote sodium excretion since adequate dietary salt restriction may be unacceptable to the patient. Bicarbonate can be given orally to control acidosis. Hyperkalaemia is uncommon, except in advanced disease; it can usually be controlled with oral ion-exchange resins containing sodium or calcium.

Hyperphosphataemia can be controlled by giving aluminium or magnesium salts by mouth. Osteodystrophy can be prevented, or treated if it does develop, by giving calcitriol or other 1α-hydroxylated derivatives of vitamin D, but care is necessary to avoid provoking hypercalcaemia.

Some limitation of dietary protein is beneficial to

Clinical features of chronic renal failure
Neurological lethargy peripheral neuropathy **Musculoskeletal** growth failure bone pain myopathy **Gastrointestinal** anorexia hiccough nausea and vomiting **Cardiovascular** anaemia hypertension pericarditis **Dermal** pruritus pallor purpura **Genitourinary** nocturia impotence

Fig.5.11 Clinical features of chronic renal failure.

reduce the formation of nitrogenous waste products, but the limitation should not usually be so severe as to cause negative nitrogen balance. In patients who are not candidates for maintenance dialysis or transplantation, however, a very low protein intake can cause considerable symptomatic improvement in the terminal stage of renal failure, and may even slow the rate of decline in renal function. It is important to maintain an adequate intake of carbohydrate and fat.

Conservative measures must be continued in patients on maintenance dialysis. Patients having a successful renal transplant are freed of such constraints, although they will require immunosuppressive therapy to prevent graft rejection.

Proteinuria and the nephrotic syndrome

The glomeruli normally filter 7–10g of protein per twenty-four hours, but almost all is reabsorbed by endocytosis and subsequently catabolized in the proximal tubule. Normal urinary protein excretion is less than 150mg/24h. Approximately half of this is Tamm–Horsfall protein, a glycoprotein secreted by tubular cells; less than 35mg is albumin.

The presence or absence of proteinuria is usually assessed using a reagent-impregnated strip (dip-stick) which is dipped into the urine. This reliably detects albumin at concentrations greater than 200mg/l and is less sensitive to other proteins. False positive (FP) results are obtained with urine that is alkaline, contaminated by various antiseptics or contains X-ray contrast media. It should be appreciated that a particular concentration of protein will be of more significance in a dilute urine than in a concentrated urine, since it represents a greater total excretion in a given period.

The mechanisms of proteinuria are summarized in Fig.5.12. Glomerular proteinuria may be sufficiently gross to cause hypoproteinaemia and oedema (the nephrotic syndrome).

Investigation of proteinuria

If a patient's urine gives a positive reaction for protein using a dip-stick, the presence of protein should be confirmed by an independent test in the laboratory. If Bence–Jones proteinuria is suspected, a specific test must be used since this protein is not detected by dip-stick. Before investigating renal function, incidental extrarenal causes of proteinuria such as fever, strenuous exercise and burns, should be excluded; such proteinuria is usually not of long-term significance.

When the presence of proteinuria has been confirmed, urinary protein excretion should be measured and simple tests of renal function performed. If these test results are normal and protein excretion is less than 500mg/24h, the patient need not be subjected to further investigation but there should be follow-up. With protein excretion in excess of this, or with abnormal test results, further investigation, such as ultrasound examination, contrast radiography or biopsy, is necessary to determine the cause.

Orthostatic proteinuria is a benign condition in which proteinuria is present only when subjects are upright. It occurs in approximately five percent of young adults and can be induced in many more if they

Mechanisms of proteinuria

Overflow

due to presence in serum
of a high concentration of a
low molecular weight protein,
which is filtered in a quantity
exceeding tubular reabsorptive
capacity, e.g. Bence–Jones protein

Glomerular

due to increased glomerular
permeability, e.g. albumin

Tubular

due to impaired or saturated
reabsorption of protein filtered by
normal glomeruli, e.g. β_2-microglobulin

Secreted

due to secretion by kidneys or
epithelium of urinary tract,
e.g. Tamm–Horsfall protein

Fig.5.12 Mechanisms of proteinuria.

glomerular proteinuria, higher molecular weight proteins are present. Electrophoresis of concentrated urine is the best technique for the detection of Bence–Jones proteinuria.

The ratio of the clearances of a high and a low molecular weight protein (such as IgG and albumin) gives an indication of the nature of glomerular damage in glomerular proteinuria. This ratio, expressed as a percentage, is known as the selectivity index (Equation 5.2).

$$(5.2) \quad \text{Selectivity index} = \frac{\text{IgG clearance}}{\text{albumin clearance}} \times 100$$

$$= \frac{U_{IgG}/P_{IgG}}{U_{alb}/P_{alb}} \times 100$$

where U and P refer to urinary and plasma concentrations.

An index of less than fifteen percent implies high selectivity and an index of greater than thirty percent implies poor selectivity because more high molecular weight proteins are passing through the nephron. Many types of glomerular damage give rise to poor selectivity, but in minimal change glomerulonephritis, a common cause of the nephrotic syndrome in childhood, the selectivity index is characteristically less than fifteen percent. This form of glomerulonephritis is often responsive to corticosteroids. The finding of a highly selective proteinuria in a child is thus not only of prognostic use, but it also obviates the need to perform a renal biopsy to determine the underlying cause of the proteinuria. However, in adults the selectivity index is not a useful investigation because minimal change glomerulonephritis is an uncommon cause of the nephrotic syndrome after childhood.

The nephrotic syndrome

Hypoproteinaemia with oedema may develop if large amounts of protein are excreted in the urine. For this to occur, proteinuria must usually exceed 5g/24h. Although the ability of the liver to synthesize protein is greater than this, much of the filtered protein is catabolized after endocytosis by renal tubular cells and is thus lost from the circulation, although it is not excreted in the urine. Conditions in which the nephrotic syndrome may occur are shown in Fig.5.13.

The amount of proteinuria is not necessarily a useful guide to the severity of renal disease; for example, in minimal change glomerulonephritis, which has a

adopt a lordotic posture. The prevalence decreases with increasing age. Orthostatic proteinuria arises as a result of an increase in the hydrostatic pressure in the renal veins, itself a result of pressure of the liver on the inferior vena cava. It is of no clinical significance and can confidently be diagnosed if a sample of urine collected immediately on rising in the morning is protein-free.

Electrophoresis of a concentrated specimen of urine may help to distinguish between the various types of proteinuria. In tubular proteinuria, for example, the predominant proteins are of low molecular weight, being filtered proteins which are not reabsorbed. In

The nephrotic syndrome		
Causes	**Clinical and biochemical features**	
	Feature	Mechanism

minimal change glomerulonephritis
membranous glomerulonephritis:
   idiopathic
   associated with carcinoma, drugs or
     infection; e.g. malaria and hepatitis B
systemic lupus erythematosus
diabetic nephropathy
other forms of glomerulonephritis
	proteinuria	

	oedema	low serum albumin secondary hyperaldosteronism
	increased susceptibility to infection	low serum immunoglobulins and complement
	thrombotic tendency	hyperfibrinogenaemia and low antithrombin III
	hyperlipidaemia	increased apolipoproteins

Fig.5.13 The nephrotic syndrome: causes and clinical and biochemical features.

good prognosis, the proteinuria may exceed that seen in patients with more aggressive glomerular lesions.

CASE HISTORY 5.4

An eight-year-old girl was admitted to hospital with generalized oedema. Her urine had become frothy and the family practitioner had found proteinuria.

Investigations

serum: sodium   130mmol/l
        potassium   3.6mmol/l
        bicarbonate   32mmol/l
        urea   2.0mmol/l
        creatinine 45μmol/l
        calcium   1.70mmol/l
        total protein   35g/l
        albumin   15g/l
        triglyceride   16mmol/l
        cholesterol   12mmol/l

24h urine protein excretion   12g
selectivity index   12%

The serum was grossly lipaemic.

Comment

The presence of proteinuria, hypoproteinaemia and oedema constitutes the nephrotic syndrome. The high selectivity of the proteinuria suggests that this is due to minimal change glomerulo-nephritis (a condition of unknown aetiology). The oedema is, in part, a result of redistribution of ECF between the vascular and interstitial compartments; secondary hyperaldosteronism, with evidence of potassium depletion is often present as a consequence.

Loss of protein is not confined to albumin. Serum concentrations of hormone-binding proteins, transferrin and antithrombin III are also reduced. On the other hand, there is usually an increase in the concentrations of high molecular weight proteins such as α_2-macroglobulin, fibrinogen and the apolipoproteins. The increase in apolipoproteins causes secondary hyper-cholesterolaemia and hypertriglyceridaemia and these may in turn cause spurious hyponatraemia. Increased serum fibrinogen (and decreased antithrombin III) can predispose to venous thrombosis, particularly in the renal veins. The hypocalcaemia is related in part to decreased protein binding and in part to renal excretion of vitamin D metabolites bound to vitamin D-binding globulin. Loss of immunoglobulins and complement components renders patients with nephrotic syndrome very susceptible to infection.

The GFR may be either low, normal or raised in patients with the nephrotic syndrome. In minimal change glomerulonephritis, it is often raised and the low urea and creatinine concentrations in this patient reflect this. The quantity of protein excreted must be judged in relation to the GFR. A decrease in excretion is usually due to a decrease in glomerular permeability but it may occur because of a decrease in GFR or, in other words, a deterioration in renal function.

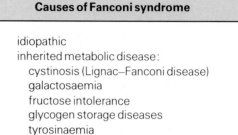

Causes of Fanconi syndrome
idiopathic
inherited metabolic disease:
cystinosis (Lignac–Fanconi disease)
galactosaemia
fructose intolerance
glycogen storage diseases
tyrosinaemia
Wilson's disease
nephrotoxins:
heavy metals
drugs
paraproteinaemia
amyloid

Fig.5.14 Causes of the Fanconi syndrome.

The clinical and biochemical features of the nephrotic syndrome are summarized in Fig.5.13. There are two aspects to management: treatment of the underlying disorder, where the disorder can be identified and treatment is possible, and treatment of the consequences of protein loss. Minimal change glomerulonephritis often responds to corticosteroids or immunosuppressive drugs, but other types of glomerulonephritis are generally much less responsive to treatment.

General measures to counteract the consequences of protein loss include a high protein, low salt diet. A high protein intake must be introduced with caution when there is concurrent renal failure. It is important not to cause too rapid a diuresis since this could lead to hypovolaemia and thus impair renal function; potassium depletion must also be avoided. Spironolactone is the diuretic of first choice but thiazides or frusemide may be necessary in addition. Prevention of infection is vital and antibiotics are often administered prophylactically. The risk of thrombosis, especially renal vein thrombosis which may cause a rapid increase in proteinuria, may warrant the prophylactic use of anticoagulants.

Renal tubular disorders

Renal tubular disorders may be congenital or acquired; they may involve single or multiple aspects of tubular function. The congenital conditions are all rare; their clinical sequelae relate to the consequences of loss of substances which are normally completely or partially reabsorbed by the tubules.

The Fanconi syndrome

This is a generalized disorder of tubular function characterized by glycosuria, aminoaciduria, phosphaturia and acidosis. It may occur secondarily to a variety of conditions (Fig.5.14). One of these is cystinosis, or Lignac–Fanconi disease, a rare inherited disease (only one in forty-thousand live births in the United Kingdom) of unknown aetiology. In this condition, cystine crystals are deposited in many body tissues, including the kidney. Affected infants fail to thrive, develop rickets and polyuria with dehydration and eventually progress to renal failure. There is no specific treatment. Cystinosis should not be confused with cystinuria, a disorder of tubular transport.

Primary Fanconi syndrome may also develop in

Urinary acidification test	
Procedure	**Results**
take blood for bicarbonate measurements, after overnight fast measure pH of freshly passed urine: if serum bicarbonate < 16 mmol/l and urine pH < 5.5, test unnecessary	normal response: urine pH < 5.2 in at least one sample if this pH is not obtained, serum bicarbonate should be measured. Test should be repeated if serum bicarbonate is not below the lower limit of normal
administer ammonium chloride (100 mg/kg body weight) orally measure pH of freshly passed urine hourly for 8 hours	renal tubular acidosis: urine pH usually ≥ 6.5

Fig.5.15 Urinary acidification test. This test should not be performed in a patient with liver disease.

young adults; it is inherited, but the nature of the defect is unknown.

Renal tubular acidosis

Proximal (type II) renal tubular acidosis, which arises as a result of impaired bicarbonate reabsorption, is a component of the Fanconi syndrome but may also occur as an isolated phenomenon. A transient form may occur in infants. Bicarbonate can be completely reabsorbed if the plasma bicarbonate concentration is low, so that patients may excrete normal amounts of acid but at the expense of systemic acidosis. Treatment consists of administering large amounts of bicarbonate, for example, 10mmol/kg body weight/day. Distal (type I or classical) renal tubular acidosis occurs more frequently. It may be either inherited or acquired, for example, secondarily to hypercalcaemia and autoimmune diseases. There is a defect in hydrogen ion excretion and the urine cannot be acidified. Consequences include osteomalacia, hypercalciuria, nephrocalcinosis, renal calculi and often hypokalaemia. Treatment involves the administration of bicarbonate in sufficient quantities to buffer normal hydrogen ion production (1–3mg/kg body weight/day) and potassium supplements.

Distal renal tubular acidosis can be diagnosed by a urinary acidification test (Fig.5.15). The diagnosis of proximal renal tubular acidosis requires determination of the renal threshold for bicarbonate.

Defects of urinary concentration

Impairment of urinary concentration is a feature of nephrogenic diabetes insipidus, a primary tubular disorder. It is also a feature of cranial diabetes insipidus and chronic renal failure and can occur with hypercalcaemia, hypokalaemia and certain drugs, notably lithium. In nephrogenic diabetes insipidus, vasopressin secretion is normal but there is a defect involving either its receptors or one of the post receptor-binding events required for its normal action. Hypercalcaemia and hypokalaemia also interfere with this cAMP-mediated pathway.

Glycosuria

Benign renal glycosuria is discussed on page 179. Renal glycosuria can also occur in association with other tubular abnormalities, for example, as part of the Fanconi syndrome.

71

Aminoaciduria

Renal aminoaciduria may occur in combination with normal plasma levels of amino acids as a result of defective tubular reabsorption, for example, Hartnup disease and cystinuria. Overflow aminoaciduria occurs secondarily to elevated plasma levels when the tubular transport mechanism is saturated, as in for instance, phenylketonuria.

Cystinuria has an incidence of about one in seven thousand live births. Defective tubular reabsorption of cystine, ornithine, arginine and lysine leads to their excretion in the urine. The loss of these amino acids would alone be of little consequence, but cystine is relatively insoluble and cystinuria predisposes the patient to renal calculus formation. The management of cystinuria is discussed on page 253.

Hypophosphataemic rickets

This condition, also known as vitamin D-resistant rickets, has a dominant X-linked pattern of inheritance. A defect in tubular phosphate reabsorption leads to severe rickets. This does not respond to treatment with vitamin D alone, even if administered in massive doses, but can be treated effectively with a combination of oral phosphate supplements and vitamin D, usually given as a 1α-hydroxylated derivative.

Hypophosphataemic rickets should not be confused with vitamin D-dependent rickets which is inherited as an autosomal recessive condition. The defect is in the 1α-hydroxylation of 25-hydroxy-cholecalciferol. This condition can be treated with 1α-hydroxylated derivatives of vitamin D alone.

Renal calculi

Pathogenesis

Renal stones or calculi can form in urine when it is supersaturated with the crystalloid components of the calculus. Factors predisposing to this, and the commoner types of calculus that occur clinically, are shown in Fig.5.16.

Hypercalciuria is present in up to twenty-five percent of patients with calcium oxalate/phosphate stones. It may be associated with hypercalcaemia, for example, due to primary hyperparathyroidism. However, many patients with calcium-containing stones are normocalcaemic and the primary abnor-

---

Renal calculi

Factors predisposing to formation

dehydration
urinary tract infection
persistently alkaline urine
hypercalciuria
hyperuricosuria
hyperoxaluria
urinary stagnation (due to obstruction)
lack of urinary inhibitors of crystallization

Composition

calcium oxalate ($\pm$ phosphate)
calcium phosphate
magnesium ammonium phosphate ('triple phosphate')
uric acid
cystine

Biochemical investigations

analysis of calculus (if available)
serum:
  calcium
  urate
  phosphate
urine:
  pH
  qualitative test for cystine
  24h excretion of calcium, oxalate and urea
  urinary acidification test

Fig.5.16 Renal calculi: composition, factors predisposing to their formation and biochemical investigations.

mality is an increase in intestinal calcium absorption.

Hyperoxaluria predisposes to renal calculus formation. The condition may be due to a rare inherited metabolic disorder (primary hyperoxaluria), but it is usually caused by increased intestinal absorption of dietary oxalate, with or without increased oxalate ingestion. This may be seen in patients with a variety of gastrointestinal disorders, in particular, with inflammatory bowel diseases and conditions associated with malabsorption. In these circumstances, free fatty acids, which are not absorbed, bind with calcium. This limits the amount of calcium available to combine with oxalate to form calcium oxalate, an insoluble substance which is normally excreted in the faeces. As a result, an increased amount of oxalate remains in solution and can be absorbed into the bloodstream.

Investigation

The history and examination may suggest an underlying cause for renal calculi, such as inadequate fluid intake. Biochemical tests that should be performed on serum and urine are shown in Fig.5.16. The single most useful test is to analyze a stone, if available. The urine must be examined for evidence of infection in all patients presenting with renal calculi. The radiographic appearance of a retained stone may be characteristic: for example, 'staghorn' calculi contain mixed phosphates and are related to chronic infection; pure uric acid stones (not containing calcium) are radiolucent as are cystine stones. An intravenous pyelogram may show a predisposing anatomical abnormality. Most calculi can be detected by ultrasound.

Management

Small calculi are often passed spontaneously. Larger calculi may require surgical removal or disintegration by ultrasound. Any urinary tract infection should be treated. The identification of the cause of renal calculus formation should make it possible to design an effective regimen to prevent further stone formation. This is particularly important in patients who form stones recurrently.

The management of cystinuria is considered on page 253. Hyperuricaemia should be treated with allopurinol (see page 269). Alkalization of the urine increases the solubility of both cystine and uric acid but may be difficult to achieve. A high fluid intake is appropriate in all patients with a tendency to form renal calculi.

If patients who form calcium stones are hypercalcaemic, the underlying cause should be treated. In the normocalcaemic majority, dietary manipulation to correct excessive intake of calcium or oxalate is appropriate. However, calcium restriction below maintenance levels is inadvisable since oxalate absorption may be increased and there may be adverse effects on the skeleton. In patients who do not respond to such measures, thiazide diuretics (which decrease urinary calcium excretion) are often very effective at preventing recurrence.

SUMMARY

The kidneys are essential for the control of extracellular fluid volume and composition, hydrogen ion homoeostasis, the excretion of waste products of metabolism and as endocrine organs, particularly in relation to calcium homoeostasis, erythropoiesis and the control of blood pressure. The best simple test of overall renal function is the serum creatinine concentration while the presence of proteinuria is a sensitive, although not specific, indicator of kidney damage.

Acute renal failure is a life-threatening condition in which there is a deterioration in renal function which is potentially reversible. It is most frequently caused by either renal hypoperfusion or exposure to nephrotoxins. When hypoperfusion is responsible, it may be possible to prevent the development of intrinsic (intrarenal) renal damage by restoration of normal perfusion. The biochemical features of acute renal failure include increases in serum urea and creatinine concentrations, hyperkalaemia, hyperphosphataemia, acidosis and fluid retention. Patients are usually oliguric and require dialysis until renal function recovers. The onset of recovery is heralded by an increase in urine production and a diuretic phase follows before normal renal function returns. During the diuretic phase there may be considerable losses of fluid and ions from the body.

In chronic, or end-stage, renal failure, renal function is irreversibly lost and patients will eventually require transplantation or long-term dialysis. This condition usually develops slowly and because the kidneys have considerable functional reserves, patients tend to present late in the course of the illness. Retention of urea, creatinine and other waste products and disturbance of sodium and water homoeostasis are characteristic but severe acidosis and hyperkalaemia are only late features of the condition. Bone disease, with hypocalcaemia and hyperphosphataemia, and anaemia result from impairment of renal endocrine function.

The nephrotic syndrome comprises proteinuria, hypoproteinaemia and oedema and can be a result of a variety of diseases affecting the glomerulus. The clinical and biochemical features stem from the loss of protein from the body. In addition to albumin, the loss of which is responsible for the oedema, the loss of other proteins leads to increased susceptibility to infection and hypercoagulability. Uraemia may or may not be present, depending on the nature of the underlying glomerular damage.

The formation of renal calculi is essentially the result of supersaturation of the urine. Factors predisposing to calculus formation include the excretion of high solute loads, for example, calcium, oxalate and urate, inadequate water intake and infection.

Conditions affecting the renal tubules alone are in general uncommon. They can be either congenital or acquired and can affect either single or multiple tubular functions. There may be either excessive loss of substances which are normally reabsorbed by the tubules, for example, phosphate, glucose and amino acids, or inadequate excretion of substances normally secreted by the tubules, for example, hydrogen ions, thus producing renal tubular acidosis.

FURTHER READING

de Wardener HE (1985) *The Kidney*. 5th edition. Edinburgh: Churchill Livingstone.

6. The Liver

INTRODUCTION

The liver is of vital importance in intermediary metabolism and in the detoxification and elimination of toxic substances (Fig.6.1). Damage to the organ may not obviously affect its activity since the liver has considerable functional reserve and, as a consequence, tests of liver function alone are insensitive indicators of liver disease. Other tests are often superior, for example, measurement of hepatic enzyme levels in the plasma (often classified incorrectly as tests of liver function).

The results of biochemical tests are frequently non-specific, since they reflect the basic pathological processes common to many conditions; they therefore provide limited diagnostic information. It is important to be aware of the limitations of biochemical tests and of the contributions made to diagnosis and management by other techniques.

The metabolic activity of the liver takes place within the parenchymal cells which constitute eighty percent of the organ mass; the liver also contains Kupffer cells of the reticuloendothelial system. Parenchymal cells are contiguous with the venous sinusoids, which carry blood from the portal vein and hepatic artery, and with the biliary canaliculi, the smallest ramifications of the biliary system (Fig.6.2). Substances destined for excretion in the bile are secreted from hepatocytes into the canaliculi, pass through the intrahepatic ducts and reach the duodenum via the common bile duct.

The most common disease processes affecting the liver are:
- Hepatitis, with damage to liver cells.
- Cirrhosis, in which increased fibrous tissue formation leads to shrinkage of the liver, decreased hepatocellular function and obstruction of bile flow.
- Tumours, most frequently secondary; for example, metastases from cancers of the large bowel, stomach and bronchus.

Extrahepatic disease may exhibit the clinical features of liver disease or may have secondary effects on the liver; for instance, obstruction to the common bile duct may cause jaundice and, if prolonged, a form of cirrhosis.

Major functions of the liver
Carbohydrate metabolism gluconeogenesis glycogen synthesis and breakdown
Fat metabolism fatty acid synthesis cholesterol synthesis and excretion lipoprotein synthesis ketogenesis bile acid synthesis 25-hydroxylation of vitamin D
Protein metabolism synthesis of plasma proteins (including clotting   factors but not immunoglobulins) urea synthesis
Hormone metabolism metabolism and excretion of steroid hormones metabolism of polypeptide hormones
Drugs and foreign compounds metabolism and excretion
Storage glycogen vitamin A vitamin B_{12} iron
Metabolism and excretion of bilirubin

Fig.6.1 Major functions of the liver.

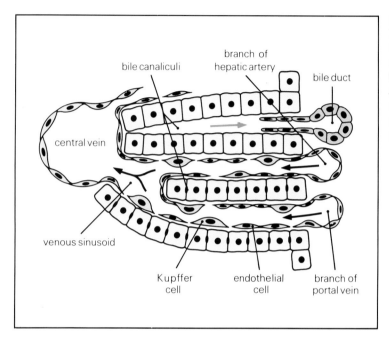

Fig.6.2. Microstructure of the liver. The liver consists of acini in which sheets of hepatocytes, one cell thick, are permeated by sinusoids carrying blood from portal venules and hepatic arterioles to the central vein. Bile is secreted from the hepatocytes into canaliculi, which drain into the bile ducts.

BILIRUBIN METABOLISM

Bilirubin is derived mainly from the haem moiety of the haemoglobin molecules liberated when effete red cells are removed from the circulation by the reticulo-endothelial system (Fig.6.3); the iron in haem is re-utilized but the tetrapyrrole ring is degraded to bilirubin. Other sources of bilirubin include myoglobin and the cytochromes.

Unconjugated bilirubin is not water-soluble; it is transported in the blood stream bound to albumin. In the liver, it is taken up by hepatocytes in a process involving specific carrier proteins. Bilirubin is then transported to the smooth endoplasmic reticulum where it undergoes conjugation, principally with glucuronic acid to form a diglucuronide; this process is catalyzed by the enzyme bilirubin-uridyl diphosphate glucuronyl transferase. Conjugated bilirubin is water-soluble and is secreted into the biliary canaliculi, even-tually reaching the small intestine via the ducts of the biliary system. Secretion into the biliary canaliculi is the rate-limiting step in bilirubin metabolism.

In the gut, bilirubin is converted by bacterial action into urobilinogen, a colourless compound. Some urobilinogen is absorbed from the gut into the portal blood; hepatic uptake of this is incomplete, and a small quantity reaches the systemic circulation and is excreted in the urine. Most of the urobilinogen in the gut is oxidized in the colon to a brown pigment, urobilin, which is excreted in the stool.

In the absence of liver disease, no bilirubin is excreted in the urine. The bilirubin present in the blood is mainly unconjugated. Since it is protein-bound, it is not filtered by the renal glomeruli. All the conjugated bilirubin is secreted into the biliary system and does not reach the blood stream.

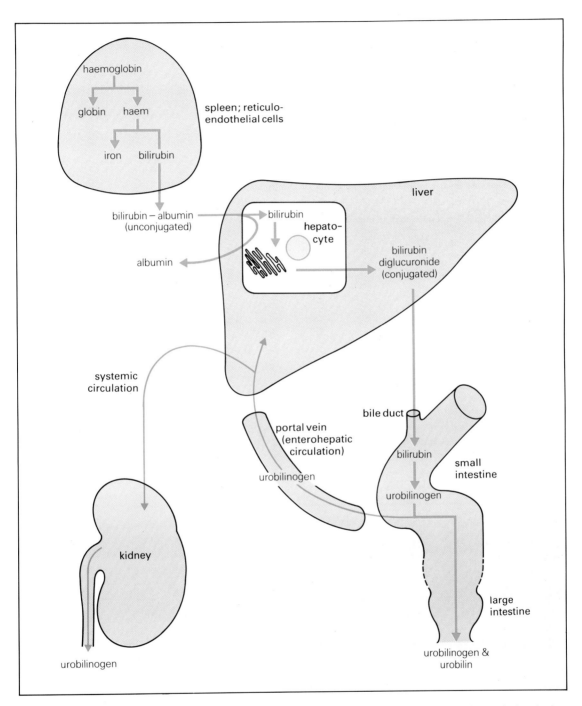

Fig.6.3 Excretion of bilirubin by the liver. Bilirubin, bound to albumin, is taken up into hepatocytes, conjugated in the smooth endoplasmic reticulum and excreted via the bile ducts into the gut, where it is converted to urobilinogen. Most of the urobilinogen is oxidized to urobilin in the colon and excreted in the stool. Some urobilinogen is absorbed from the small intestine and enters the enterohepatic circulation; while most is excreted in the bile, some reaches the systemic circulation and is excreted in the urine.

Major causes of jaundice
Pre-hepatic
haemolysis
ineffective erythropoiesis
Hepatic
pre-microsomal:
Gilbert's syndrome
drugs, e.g. rifampicin, which interfere with
bilirubin uptake
microsomal:
prematurity
hepatitis, e.g. viral or drug-induced
drugs, e.g. rifampicin
Crigler–Najjar syndrome
post-microsomal:
impaired excretion:
hepatitis
drugs, e.g. methyltestosterone and rifampicin
Dubin–Johnson syndrome
intrahepatic obstruction:
hepatitis
cirrhosis
infiltrations, e.g. lymphoma and amyloid
biliary atresia
tumours
Post-hepatic
gallstones
stricture
carcinoma of pancreas or biliary tree

Fig.6.4 Classification and major causes of jaundice. In hepatitis, bilirubin metabolism may be affected at various steps. The jaundice is usually due to conjugated bilirubin.

Jaundice, the yellow discoloration of tissues due to bilirubin deposition, is a frequent feature of liver disease. Clinical jaundice may not be present unless the serum bilirubin concentration is more than two and a half times the upper limit of normal, that is, more than 50μmol/l. Hyperbilirubinaemia can be caused by increased production of bilirubin, impaired metabolism, decreased excretion or a combination of these. Causes of jaundice are listed in Fig.6.4.

BIOCHEMICAL ASSESSMENT OF LIVER FUNCTION

Bilirubin

Hyperbilirubinaemia is not always present in patients with liver disease nor is it exclusively associated with liver disease. For example, it is not usually present in patients with well-compensated cirrhosis but it is a common feature of advanced pancreatic carcinoma.

Unconjugated hyperbilirubinaemia

When an excess of bilirubin is unconjugated the concentration in adults rarely exceeds 100μmol/l. In the absence of liver disease, unconjugated hyperbilirubinaemia is most often due to either haemolysis or to Gilbert's syndrome, an inherited abnormality of bilirubin metabolism.

In haemolysis, hyperbilirubinaemia is due to an increased production of bilirubin which exceeds the capacity of the liver to remove and conjugate the pigment. Nevertheless, more bilirubin is excreted in the bile, the amount of urobilinogen entering the entero-hepatic circulation is increased and urinary urobilinogen is increased. The laboratory findings in haemolytic (pre-hepatic) jaundice are summarized in Fig.6.5.

Activity of the hepatic conjugating enzymes is usually low at birth but increases rapidly thereafter; the transient 'physiological' jaundice of the newborn reflects this. With excessive haemolysis, as in Rhesus incompatibility, or a lack of enzyme activity, as occurs in prematurity and in the Crigler–Najjar syndrome, there may be a massive rise in the serum concentration of unconjugated bilirubin. If bilirubin levels exceed approximately 340μmol/l, its uptake into the brain may cause severe brain damage (kernicterus).

Conjugated hyperbilirubinaemia

This condition is due to leakage of bilirubin from either hepatocytes or the biliary system into the blood stream when its normal route of excretion is blocked. The water-soluble conjugated bilirubin entering the systemic circulation is excreted in the urine, giving it a deep orange-brown colour. In complete biliary obstruction, no bilirubin reaches the gut, no urobilin is formed and the stools are pale in colour. The differential diagnosis of jaundice due to conjugated bilirubin is considered on pages 84 and 85.

Laboratory findings in haemolytic jaundice	
serum bilirubin	unconjugated rarely >100μmol/l except in neonates
serum enzymes	aspartate transaminase and hydroxybutyrate dehydrogensase slightly increased
serum haptoglobins	decreased
urine urobilinogen	increased
peripheral blood	increased reticulocytes decreased haemoglobin possible evidence of haemolysis on blood film

Fig.6.5 Laboratory findings in haemolytic jaundice.

The separate measurement of conjugated and unconjugated bilirubin concentrations is useful in the diagnosis of neonatal jaundice where there may be some doubt as to the relative contributions of defective conjugation and other causes; it is less often required in adults. If the serum bilirubin concentration is less than 100μmol/l and other tests of liver function are normal, it can be inferred that the raised levels are due to the unconjugated form of the pigment. The urine can be tested to confirm this, since with unconjugated hyperbilirubinaemia there is no bilirubin in the urine.

Serum enzymes

Enzymes used in the assessment of hepatic function include aspartate and alanine transaminases, alkaline phosphatase and γ-glutamyl transpeptidase, as discussed in detail in *Chapter 17*. In general, these enzymes are not specific indicators of liver dysfunction. The hepatic isoenzyme of alkaline phosphatase is an exception, and alanine transaminase is more specific to the liver than aspartate transaminase.

Increased transaminase levels reflect cell damage; serum levels are often twenty times the upper limit of normal (ULN) in patients with hepatitis. In cholestasis, serum alkaline phosphatase activity is increased. This is a result of enzyme induction, itself a consequence of cholestasis although the mechanism involved is unknown. In severe obstructive jaundice, the serum alkaline phosphatase level may be up to ten times the ULN.

However, these enzyme changes do not always occur in isolation. In primarily cholestatic disease there may be secondary hepatocellular damage and increased serum transaminase levels, while cholestasis frequently occurs in hepatocellular disease. Increased γ-glutamyl transpeptidase is found in both cholestasis and hepatocellular damage; this enzyme is very sensitive but non-specific. Thus, although certain patterns of enzyme levels are frequently observed in various types of liver disease, they are not reliably diagnostic.

Serum enzyme levels are very useful in following the progress of liver disease once the diagnosis has been made. Falling transaminase levels suggest a decrease in hepatocellular damage and falling alkaline phosphatase levels indicate a resolution of cholestasis. However, in fulminant hepatic failure, a decrease in transaminase levels may be a poor prognostic sign, reflecting almost complete destruction of parenchymal cells.

Serum proteins of diagnostic value		
Protein	**Condition**	**Change in concentration**
albumin	chronic liver disease	↓
α_1-antitrypsin	cirrhosis due to α_1-antitrypsin deficiency	↓
caeruloplasmin	Wilson's disease	↓
α-fetoprotein	marker for hepatocellular carcinoma	greatly ↑
transferrin	haemochromatosis	normal but 100% saturated with iron

Fig.6.6 Serum proteins of diagnostic value in liver disease.

Serum proteins

Albumin is synthesized in the liver and its concentration in the serum is in part a reflection of the functional capacity of the organ. Serum albumin concentration is decreased in chronic liver disease, but tends to be normal in the early stages of acute hepatitis due to its long half-life of approximately twenty days. There are, however, many other causes of hypoalbuminaemia as discussed on pages 210 & 211.

The prothrombin time is a test of plasma clotting activity and reflects the activity of vitamin K-dependent clotting factors synthesized by the liver, of which factor VII has the shortest half-life of four to six hours. An increase in the prothrombin time is often an early feature of acute liver disease, but a prolonged prothrombin time may also reflect vitamin K deficiency. The cause can be determined by administering the vitamin parenterally; in the absence of liver disease, the prothrombin time should return to normal within eighteen hours.

A polyclonal increase in immunoglobulins is frequently associated with cirrhosis, particularly when the disease is autoimmune in origin, and may cause the total serum protein to be normal, or even increased, in spite of a low albumin concentration.

Serum protein electrophoresis is of little value in the diagnosis of liver disease; typical patterns may be seen in certain hepatic disorders, such as fusion of the β and γ bands due to an increase in IgA in alcoholic cirrhosis, but they are neither specific nor invariably present.

The measurement of individual immunoglobulins is also of little diagnostic value, although increases in specific auto-antibodies may point to a diagnosis, for example, anti-smooth muscle antibody levels are increased in more than seventy percent of patients with chronic active hepatitis. Diagnostically useful changes in the concentration of other serum proteins in liver disease are listed in Fig.6.6.

LIVER DISEASE

Hepatitis

Acute hepatitis is most frequently caused by infectious agents, particularly viruses, and toxins. Patients may present with jaundice but this is often not apparent early in the course of the disease.

CASE HISTORY 6.1

A twenty-year-old student developed a flu-like illness with loss of appetite, nausea and pain in the right hypochondrium. On examination the liver was just palpable and was tender. Two days later he developed jaundice, his urine became darker in colour and his stools became pale.

Biochemical changes during acute hepatitis		
	Pre-icteric	**Icteric**
serum bilirubin	slight ↑	large ↑
serum transaminases	large ↑	↑
serum alkaline phosphatase	normal	slight ↑
urinary bilirubin	↑	↑
urinary urobilinogen	↑	absent

Fig.6.7   Biochemical changes during acute hepatitis.

Investigations

	on presentation	one week later
serum:		
bilirubin	38μmol/l	230μmol/l
albumin	40g/l	38g/l
aspartate		
transaminase	450iu/l	365iu/l
alkaline		
phosphatase	70iu/l	150iu/l
γ-glutamyl		
transpeptidase	60iu/l	135iu/l

urine:		
bilirubin	positive	positive
urobilinogen	positive	negative

Comment

The first set of results is characteristic of early hepatitis, with a raised transaminase reflecting cell damage. This usually precedes the rise in bilirubin and the development of jaundice. Impairment of the hepatic secretion of conjugated bilirubin and of urobilinogen uptake from the portal blood, causes both these substances to be excreted in the urine.

The second set of results shows the expected high serum bilirubin but with a fall in transaminase as the phase of maximum cellular damage has passed. An increase in alkaline phosphatase, usually of not more than three times the ULN, is common at this stage. In hepatitis, the bilirubin in serum is both conjugated and unconjugated, with the former predominating. Conjugated bilirubin is excreted in the urine and the pale stool reflects decreased biliary excretion. The serum albumin has remained normal in this acute illness.

Early in the course of acute hepatitis, bilirubin and urobilinogen are usually readily detectable by a simple dip-stick technique. For as long as the serum bilirubin is raised, bilirubin continues to be excreted in the urine. Urobilinogen may disappear from the urine at the height of the jaundice, because no bilirubin reaches the gut, but it reappears as the hepatitis resolves and biliary excretion returns to normal. These changes (Fig.6.7) are of no practical value in the management of hepatitis, but the detection of bilirubin in the urine is a simple and valuable diagnostic pointer to hepatitis in the pre-icteric stage of the illness.

The viruses primarily associated with hepatitis are hepatitis A, B and 'non-A, non-B', but many others, such as the Epstein–Barr virus and cytomegalovirus, can cause the disease. Many toxins and drugs can also cause acute hepatitis, including alcohol, paracetamol and carbon tetrachloride.

Most cases of viral hepatitis resolve completely. In severe cases, usually due to hepatitis B or 'non-A, non-B', hepatic failure may develop, but most patients who survive the acute illness eventually recover completely. In some cases of hepatitis B infection, persistent antigenaemia occurs and may be associated with the development of chronic liver disease; infection with hepatitis A never leads to chronic disease.

Cirrhosis

Causes of cirrhosis include chronic excessive alcohol intake, autoimmune disease, persistence of the hepatitis B virus and various inherited metabolic diseases, such as Wilson's disease, haemochromatosis and α_1-antitrypsin deficiency.

Due to the great functional capacity of the liver, metabolic and clinical abnormalities may not become apparent until late in the course of the disease; until this time the cirrhosis is said to be 'compensated'. There are no reliable biochemical tests to diagnose subclinical disease; the bromsulphthalein retention test has been used in the past, but is potentially dangerous and is now rarely used.

CASE HISTORY 6.2

A middle-aged female publican was admitted to hospital following a haematemesis. Endoscopy revealed the presence of oesophageal varices. The only biochemical abnormality was an elevated γ-glutamyl transpeptidase (245iu/l). Her varices were treated by sclerotherapy and no further bleeding occurred. The patient was told to abstain from alcohol. She was readmitted one year later, jaundiced, drowsy and with the clinical signs of chronic liver disease.

Investigations

serum:

albumin	25g/l
bilirubin	260μmol/l
alkaline phosphatase	315iu/l
aspartate transaminase	134iu/l
γ-glutamyl transpeptidase	360iu/l

Comment

The patient had continued to drink and the resulting liver damage eventually affected hepatic function. The serum albumin is decreased, serum bilirubin elevated and enzyme changes are consistent with cirrhosis and active liver cell damage; the prothrombin time was also prolonged.

Hepatic decompensation may be precipitated in chronic liver disease by sepsis, bleeding into the gut, for example, from varices, erosions and ulcers, and by various drugs, including diuretics. Diuretics may be given to treat ascites, a common feature of chronic liver disease, but must be used with great caution. Several factors probably contribute to the development of ascites, including hypoalbuminaemia, portal venous obstruction and increased hepatic lymph production.

Encephalopathy, characterized by a decrease in consciousness and impairment of higher functions, is often present in decompensated cirrhosis and may also be a feature of severe acute hepatitis. Substances implicated in encephalopathy include ammonia, which accumulates when urea synthesis is impaired, and false neurotransmitters, such as octopamine and β-phenylethanolamine. These false neurotransmitters are derived from the amino acids tyrosine and phenylalanine respectively, by bacterial decarboxylation in the gut, and are normally detoxified in the liver.

Treatment of hepatic encephalopathy involves: appropriate management of any precipitating factors such as gastrointestinal haemorrhage; restriction of dietary protein intake; and the provision of enemas or laxatives, for example, lactulose, to empty the bowels of nitrogen-containing material. Neomycin, a non-absorbable antibiotic, is used to sterilize the gut in order to reduce the production of toxins by bacteria. An adequate calorie intake is essential and fluid and electrolyte balance must be maintained. If ascites is present, sodium restriction is essential.

Alcohol is a common cause of liver disease. This either may take the form of an acute hepatitis, particularly after a bout of heavy drinking in patients with a history of chronic excessive alcohol ingestion, or may present as established cirrhosis. The most important aspect of management, apart from general supportive measures and treatment of any complications, is to persuade the patient to abstain totally from alcohol. If this can be achieved, the prognosis in alcoholic cirrhosis is better than in cirrhosis due to other causes.

CASE HISTORY 6.3

A forty-year-old woman presented with jaundice. There was no history of contact with hepatitis, recent foreign travel, injections or transfusions. She did not drink alcohol. She had been well in the past but had suffered from increasingly intense pruritus during the previous eighteen months.

Investigations

serum:

total protein	85g/l
albumin	28g/l
bilirubin	340μmol/l
alkaline phosphatase	522iu/l
aspartate transaminase	98iu/l
γ-glutamyl transpeptidase	242iu/l

Comment

The very high alkaline phosphatase indicates a cholestatic jaundice; the low albumin is consistent with chronic liver disease. The clue to the diagnosis is the high total protein, implying a serum globulin level of 57g/l. This is often seen in autoimmune liver disease. Further investigations revealed a high titre of antimitochondrial antibodies, characteristic of primary biliary cirrhosis. This diagnosis was confirmed by histological examination of tissue obtained by percutaneous liver biopsy.

Pruritus in chronic liver disease is due to the accumulation of bile salts. Measurement of serum bile salts has been suggested as a sensitive test of hepatocellular function but has not been adopted routinely.

Once established, hepatic cirrhosis is irreversible. If possible, any underlying cause should be treated appropriately. Specific complications, including ascites, bleeding, for example, from oesophageal varices resulting from portal hypertension, and malabsorption, may also be amenable to treatment.

Causes of death include hepatic encephalopathy, uncontrollable bleeding and septicaemia.

Tumours and infiltrations

The liver is a common site for tumour metastasis. Primary liver tumours are rare in the Western world but occur frequently in other geographical areas. Primary tumours are associated with cirrhosis, persistence of serological markers for hepatitis B and various carcinogens, including aflatoxins. Serum α-fetoprotein is elevated at diagnosis in approximately seventy percent of patients with primary hepatocellular carcinomas and is a valuable marker for this tumour. Infiltrative conditions which can affect the liver include lymphomas and amyloidosis. Patients with such conditions, and with intrahepatic tumours, are often not jaundiced.

CASE HISTORY 6.4

An elderly woman consulted her general practitioner because of weight loss and constipation. She had lost approximately 8kg in weight in two months and had lost her appetite. She had previously opened her bowels daily but had recently had intervals of several days between movements and had passed only small amounts of stool on each occasion. On examination she was anaemic and had obviously lost weight. The liver was enlarged and had an irregular edge; a mass was palpable in the right iliac fossa.

Investigations

serum:

albumin	30g/l
alkaline phosphatase	314iu/l
bilirubin, aspartate transaminase and γ-glutamyl transpeptidase	normal
stool occult blood	positive

A barium enema revealed a carcinoma of the caecum; an isotopic liver scan showed multiple filling defects characteristic of tumour deposits.

Comment

An increase in serum alkaline phosphatase in a patient with carcinoma could be due to metastasis in bone or liver. When the source of the enzyme is not obvious clinically, it can usually be inferred from isoenzyme studies. With hepatic metastases there is often no increase in serum bilirubin levels unless lymph nodes at the porta hepatis are involved and obstruct the major bile ducts. Although tumour deposits within the liver can cause obstruction of small bile ducts, which is reflected by the increase in alkaline phosphatase, bilirubin leaking into the blood stream can be taken up and excreted by parts of the liver not affected by tumour. Thus, there is little or no increase in the serum bilirubin concentration.

Hypoalbuminaemia is common in malignant disease and is usually multifactorial. Poor nutrition, increased catabolism (cancer cachexia) and replacement of normal hepatic tissue by tumour are all possible contributory causes in this case.

Carcinoma of the caecum is often clinically silent and may not present until extensive secondary spread has occurred.

Cholestasis and jaundice

In a patient presenting with jaundice due to excess conjugated bilirubin, the possible diagnoses include intrinsic hepatocellular diseases and both intra- and extrahepatic cholestasis (see Fig.6.4). Valuable diagnostic information may be provided by the history and examination. Biochemical tests can also give valuable information; for example, a very high serum transaminase level suggests the presence of hepatocellular damage while a very high alkaline phosphatase level suggests cholestasis.

It is often not possible to distinguish reliably between intra- and extrahepatic cholestasis from the results of biochemical tests alone. Other useful diagnostic techniques for the investigation of patients with cholestasis include ultrasound scanning, isotopic imaging, histological examination of tissue obtained by percutaneous biopsy and specialized radiological investigations such as either percutaneous or retrograde cholangiography (Fig.6.8). The simple biochemical tests are cheap and non-invasive; their results, used in conjunction with the clinical findings, may suggest a diagnosis but this is often made, and usually confirmed, by other means.

CASE HISTORY 6.5

A retired publican presented to his family practitioner with a three-month history of epigastric pain radiating into the back and not related to meal-times. He was given antacids but returned one month later with more severe pain and weight loss. Over the past week his urine had become dark in colour and his stools pale. He had also become jaundiced. On examination, apart from the jaundice and signs of recent weight loss, no abnormality was found.

Investigations

serum:

total protein	72g/l
albumin	40g/l
bilirubin	380μmol/l
alkaline phosphatase	510iu/l
aspartate transaminase	80iu/l
γ-glutamyl transpeptidase	115iu/l

Ultrasound examination demonstrated the presence of dilated bile ducts.

A barium meal and follow-through revealed indentation of the second part of the duodenum by an extrinsic mass, thought to be a carcinoma of the head of the pancreas.

A computerized tomogram of the abdomen also suggested the presence of tumour within the pancreas and this was confirmed at laparotomy.

Comment

The results of the biochemical tests suggest that the jaundice was due to biliary obstruction and militate against, although do not exclude, the presence of liver disease. The clinical features are very suggestive of a carcinoma of the head of the pancreas obstucting the common bile duct as it enters the duodenum. However, the biochemical results, although compatible with this diagnosis,

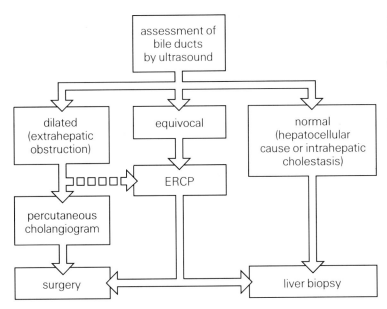

Fig.6.8 Procedures for the investigation of cholestatic jaundice. Endoscopic retrograde cholangio-pancreatography (ERCP) involves the introduction of X-ray contrast medium directly into the common bile duct using an endoscope. Percutaneous cholangio-graphy is only performed immediately before laparotomy.

could also be caused by either a calculus obstructing the common bile duct, a metastatic tumour involving lymph nodes at the hilum of the liver or intrahepatic cholestasis.

Pancreatic carcinoma often presents with pain and cholestatic jaundice; there are, as yet, no reliable biochemical tests to aid in the diagnosis of this tumour.

Inherited abnormalities of bilirubin metabolism

There are four conditions in which jaundice is caused by an inherited abnormality of bilirubin metabolism: Gilbert's, Crigler–Najjar, Dubin–Johnson and Rotor syndromes. Their characteristics are summarized in Fig. 6.9. Gilbert's syndrome affects two to five percent of the population but the others are rare.

CASE HISTORY 6.6

A medical student recovering from an attack of influenza was noticed to be slightly jaundiced. Worried that he might have hepatitis, the student had some blood taken for biochemical tests.

Investigations

serum:
bilirubin	60μmol/l
alkaline phosphatase	74iu/l
aspartate transaminase	35iu/l
haemoglobin	16g/dl
reticulocytes	1%
urine bilirubin	negative

Comment

The negative urine bilirubin indicates that the excess bilirubin in the serum is unconjugated. There is no evidence of hepatocellular damage and the normal haemoglobin and reticulocyte count indicate that haemolysis cannot be the cause of the raised bilirubin. By elimination, the diagnosis is Gilbert's syndrome.

In this condition, there is reduced activity of UDP-glucuronyl transferase (the enzyme responsible for the conjugation of bilirubin) and often defective uptake of bilirubin into the liver cells. In cases where there is a family history, the pattern of inheritance is characteristic of an autosomal dominant, single gene defect.

Inherited disorders of bilirubin metabolism		
Syndrome	**Defect**	**Clinical features**
Gilbert's	decreased conjugation of bilirubin and decreased uptake in some cases	mild, fluctuant unconjugated hyperbilirubinaemia which increases on fasting normal biopsy normal lifespan

Crigler–Najjar	Type 1
	   absence of conjugating
	     enzyme
	
	Type 2
	   partial defect of
	     conjugating enzyme
		severe unconjugated hyperbilirubinaemia
		early death due to kernicterus
		
		severe unconjugated hyperbilirubinaemia, but may respond to phenobarbitone and phototherapy
		possible survival to adulthood

Dubin–Johnson	decreased hepatic excretion of bilirubin	mild, fluctuant conjugated hyperbilirubinaemia hepatic pigment deposition (melanin) increased coproporphyrin I/III ratio in urine normal lifespan
Rotor	unknown	similar to Dubin–Johnson but no hepatic pigmentation

Fig.6.9 Inherited disorders of bilirubin metabolism. The precise nature of the metabolic defect in Dubin–Johnson syndrome is unknown.

The jaundice of Gilbert's syndrome is typically mild and present only intermittently. It is often noticed after an infection or a period of decreased food intake, possibly because free fatty acids compete with bilirubin for uptake into liver cells. There may be mild malaise and hepatic tenderness but there are no other abnormal physical signs. The liver is histologically normal and there are no sequelae.

When suspected, the diagnosis should be confirmed and documented to avoid unnecessary investigations in the future. An increase in serum bilirubin concentration of $20\mu mol/l$ in response to either a 400kcal food intake over forty-eight hours or an infusion of nicotinic acid (50mg intravenously over thirty seconds, with blood samples taken every half hour for two hours, then hourly for three hours) is characteristic.

Uncommon liver diseases

Wilson's disease is an inherited abnormality (autosomal recessive) of copper metabolism, characterized by decreased biliary excretion of copper and by decreased incorporation of copper into caeruloplasmin, a serum protein. Copper is deposited in the liver, the basal ganglia of the brain and the cornea of the eye. Patients with Wilson's disease may present either in childhood with a fulminating hepatitis, accompanied in many cases by haemolysis and a renal tubular defect, or as young adults with cirrhosis or manifestations of disease of the basal ganglia.

The biochemical features of Wilson's disease are a reduced serum caeruloplasmin level, low-normal or low serum copper (with increased binding to albumin) and increased urinary copper excretion. The decrease in caeruloplasmin is not unique to Wilson's disease, but is also seen in chronic hepatitis and malnutrition. The definitive diagnostic test is the demonstration of a high copper content in liver tissue obtained by biopsy; increased hepatic copper deposition is also seen to a lesser extent in primary biliary cirrhosis and in neonatal biliary atresia but these conditions have other distinguishing features.

Wilson's disease is treated with penicillamine which chelates copper and increases its urinary excretion; in chronic cases this often halts the progress of the disease. When patients present with fulminant hepatitis the prognosis is very poor, but some cases have been treated successfully by liver transplantation; since the genetic defect is expressed in the liver, transplantation effectively cures the disease.

Haemochromatosis (see pages 262 & 263), an inherited disorder characterized by excessive iron uptake from the gut and by iron deposition in the tissues, can affect many organs including the liver.

α_1-antitrypsin deficiency (see page 261), an inherited condition characterized either by the absence of this protein from the serum or by the presence of an abnormal protein, is another rare cause of cirrhosis.

Fulminant hepatic failure

This clinical syndrome develops when there is massive liver cell necrosis. Such severe destruction of liver tissue is fortunately rare, but it is particularly associated with viral hepatitis and paracetamol poisoning. The underlying hepatic lesion is usually potentially reversible since the liver has a considerable capacity for regeneration, but the metabolic disturbance is profound and the prognosis poor; fulminant hepatic failure is often accompanied by renal failure.

Metabolic features of fulminant hepatic failure include severe hyponatraemia, hypocalcaemia and hypoglycaemia. Hydrogen ion homoeostasis is also disturbed. Lactic acidosis may develop as a result of the failure of hepatic gluconeogenesis from lactate, but may be masked by a respiratory alkalosis caused by toxic stimulation of the respiratory centre. Generalized depression of the brain stem may lead to respiratory arrest. In some cases, a metabolic alkalosis predominates; this is in part related to excessive urinary potassium loss, due to intracellular potassium depletion and secondary aldosteronism, and in part to the accumulation of basic substances, such as ammonia, in the blood.

Despite the fact that renal failure may also be present, the serum urea concentration is often low, reflecting decreased hepatic synthesis. The serum creatinine concentration is a more reliable guide both to renal function and to whether the patient should be haemodialyzed. The prothrombin time is greatly prolonged as a result of impaired hepatic synthesis of clotting factors and bleeding is a common clinical problem.

Management

Management involves support of vital functions and correction of the metabolic imbalances. Respiratory failure may necessitate artificial ventilation and haemodialysis may be necessary if renal failure occurs. Close cooperation between the laboratory and clinical staff is vital in the management of fulminant hepatic failure. Despite the development of improved methods of artificial liver support, the prognosis remains poor.

Gallstones

Gallstones are usually composed primarily of cholesterol with varying amounts of bilirubin and calcium salts. Cholesterol is maintained in solution in bile by virtue of the surface-active properties of bile salts and lecithin, but while a change in the proportion of these components may predispose to stone formation, other factors are also involved. Stones consisting primarily of bilirubin diglucuronide may develop in patients with chronic haemolytic anaemias.

Gallstones may be clinically silent. They can, however, cause biliary obstruction and predispose to cholecystitis, cholangitis and pancreatitis. Biochemical tests may be of value in the management of these conditions, but the analysis of biliary calculi is of no importance in the routine diagnosis or surgical management of patients with gallstones.

Drugs and the liver

The liver plays a central role in the metabolism of many drugs, converting them to polar, water-soluble metabolites which can be excreted in bile and urine. The enzymes involved are located in the smooth endoplasmic reticulum of the hepatocytes. Metabolism usually involves two types of reaction: phase 1 metabolism, for example, oxidation or demethylation by cytochrome P450-linked enzymes; and phase 2 metabolism in which phase 1 metabolites are conjugated with polar molecules, for example, glucuronic acid or glutathione.

Drug-induced liver damage may be predictable, arising when a toxic metabolite is produced by a phase 1 reaction at a rate which exceeds the detoxicating capacity of the phase 2 reaction as occurs, for example, in a paracetamol overdose. However, many drugs may cause toxic effects when used in therapeutic doses (Fig.6.10); this response (idiosyncratic hepatotoxicity) is unpredictable and is independent of the dose of the drug administered. Some idiosyncratic reactions to drugs, such as halothane-induced liver damage, have an immunological basis; the binding of a metabolite to a liver cell protein alters its antigenicity and provokes an immune response.

Some drugs are associated with the development of cholestasis; this may be an idiosyncratic response, as is the case with chlorpromazine, and additionally there is often evidence of liver cell damage. Other drugs, for instance, 17α-alkyl-substituted steroids, including some anabolic steroids, predictably cause cholestasis without hepatocellular damage when administered in high doses.

Minor degrees of hepatic dysfunction occur relatively frequently as results of idiosyncratic responses, but overt hepatotoxicity is fortunately rare.

Drugs causing jaundice
Dose-dependent hepatotoxicity
paracetamol (in overdose) salicylates (high doses only) tetracyclines (high doses only)
Idiosyncratic hepatotoxicity
isoniazid halothane methyldopa rifampicin
Dose-dependent cholestasis
methyltestosterone
Idiosyncratic cholestatic hepatitis
chlorpromazine erythromycin estolate

Fig.6.10 Drugs that cause jaundice. In addition, rifampicin also impairs bilirubin metabolism and induces hepatic enzymes involved in drug metabolism.

SUMMARY

The liver has a central role in intermediary metabolism and is also responsible for: detoxification of many foreign compounds; deamination of amino acids and synthesis of urea; synthesis and excretion of bile; metabolism of some hormones; synthesis of plasma proteins; and storage of certain vitamins.

Because of its considerable functional reserve, biochemical tests tend to be insensitive indicators of hepatic function, although they can be highly sensitive indicators of damage to the liver. The results of biochemical tests often indicate the nature of a liver disease, but less often indicate a specific diagnosis.

The most frequently performed biochemical tests are the measurement of serum bilirubin and serum albumin concentrations and the measurement of the activities of transaminases, alkaline phosphatase and γ-glutamyl transpeptidase in the serum.

A raised serum bilirubin concentration is a frequent but not invariable finding in patients with liver disease. However, conjugated hyperbiliribinaemia can result from extrahepatic biliary obstruction and a mild, unconjugated hyperbilirubinaemia is often a result of haemolysis. Greatly increased serum transaminase activities are characteristic of hepatocellular damage; greatly increased alkaline phosphatase activity is characteristic of biliary obstruction. However, transaminases may be increased irrespective of either the nature or cause of hepatocellular damage and alkaline phosphatase may be increased with both intra- and extrahepatic obstruction. In many patients with liver disease, moderate increases in both enzymes are observed. Further, changes in neither of these enzymes is specific to liver disease.

Serum γ-glutamyl transpeptidase activity is frequently increased in liver disease but an isolated increase may indicate excessive alcohol consumption; the finding of an increase in γ-glutamyl transpeptidase in a patient with an increased serum alkaline phosphatase level implies a hepatic origin for the latter. Albumin is synthesized by the liver but because of its long plasma half-life, serum albumin concentration tends to be decreased only in chronic liver disease. Many other factors can also affect albumin concentration. Blood clotting factors are synthesized in the liver and the prothrombin time provides a sensitive and rapidly responsive index of hepatic synthetic capacity.

Serum protein electrophoresis has no role in the investigation of liver disease; serum immunoglobulins are of limited value but serological tests for specific auto-antibodies are valuable in the differential diagnosis of chronic liver disease. Other tests which may be useful in specific liver diseases include the measurement of α-fetoprotein (liver cancer), α_1-antitrypsin (α_1-antitrypsin deficiency) and copper and caeruloplasmin (Wilson's disease).

Biochemical tests which reflect liver damage and function are in general cheap and simple to perform but they often do not provide a precise diagnosis. They are, however, invaluable in monitoring the course of liver disease and the response of patients to treatment.

FURTHER READING

Sherlock S (1985) *Diseases of the Liver and Biliary System*. 7th edition. Oxford: Blackwell Scientific.

7. The Gastrointestinal Tract

Gastrin	
Functions	**Control of secretion**

stimulation of:
    gastric acid secretion
    pepsin secretion
    gastric motility
    growth of gastric mucosa
	stimulated by:
	    increased vagal discharge
	    gastric distension
	    food, particularly amino
	        acids and peptides, in
	        stomach
	    calcium in blood
	
	inhibited by:
	    gastric acidity
	    gastrointestinal hormones,
	        e.g. secretin

Fig.7.1 Gastrin: functions and control of secretion.

INTRODUCTION

The digestion and absorption of food is a complex process which depends upon the integrated activity of the organs of the alimentary tract. This chapter covers those disorders of the stomach, pancreas and intestine for which biochemical tests are of particular importance in diagnosis and management.

THE STOMACH

In the stomach, food mixes with acidic gastric juice, which contains the proenzyme of pepsin, and with intrinsic factor, essential for the absorption of vitamin B_{12}. Secretion of gastric juice is under the combined control of the vagus nerve and the hormone, gastrin.

Gastrin is secreted by G-cells in the antrum of the stomach itself and has several physiological functions (Fig.7.1). It is a polypeptide hormone, present in the blood stream mainly in two forms: G-17 and G-34 containing seventeen and thirty-four amino acids respectively. Other gastrin molecules have been identified in the blood, but the physiological significance of this heterogeneity is unknown. All the variants have an identical C-terminal amino acid sequence.

Disorders and investigation of gastric function

Biochemical tests are of limited use in the diagnosis of gastric disorders; the stomach can be directly inspected by endoscopy and contrast radiography can also provide valuable information. Biochemical tests are used principally to investigate conditions in which gastric acid secretion is either excessive or inadequate.

Excessive gastric acid secretion is an important factor in the pathogenesis of duodenal, though not of gastric ulcers. The management of both types of peptic ulceration has changed radically since the introduction of H_2-blockers (antagonists of H_2-histamine receptors). These drugs inhibit gastric acid secretion and as a result of their use surgical procedures are now required far less frequently.

Maximal gastric acid secretion can be measured by the pentagastrin test (Fig.7.2); pentagastrin is a synthetic analogue of gastrin. Acid secretion tends to be higher than normal in patients with duodenal ulceration and low in patients with gastric ulceration, but these findings are inconsistent and of no diagnostic use.

In achlorhydria, basal gastric acid secretion is low or completely undetectable and there is no response to pentagastrin. This condition is most frequently seen in

Pentagastrin test	
Procedure	**Results**
fast patient overnight	resting juice: normally <50ml with a low acid content; large volume suggests gastric stasis
pass nasogastric tube under fluoroscopic control into stomach	
position tip in antrum	basal acid secretion: normally <5mmol/h; very high levels are characteristic of the Zollinger–Ellison syndrome
aspirate gastric juice (resting juice)	
aspirate stomach every 15 min, or continuously, for 1 hour (basal secretion)	stimulated acid secretion: normally <45mmol/h in males and <35mmol/h in females; pH does not fall below 7.0 in achlorhydria
combine samples for analysis	
give pentagastrin 6μg/kg body weight i.m.	
aspirate stomach every 15 min, or continuously, for 1 hour (stimulated secretion)	
analyze samples separately	
measure pH of all samples	
if acidic, titrate with sodium hydroxide and calculate acid secretion	

Fig.7.2 Pentagastrin test.

patients with atrophic gastritis, but is also present in pernicious anaemia and in association with gastric carcinoma.

In practice, pentagastrin tests are now most often performed to exclude achlorhydria as a cause of increased gastrin secretion in patients with peptic ulceration. Gastric acid secretion is also stimulated by stress, this response being mediated via the vagus nerve. The secretion of acid in response to stress induced by hypoglycaemia can be used to assess the completeness of a vagotomy, for example, if a patient thus treated develops recurrent ulceration.

Gastrin can be measured in the plasma by radio-immunoassay. As gastrin is very labile, the blood must be mixed with aprotinin, a protease inhibitor, immediately after venesection to prevent degradation. The usual indication for the measurement of gastrin in serum is atypical peptic ulceration, for example, duodenal ulcers resistant to medical treatment, recurrent duodenal ulcers after surgery, multiple duodenal ulcers and jejunal ulcers. A further indication is the presence of excessive gastric secretion detected endoscopically or radiologically. Atypical peptic ulceration

is a feature of Zollinger–Ellison syndrome, a rare condition in which hypergastrinaemia is caused by either a tumour (gastrinoma) of the endocrine pancreas or, less frequently, of the gastric G-cells. Approximately sixty percent of these tumours are malignant and in approximately twenty percent of cases they occur as part of a pluriglandular syndrome. Patients with Zollinger–Ellison syndrome may have gastric acid secretion rates in excess of 100mmol/h. They usually present with recurrent or atypical peptic ulceration and sometimes have steatorrhoea due to inhibition of pancreatic lipase by the excessive gastric acid.

Zollinger–Ellison syndrome is treated by surgical removal of the tumour, where possible. Vagotomy and long-term treatment with H_2-receptor antagonists may also be necessary and may be the only possible treatment if the tumour cannot be resected.

Other causes of hypergastrinaemia are described in Fig.7.3. The response to an intravenous bolus of secretin (1–2 u/kg body weight) and to a protein-rich meal may be useful in distinguishing between some of them and is also shown in Fig.7.3.

Causes of hypergastrinaemia			
Disorder	**Gastric acid secretion**	**Gastrin response to:** secretin	protein meal
Zollinger–Ellison syndrome	greatly ↑	↑	normal
hypersecretion of gastrin by antral G-cells	greatly ↑	↓	greatly ↑
pernicious anaemia	↓	↓	↑
post vagotomy	↓	↓	↑
chronic renal failure	variable	↓	normal

Fig.7.3 Some causes of hypergastrinaemia.

THE PANCREAS

The pancreas is an essential endocrine organ producing insulin, glucagon, pancreatic polypeptide and other hormones; its endocrine functions are discussed in *Chapter 12*. The exocrine secretion of the pancreas is an alkaline, bicarbonate-rich juice containing various enzymes essential for normal digestion: the proenzyme forms of the proteases, trypsin, chymotrypsin and carboxypeptidase; the lipolytic enzyme, lipase, and co-lipase; and amylase.

The secretion of pancreatic juice is primarily under the control of two hormones secreted by the small intestine: secretin, a twenty-seven amino acid polypeptide, which stimulates the secretion of an alkaline fluid, and cholecystokinin (CCK), which stimulates the secretion of pancreatic enzymes. Like gastrin, CCK is a heterogeneous hormone; the predominant form in the gut is a thirty-three amino acid polypeptide, but an eight amino acid form is present in some parts of the central nervous system and may function as a neurotransmitter. Both secretin and CCK are secreted in response to the presence of acid in the duodenum.

Investigation of pancreatic function

Tests of pancreatic function fall into two groups: direct and indirect, according to whether or not it is necessary to intubate the patient to obtain a sample of pancreatic juice. Tests involving intubation are unpleasant for the patient and the procedure itself requires considerable skill and time. Many tests have been developed and two particularly useful techniques for each group are described.

Direct tests

SECRETIN–CHOLECYSTOKININ TEST
In this test (Fig.7.4.) a double-lumen tube is used, with one orifice in the stomach and the other in the duodenum near the orifice of the pancreatic duct; this allows separate removal of gastric and duodenal juices and prevents the gastric juice from contaminating that in the duodenum. The volume of juice, the bicarbonate output and tryptic activity are measured.

A decrease in bicarbonate secretion is characteristic of chronic pancreatic insufficiency; enzyme activity is usually also reduced. Abnormal results are seen in some cases of pancreatic cancer, particularly when the tumour is in the head of the pancreas, in which enzyme secretion tends to be affected to a greater extent than bicarbonate. The range of normal responses is very wide; Fig.7.4 gives typical lower limits of normal.

Secretin – cholecystokinin test	
Procedure	**Results**

fast patient overnight
 	
intubate with a double-lumen tube
 	
aspirate throughout test but discard all gastric juice
 	
keep all samples on ice
 	
0 minutes:
  discard resting duodenal juice
  give secretin i.v.
  collect duodenal juice for three 10 min periods
    over 30 min
 	
30 minutes:
  give CCK i.v.
  collect duodenal juice as before
 	
analyze samples immediately for bicarbonate and tryptic activity and measure volume of juice
	lower limits of normal:
	  total volume of aspirate in 60 min 150ml
	
	  peak bicarbonate concentration 60mmol/l
	
	  total bicarbonate output
	    after secretin 5.7mmol
	
	  peak tryptic activity 30iu/ml
	
	  total tryptic activity after CCK 2200iu

Fig.7.4 Secretin–cholecystokinin test. The doses of secretin and CCK depend upon the brand of hormone used. The values for tryptic activity are guides only; the lower limit of normal for enzyme secretion should be determined by individual laboratories.

Lundh test	
Procedure	**Results**
fast patient overnight intubate with a single-lumen tube aspirate resting duodenal juice and discard give test meal aspirate duodenal juice for four 30 min periods chill samples on ice analyze samples for tryptic activity immediately	normal tryptic activity >25 iu/ml in at least one sample

Fig.7.5 Lundh test. The value given for normal tryptic activity is a guide only; each laboratory should determine its own normal value. The test meal consists of 18g corn oil, 40g glucose and 15g casilan in 300ml water.

LUNDH TEST

In this test (Fig.7.5) pancreatic secretion is stimulated physiologically by giving a test meal containing corn oil, skimmed milk powder and dextrose. A single-lumen tube is used and tryptic activity is measured in pooled duodenal juice collected over two hours. If properly performed, the test has the same sensitivity as the secretin–cholecystokinin test for the detection of pancreatic insufficiency.

Fluorescein dilaurate test	
Procedure	**Results**

day 1 (test):
    give 0.5mmol fluorescein dilaurate orally
    ensure adequate fluid intake
    collect urine for 10h
    measure amount of fluorescein excreted
 	
day 2 (control):
    give 0.5mmol fluorescein orally
    follow same procedure as day 1
	$\dfrac{\text{fluorescein excreted on day 1}}{\text{fluorescein excreted on day 2}} \times 100 = \dfrac{T}{K}\,\text{index}$
	
	normal pancreatic function: T/K index >30%
	
	pancreatic insufficiency: T/K index <20%

Fig.7.6   Fluorescein dilaurate test.

OTHER DIRECT TESTS

Pure pancreatic juice can be collected by endoscopic cannulation of the pancreatic duct, but the measurement of enzymes and bicarbonate in this fluid does not appear to offer any advantage over the other tests described. The concentration of lactoferrin, an iron-containing glycoprotein, appears to be increased in patients with chronic pancreatitis; similar results are obtained if the protein is measured in duodenal juice.

Indirect tests

FLUORESCEIN DILAURATE TEST

This test (Fig.7.6) is based on the principle that fluorescein dilaurate, administered orally, is hydrolyzed in the gut by pancreatic esterase. The fluorescein released is absorbed from the gut and excreted in the urine, where it can be measured. However, since pancreatic esterase is dependent on bile salts for its activity, the test effectively assesses combined pancreaticobiliary function; results may wrongly indicate pancreatic unsufficiency if bile-salt secretion is defective.

The possibility of any defect in either intestinal absorption or renal excretion is eliminated by comparing the excretion of fluorescein after the test with the excretion of non-esterified fluorescein given on a different day. Although not specific, this is a sensitive, simple and cheap test for assessing pancreatic function.

[14]C–PABA EXCRETION TEST

This test (Fig.7.7) utilizes a similar principle to the fluorescein dilaurate test, although a different enzyme is involved; the synthetic peptide, N-benzoyl-L-tyrosyl-p-aminobenzoic acid (BT-PABA) is hydrolyzed to p-aminobenzoic acid(PABA)by chymotrypsin. PABA is absorbed from the gut and excreted unchanged in the urine. To eliminate extrapancreatic factors, the BT-PABA is given with a tracer quantity of ^{14}C-labelled PABA and the amounts of PABA and ^{14}C excreted are measured and expressed as a ratio of the doses given.

This test is sensitive and highly specific for detecting pancreatic insufficiency, but has the disadvantage that facilities for counting β-emission are required. It is not suitable for use in pregnancy or in renal failure and some drugs, including paracetamol and sulphonamides, interfere with the measurement of PABA.

OTHER INDIRECT TESTS

The use of breath tests is described later in the section on tests of intestinal function. The measurement of serum amylase activity, useful in the diagnosis of acute pancreatitis, is of no value in the assessment of pancreatic function.

The most sensitive non-biochemical test for investigating pancreatic function is pancreatic scanning after administration of ^{75}Se-selenomethionine. The isotope is incorporated into pancreatic enzymes and its presence can be demonstrated by pancreatic scintiscan-

¹⁴C – PABA excretion test	
Procedure	**Results**

fast patient overnight
 	
give 0.5g BT–PABA, 5μCi ^{14}C–PABA and 25g casein in 500ml water orally
 	
collect urine for 6h
 	
analyze urine for PABA and ^{14}C content
	$\dfrac{(\text{PABA excreted/PABA ingested})}{(^{14}\text{C excreted}/^{14}\text{C ingested})} \times 100 = \text{PABA}/^{14}\text{C index}$
	
	normal pancreatic function:
	   PABA/^{14}C index >0.76
	
	pancreatic insufficiency:
	   PABA/^{14}C index <0.76

Fig.7.7 ^{14}C–PABA excretion test. Casein is used as a competitive inhibitor of chymotrypsin to increase the sensitivity of the test.

ning. Although the technique is sensitive, false positive results may occur; low pancreatic uptake of the iso-tope also occurs in some extrapancreatic conditions.

Disorders of pancreatic function

Disorders of the exocrine pancreas include acute and chronic pancreatitis, pancreatic carcinoma and cystic fibrosis. Clinical evidence of impaired exocrine func-tion is usually only seen in advanced disease. Endo-crine function tends to be well-preserved in all these conditions, although glucose intolerance may develop in severe or advanced disease. Diabetes mellitus and other conditions affecting only the endocrine pancreas are discussed in *Chapters 12 and 13*.

Other gastrointestinal disorders may mimic the effects of pancreatic disease; for example, enzymes may be destroyed in Zollinger–Ellison syndrome or be rendered less effective by an insufficient concentration of bile salts, while the control of pancreatic exocrine secretion may be affected by previous gastric surgery.

Acute pancreatitis

This condition presents as an acute abdomen with severe pain and a variable degree of shock. Causes include excessive alcohol ingestion, gallstones and vir-al infection. The pancreas becomes acutely inflamed and, in severe cases, haemorrhagic.

CASE HISTORY 7.1

A fifty-three-year-old man, who admitted to a heavy alcohol intake over many years, developed severe abdominal pain which radiated through to the back. The pain had started quite suddenly, eighteen hours before admission to hospital. He had no previous history of gastrointestinal disease. On examination, the patient was mildly shocked and his abdomen was tender in the epigastric region with slight guarding. There was no evidence of either intestinal obstruction or perforation of a viscus on radiographic examination. Blood was taken for urgent biochemical investigation.

Investigations

serum:	urea	10mmol/l
	creatinine	90μmol/l
	calcium	2.10mmol/l
	albumin	30g/l
	glucose	12mmol/l
	amylase	5000iu/l

Comment

The level of amylase in the serum cannot be used alone to diagnose pancreatitis (see *Chapter 17*) and it is necessary to consider all the available evidence. In this case, the history is suggestive of pancreatitis and the clinical findings, although non-specific, are consistent with this diagnosis. The radiological findings militate against, but do not exclude, intestinal obstruction and perforation, two important differential diagnoses.

The finding of a very high amylase concentration strongly supports a diagnosis of acute pancreatitis. The slightly raised urea, with normal creatinine, can be explained by renal hypoperfusion due to shock. Loss of protein-rich exudate into the peritoneal cavity commonly causes a fall in serum albumin concentration and contributes to the hypocalcaemia which is often present, especially in severe cases of acute pancreatitis. The formation of insoluble calcium salts of fatty acids, released within and around the inflamed pancreas by pancreatic lipase, may also contribute to hypocalcaemia and hormonal disturbances, for example, glucagon-stimulated release of calcitonin, have also been implicated. Hyperglycaemia may occur, but is usually transient.

In severe pancreatitis, methaemalbumin may be detectable in the serum, but this finding is not sufficiently consistent to be of diagnostic value. The serum of patients with pancreatitis may be lipaemic and there may be a mild increase in bilirubin and alkaline phosphatase.

The management of acute pancreatitis is essentially conservative. The gut is 'rested' by nasogastric aspiration and fluid, electrolyte and protein losses are replaced intravenously; parenteral nutrition may be required if the condition does not settle within a short period. Pain is controlled with appropriate analgesics; opiates may exacerbate the condition and should be avoided.

Chronic pancreatitis

Chronic pancreatitis is an uncommon condition which usually presents with malabsorption and occasionally with impaired glucose tolerance. The malabsorption is due to impaired digestion of foodstuffs, but there is considerable functional reserve and pancreatic lipase output must be reduced to only ten percent of normal before steatorrhoea is produced. Such a reduction only occurs in extensive disease or if the main pancreatic duct is obstructed. Alcohol is an important aetiological factor and there may be a history of recurrent acute pancreatitis.

The treatment of chronic pancreatitis involves treatment of the underlying cause, if known, and, since damage to the organ is irreversible, long-term treatment of its consequences by, for example, prevention of malabsorption by the addition of pancreatic extracts to food.

Carcinoma of the pancreas

Pancreatic carcinoma may be difficult to diagnose (see *Case History 6.5*). Presentation often occurs as a result of metastases rather than as a direct effect of the primary tumour. Other presentations include obstructive jaundice, when a tumour in the head of the pancreas obstructs the common bile duct, and malabsorption. Biochemical tests of pancreatic function are rarely of any use in diagnosis and other techniques, such as contrast radiography, are far more powerful diagnostic tools.

Pancreatic carcinoma is therefore usually diagnosed late, by which time metastases are often present and only palliative surgical procedures are feasible. However, even in advanced disease, the demonstration of a rise in serum amylase after administration of CCK suggests that surgical drainage of the pancreatic duct may be beneficial in relieving pain.

Cystic fibrosis

This is an inherited condition in which increased viscosity of pancreatic exocrine secretions results in obstruction of the pancreatic ducts and eventual fibrosis of the gland. The resultant malabsorption can be prevented by adding pancreatic extract to the food. Cystic fibrosis also affects mucus secretion in the bronchi, predisposing to recurrent respiratory infections and bronchiectasis, and biliary secretion, leading in some cases to cirrhosis. The diagnosis of cystic fibrosis is considered in *Chapter 18*.

Xylose absorption test	
Procedure	**Results**

fast patient overnight; bladder must be emptied before test
 	
0 minutes:
  give 5g xylose in water
  collect all urine passed until end of test
  patient should drink at least 500ml of water
    over the next 2h
 	
1 hour:
  draw blood and determine xylose level
 	
5 hours:
  collect final urine
  analyze urine for xylose
	normal serum xylose at 60min >1.3mmol/l
	
	normal urine xylose excretion >7.0mmol/5h

Fig.7.8   Xylose absorption test. Urinary xylose excretion can be measured over two hours (normal excretion >4.0mmol) but this is less reliable.

THE SMALL INTESTINE

The small intestine is the site of absorption of all nutrients; most of this absorption takes place in the duodenum and jejunum, but vitamin B_{12} and bile salts are absorbed in the terminal ileum. Approximately eight litres of fluid enters the gut every twenty-four hours. This is derived from ingested food and water and from the digestive juices, including those secreted by the small intestine itself. Most of this fluid, and the salts it contains, is reabsorbed in the jejunum, ileum and large intestine.

Investigation of intestinal function

Disorders of the small intestine can lead to malabsorption of nutrients. Many biochemical tests are available to monitor the absorption of nutrients and these are widely used in the investigation of patients with suspected malabsorption.

Tests of carbohydrate absorption

The most widely used test of carbohydrate absorption is the xylose absorption test (Fig.7.8). Xylose, a plant sugar, is absorbed from the jejunum without prior digestion. It is not metabolized in the body and is excreted unchanged in the urine where it can be measured. An accurately timed urine collection is essential.

Misleadingly low results are obtained if the glomerular filtration rate is decreased, as occurs in renal failure and many normal elderly people. Other factors which can produce misleading results include delayed gastric emptying, oedema, obesity and metabolism of xylose by bacteria in the gut lumen if there is bacterial overgrowth.

An alternative approach is to measure serum xylose concentration sixty minutes after the xylose is administered. This test is cheap and simple to perform. False positive results do occur, but a normal result reliably excludes proximal small bowel disease. Xylose absorption is usually normal in patients with small bowel disease affecting the ileum alone and in patients with pancreatic malabsorption.

Impaired absorption of glucose may produce a 'flat' response in a glucose tolerance test. However, the number of other factors involved in determining the response to a glucose load is such that this test is of no practical value in the diagnosis of malabsorption.

Intestinal disaccharidase deficiency may be diagnosed by administering the appropriate disaccharide orally and measuring the blood glucose response. To increase sensitivity, the test is performed with the disaccharide (50g) and then with the equivalent quantities (25g each) of the constituent monosaccharides. The commonest of these disorders is lactase deficiency, which may be congenital or acquired; it often occurs transiently when there is damage to gut mucosa, such as after gastroenteritis. Lactose itself cannot be absorbed and in lactase deficiency lactose reaches the colon and undergoes bacterial fermentation; hydrogen, a byproduct of this process, can be measured in expired air by gas chromatography. Breath hydrogen excretion is also high in patients with bacterial overgrowth in the small intestine. The definitive test for disaccharidase deficiencies is measurement of the appropriate enzyme in a biopsy sample.

Tests of amino acid absorption

Tests of amino acid absorption from the gut are only used as research procedures. Generalized malabsorption of amino acids occurs only with extensive small bowel disease. Malabsorption of specific amino acids occurs in certain inherited metabolic disorders; for example, deficiency of tryptophan may occur in Hartnup disease, an inherited disorder of the renal transport of neutral amino acids. In cystinuria there is impaired transport of the dibasic amino acids, lysine, cystine, ornithine and arginine, but this condition is not associated with a deficiency syndrome.

Loss of protein from the gut in a protein-losing enteropathy can be assessed by measuring faecal radioactivity after parenteral administration of isotopically labelled proteins or polyvinylpyrrolidine. Such investigations are not commonly performed, however, since the cause of any hypoproteinaemia is usually obvious in such conditions.

Tests of fat absorption

FAECAL FAT TEST

Fat absorption has traditionally been assessed by measuring the excretion of fat in faeces. After digestion, dietary fat is normally absorbed completely in the small intestine; a small quantity of fat (<18mmol/24h) is excreted in faeces but this is derived from enterocytes.

With malabsorption of fat, its excretion in the faeces is increased. However, a major problem with its measurement is the need to obtain an accurately timed faecal collection. Collections should preferably be made for five consecutive days, although, for practical reasons, three day collections are often used.

Accuracy of timing can be improved by using a non-absorbable coloured marker such as carmine. This is administered orally and faecal collection is started when marker appears in the stool; a second marker is given 120 (or 72) hours after the first, and collection is terminated when this appears.

This test is unpleasant for all concerned and is only of value if carried out correctly. Dietetic guidance should be sought to ensure that the patient consumes 90–100g fat per day for forty-eight hours before and during the period of collection; if less fat is ingested, minor degrees of malabsorption may be missed. In severe malabsorption with obvious steatorrhoea, quantifying the faecal fat excretion adds nothing to the diagnosis.

^{14}C–TRIOLEIN BREATH TEST

Due to the unpleasantness of the faecal fat test and its impracticality as an outpatient procedure, there has been considerable enthusiasm for the development of alternative tests. The ^{14}C–triolein breath test is probably the most reliable alternative (Fig.7.9). Facilities for counting β-radiation are required, but the test takes only a few hours and can be performed on a day-ward. The test is based on the principle that when ^{14}C-labelled triglyceride is taken orally, digested and absorbed, some of the label appears in the breath as ^{14}C-labelled carbon dioxide. Since it is the specific activity of the expired carbon dioxide that is measured, a constant rate of production of carbon dioxide from all other sources must be assumed; patients must be fasting and must rest throughout the test.

The ^{14}C–triolein breath test is not reliable in patients with diabetes, obesity, thyroid disease or chronic respiratory insufficiency and is not suitable for use in pregnancy. Properly performed, however, it is a sensitive test for fat malabsorption and results correlate well to those of faecal fat excretion. However, neither test differentiates between the different causes of fat malabsorption (see page 101).

Modifications of this breath test, in which the absorption of labelled triglyceride and labelled free fatty acids is compared, have been described but are not reliable. In practice, the ^{14}C–triolein breath test is

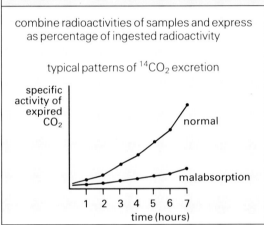

^{14}C – triolein breath test
Procedure
fast patient overnight collect basal sample of expired CO_2 (1mmol) give 10μCi ^{14}C – triolein in 60g fat meal collect 1mmol samples of expired CO_2 hourly for 7h measure radioactivity of CO_2 samples
Results
combine radioactivities of samples and express as percentage of ingested radioactivity typical patterns of $^{14}CO_2$ excretion

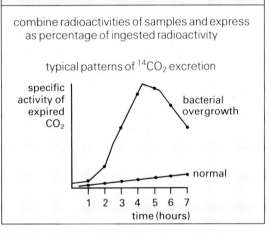

^{14}C – glycocholic acid breath test
Procedure
fast patient overnight collect basal sample of expired CO_2 (1mmol) give 10μCi ^{14}C – glycocholic acid in test meal collect 1mmol samples of expired CO_2 hourly for 7h measure radioactivity of CO_2 samples
Results
combine radioactivities of samples and express as percentage of ingested radioactivity typical patterns of $^{14}CO_2$ excretion

Fig.7.9 ^{14}C–triolein breath test. CO_2 samples are collected by bubbling expired air into vials containing 1mmol hyamine which reacts with the CO_2. An indicator is used which changes colour when the reaction is complete. Individual laboratories should determine their own lower limit of normal for $^{14}CO_2$ excretion.

Fig.7.10 ^{14}C–glycocholic acid breath test. CO_2 is collected as in the ^{14}C–triolein test. The test meal can be the same as that used in the Lundh test. (see Fig.7.5).

best used to diagnose fat malabsorption in doubtful cases, while other tests are used to define the cause.

Tests for bacterial overgrowth

Bacterial overgrowth of the small intestine can occur in a number of conditions, particularly when there is stasis of gut contents, for example, in jejunal diverticulosis. Bacterial deconjugation of bile salts leads to malabsorption of fat and failure of mixed micelle formation.

The most reliable diagnostic test for bacterial overgrowth is the aspiration and culture of duodenal con-

tents. However, this method has disadvantages: it is an invasive procedure and the cultures are sometimes negative when other evidence of bacterial overgrowth is overwhelming.

The measurement of urinary indicans (products of the bacterial metabolism of tryptophan) is widely used to screen for bacterial overgrowth, but results correlate poorly with those of duodenal aspiration.

A breath test using a radioactive bile acid (^{14}C–glycocholic acid) can be used as an indirect test for the presence of bacterial overgrowth in the small intestine (Fig.7.10); the procedure is similar to that for the ^{14}C–triolein breath test. Normally, bile salts are absorbed in the terminal ileum, some metabolism takes place and small quantities of ^{14}C–carbon dioxide appear in the expired air after several hours. With bacterial overgrowth in the proximal small bowel, deconjugation occurs and ^{14}C–glycine is released; this is rapidly absorbed and metabolized to carbon dioxide, causing the specific activity of expired carbon dioxide to rise after a short period. Deconjugation and release of isotopically labelled glycine can also occur if, as a result of malabsorption, bile salts reach the colon which has a rich bacterial flora. This occurs with diseases affecting the terminal ileum, where bile salts are normally absorbed, for example, Crohn's disease.

Theoretically, these two causes of increased isotope excretion should be distinguishable on the basis of faecal radioactivity, which should be low with bacterial overgrowth in the small gut and increased with ileal disease. In practice, this is not always the case and it is necessary to combine the bile acid breath test with other tests, such as the Schilling test (see following), to obtain reliable results.

Tests of terminal ileal function

The use of the bile acid breath test in this context has already been discussed. The Schilling test is used to assess the absorption of vitamin B_{12}, particularly in patients with suspected pernicious anaemia. Abnormal results are seen in many patients with disease of the terminal ileum, but false negatives do occur and abnormal results may be seen in patients with bacterial overgrowth of the small intestine. The Schilling test is usually performed by a haematologist.

Non-biochemical tests of intestinal function

The mucosa of the small intestine can be biopsied using a Crosby capsule; this is the definitive procedure for the diagnosis of coeliac disease (gluten-induced enteropathy, see *Case History 7.3*). The diagnosis of disaccharidase deficiencies can also be confirmed by measuring the enzyme in an intestinal biopsy.

Characteristic radiographical appearances are seen in patients with certain intestinal diseases, for example, Crohn's disease (see *Case History 7.4*). Biochemical tests, however, continue to be used for diagnosis of the malabsorption syndrome.

DISORDERS OF INTESTINAL FUNCTION

Malabsorption

The term malabsorption strictly refers to impaired absorption of the products of digestion, whilst maldigestion is failure of digestion which may be responsible for non-absorption of nutrients, for example, with pancreatic insufficiency. In practice, since the resultant clinical syndromes are basically the same, the term malabsorption is commonly used to encompass both disorders.

The clinical features of malabsorption are varied and stem from either deficiency of nutrients or retention of nutrients within the bowel lumen. The clinical features and common causes of malabsorption are shown in Fig.7.11.

More than one mechanism can be responsible for malabsorption in individual cases. After gastric surgery, for example, impaired mixing of food with digestive juices, decreased stimuli to their secretion, rapid transit and bacterial colonization of a blind afferent loop may all contribute to malabsorption.

Investigations are required for two purposes: to diagnose malabsorption and to determine its cause. Simple tests, for example, haemoglobin, prothrombin time, serum albumin, calcium, phosphate and alkaline phosphatase, should be performed first; if the results of these are normal, malabsorption is very unlikely and further expensive and invasive tests can be avoided.

CASE HISTORY 7.2

A middle-aged publican presented with flatulence and abdominal distension. On questioning, he admitted to weight loss and to passing frequent, bulky, foul-smelling bowel motions which were difficult to flush away.

Investigations

serum:		
	calcium	2.10mmol/l
	phosphate	0.70mmol/l
	glucose (fasting)	12mmol/l
	alkaline phosphatase	264iu/l
	albumin	40g/l

Malabsorption	
Clinical features	**Causes**

Retention of non-absorbed nutrients
 	
diarrhoea, steatorrhoea
abdominal discomfort and distention
flatulence
 	
Decreased absorption of nutrients
 	
anaemia (iron, folate and vitamin B_{12} deficiency)
osteomalacia and rickets (vitamin D deficiency)
oedema (hypoalbuminaemia)
bleeding tendency (vitamin K deficiency)
weight loss; growth failure in children
	pancreatic enzyme deficiency, e.g. chronic
	   pancreatitis and cystic fibrosis
	bile-salt deficiency, e.g. biliary obstruction and
	   hepatic disease
	intestinal, e.g. coeliac disease, tropical sprue,
	   Crohn's disease and partial resection
	bacterial overgrowth, e.g. gastric surgery,
	   internal fistulae, strictures and jejunal
	   diverticulosis

Fig.7.11 Malabsorption: clinical features and common causes. There may be several reasons for the development of malabsorption in individual cases.

A plain abdominal radiograph revealed pancreatic calcification.

Comment

The clinical features are characteristic of malabsorption (Fig.7.11). The patient is hypocalcaemic and hypophosphataemic with a raised alkaline phosphatase due to vitamin D deficiency with secondary hyperparathyroidism. With gross steatorrhoea, further investigations to establish that the patient has malabsorption are not necessary, but the cause must be determined.

The presence of pancreatic calcification is very suggestive of alcohol-induced chronic pancreatitis. The raised fasting level of glucose, indicating glucose intolerance, is compatible with chronic pancreatitis and no further investigations of pancreatic function were performed. The patient was given pancreatic extract to add to his food and the symptoms regressed. This therapeutic response provides further confirmation of the diagnosis of pancreatic insufficiency.

CASE HISTORY 7.3

A three-year-old boy was referred for the investigation of failure to thrive; he was below the third centile for height and the tenth for weight, although both parents were tall. The boy had frequent diarrhoea and did not appear to enjoy his food. On examination, he was anaemic and had abdominal distension; there was obvious wasting of the muscles of the limbs, buttocks and shoulder girdle.

Investigations

serum:	albumin	30g/l
	xylose (1h after 5g orally)	0.5mmol/l
haemoglobin		9.7g/dl

A jejunal biopsy showed total villous atrophy. A blood film showed hypochromic, microcytic red cells.

Comment

There are many causes of growth failure. In this case, the history and findings on examination suggest a gastrointestinal disorder. Hypoproteinaemia and a hypochromic, microcytic anaemia, characteristic of iron deficiency, are common in patients with malabsorption. The grossly abnormal xylose absorption (see Fig.7.8) indicates an intestinal lesion. The biopsy appearance is characteristic of coeliac disease, or gluten-induced enteropathy.

In this condition, damage to the small intestine occurs on exposure to gluten, a protein in wheat flour; it varies considerably in severity. It may present either in infancy with severe failure to thrive and gross steatorrhoea, in childhood or not until adult life. Growth failure is almost invariable in children. Complete withdrawal of wheat protein from the diet results in the regrowth of intestinal villi and resolution of symptoms. Secondary lactose intolerance is common and may persist for a period after treatment has been started.

CASE HISTORY 7.4

A thirty-five-year-old woman was referred for the investigation of diarrhoea and abdominal pain. She had lost weight and was clinically anaemic. She had had two previous episodes of the same symptoms, lasting for several weeks on each occasion in the preceding two years, but had not sought medical advice.

Investigations

serum albumin	28g/l
haemoglobin	8.5g/dl
red cell volume	110fl

A ^{14}C–triolein breath test was performed; the excretion of ^{14}C–carbon dioxide was very low and for a few hours after having had the fat meal the woman experienced abdominal discomfort and distension. A barium meal and follow-through revealed narrowing and ulceration of the terminal ileum, with an ileo-ileal fistula.

Comment

Weight loss is a common feature of gastrointestinal disease, even without malabsorption, and the breath test was used essentially to screen for possible malabsorption.

The patient did not have steatorrhoea; this was ascribed to her habitual low fat diet, prescribed for familial hypercholesterolaemia some years before. The development of symptoms when she was challenged with fat suggests that her symptoms might have been more florid if she had had a normal fat intake. There was nothing in the history specifically to suggest a biliary or pancreatic disorder and the diagnosis was made radiologically.

The radiographical appearances are typical of Crohn's disease, an inflammatory disease of the gut in which ulceration and fibrosis occur and may lead to the formation of strictures and fistulae. Although the condition can affect any part of the gut, the ileum is most often involved. The course is often one of remission and exacerbations.

In the acute illness, sulphasalazine, steroids and azathioprine may be used and nutritional support is often necessary. Surgery may be required for intestinal obstruction and fistulae, or if medical treatment fails. Malabsorption in Crohn's disease may be due to either damage to the ileum or bacterial overgrowth of a stagnant loop, a possible consequence of internal fistula formation.

Other intestinal disorders

Given the amount of fluid that enters the gut each day there is considerable potential for fluid and electrolyte depletion in situations of impaired reabsorption. Dehydration can complicate prolonged vomiting and diarrhoea, and enterocutaneous fistulae. Potassium depletion is also frequently associated with excessive loss of fluid from the gastrointestinal tract.

In some cases, there is increased secretion of fluid

Gastrointestinal hormones		
Hormone	**Location**	**Function**
gastrin	gastric antrum	stimulates gastric acid secretion (see Fig. 7.1)
CCK	duodenum, jejunum	stimulates pancreatic enzyme secretion and gallbladder contraction
secretin	duodenum, jejunum	stimulates pancreatic bicarbonate secretion
pancreatic polypeptide (PP)	pancreas	inhibits exocrine pancreatic secretion
gastric inhibitory polypeptide (GIP)	duodenum, jejunum	releases insulin in response to glucose and inhibits gastric acid secretion
VIP	entire GI tract	? neurotransmitter and regulates GI motility and secretion
motilin	duodenum, jejunum	stimulates GI motility

Fig. 7.12 Gastrointestinal hormones: locations and functions. ? = exact function unknown.

into the gut; for example, in cholera, massive fluid loss can occur very rapidly. Secretory diarrhoea also occurs with villous adenomata of the rectum, with tumours which secrete large volumes of potassium-rich mucus and with tumours secreting vasoactive intestinal polypeptide (VIP) which cause profuse watery diarrhoea, the Werner–Morrison syndrome.

GASTROINTESTINAL HORMONES

The functions of gastrin, secretin, CCK, insulin and glucagon have been well understood for some time. In recent years, a number of other gastrointestinal polypeptide hormones have been discovered (Fig. 7.12). Although many of their properties are known, their exact physiological functions are incompletely understood.

At present, assays for these hormones are available only in specialized laboratories and the indication for measuring them for diagnostic purposes is largely confined to cases of suspected hormone-secreting tumours, for example, in the Werner–Morrison syndrome.

SUMMARY

The gastrointestinal tract is responsible for the digestion and absorption of food. This process also depends upon normal hepatic and pancreatic function and is controlled by both neural and humoral mechanisms.

The malabsorption syndrome can be a result of intestinal, pancreatic or hepatic dysfunction. Significant malabsorption can readily be excluded by simple tests on blood or serum. If malabsorption is obvious clinically, for instance, because of weight loss and gross steatorrhoea, tests for malabsorption add nothing to the diagnosis. Such tests, for example, the ^{14}C–triolein breath test, should be used only in doubtful cases. Once a diagnosis of malabsorption has been made, investigations are required to determine the cause if this is not obvious clinically. Biochemical, histological and radiological data may all be useful in this context.

Formal assessment of gastric acid secretion is now seldom required, but measurement of the hormone, gastrin, which stimulates gastric acid secretion, is valuable in patients with atypical peptic ulceration, since this may be due to a gastrin-secreting tumour. Many other gut hormones have been described. The measurement of some of them may similarly be useful in the investigation of patients suspected of having a hormone-secreting tumour.

Chronic pancreatitis usually presents with malabsorption but acute pancreatitis typically presents as an acute abdomen. Increased serum amylase activity is characteristic of, though not specific to, acute pancreatitis. In the acute condition, shock, renal failure, hypocalcaemia and hyperglycaemia may be complicating factors.

Approximately eight litres of fluid are secreted into the gut each day, the great majority of which is reabsorbed. The loss of fluid and salts from the gut because of vomiting, diarrhoea or a fistula can lead to severe salt and water depletion.

FURTHER READING

Bouchier IAD, Allan RN, Hodgson HJF & Keighley MRB (1984) *Textbook of Gastroenterology.* London: Bailliere Tindall.

Sleisenger MH (ed.) (1983) Malabsorption and nutritional support. *Clinics in Gastroenterology,* **12**, 323–613.

8. The Hypothalamus and Pituitary Gland

INTRODUCTION

The pituitary gland consists of two parts, the anterior pituitary, or adenohypophysis, and the posterior pituitary, or neurohypophysis. Though very closely related anatomically, they are embryologically and functionally quite distinct. The anterior pituitary is comprised primarily of glandular tissue, while the posterior pituitary is of neural origin. The pituitary gland is situated at the base of the brain, in close relation to the hypothalamus (Fig.8.1) which has an essential role in the regulation of pituitary function.

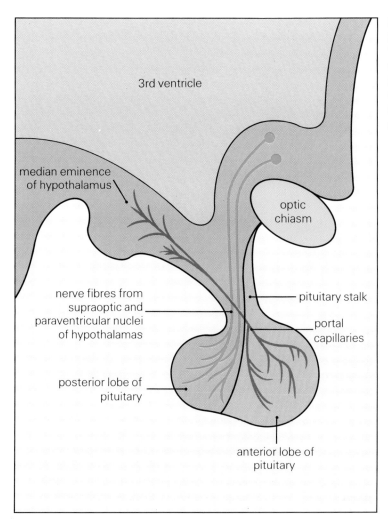

3rd ventricle

median eminence
of hypothalamus

optic
chiasm

nerve fibres from
supraoptic and
paraventricular nuclei
of hypothalamas

pituitary stalk

portal
capillaries

posterior lobe of
pituitary

anterior lobe of
pituitary

Fig.8.1 Anatomical relationship of the pituitary gland and hypothalamus. The portal blood vessels, through which hypothalamic hormones reach the anterior pituitary, and nerve fibres which transport hypothalamic hormones to the posterior pituitary are shown.

Anterior pituitary hormones		
Hormone	**Target organ**	**Action**
growth hormone (GH)	liver	somatomedin synthesis, hence growth stimulation
	others	metabolic regulation
prolactin	breast	lactation
thyroid-stimulating hormone (TSH)	thyroid	thyroid hormone synthesis and release
follicle-stimulating hormone (FSH)	ovary	oestrogen synthesis oogenesis
	testis	spermatogenesis
luteinizing hormone (LH)	ovary	ovulation corpus luteum, hence progesterone, production
	testis	testosterone synthesis
adrenocorticotrophic hormone (ACTH)	adrenal cortex	glucocorticoid synthesis and release
	skin	pigmentation
β-lipotrophin		precursor of endorphins

Fig.8.2 Anterior pituitary hormones and their actions. All the actions shown are stimulatory; trophic hormones stimulate both synthesis and release of hormones by their target organs.

ANTERIOR PITUITARY HORMONES

The anterior pituitary secretes several hormones, some of which are trophic, that is, they stimulate the activity of other endocrine glands (Fig.8.2). The secretion of hormones by the anterior pituitary is controlled by hormones secreted by the hypothalamus which reach the pituitary through a system of portal blood vessels. The secretion of the hypothalamic hormones is influenced by higher centres in the brain and the secretion of both hypothalamic and pituitary hormones is regulated by feedback from the hormones whose production they stimulate in target organs.

Growth hormone

Growth hormone (GH) is a 191-amino acid polypeptide hormone. Its release is controlled by two hypothalamic hormones, growth hormone releasing hormone (GHRH) and growth hormone release inhibiting hormone (somatostatin). It is essential for normal growth, although it acts indirectly by causing the liver to produce various polypeptide growth factors known as somatomedins. GH has a number of metabolic effects which are summarized in Fig.8.3.

The concentration of GH in the blood varies widely through the day and at certain times it may be undetectable by presently available radioimmunoassays. Physiological secretion occurs in sporadic bursts, lasting for one to two hours and mainly at night. Secretion can be stimulated by stress, exercise, a fall in blood glucose concentration, fasting and certain amino acids. Such stimuli are employed in provocative tests for diagnosing GH deficiency, particularly in children. GH secretion is inhibited by a rise in blood glucose and this effect provides the rationale for the use of the oral glucose tolerance test in the diagnosis of excessive

---

Metabolic actions of growth hormone
increases lipolysis (hence ketogenic)
increases hepatic glucose production and decreases tissue glucose uptake (hence diabetogenic)
increases protein synthesis (hence anabolic)

Fig.8.3 Metabolic actions of GH.

GH secretion. Excessive secretion due to a pituitary tumour causes gigantism in children and acromegaly in adults; a deficiency causes growth retardation in children but is usually clinically silent in adults.

Somatostatin, the fourteen amino acid hypothalamic peptide which inhibits GH secretion has many other actions both within the hypothalamic–pituitary axis and elsewhere. For example, it inhibits the release of thyroid-stimulating hormone (TSH) in response to thyrotrophin releasing hormone (TRH) and it is present in the gut and pancreatic islets where it inhibits the secretion of many gastrointestinal hormones. The physiological significance of these actions is poorly understood. Rare somatostatin-secreting tumours of the pancreas have been described.

Prolactin

The principal physiological action of prolactin is to initiate and sustain lactation. There is no known hypothalamic prolactin-releasing factor, but instead its secretion is controlled by an inhibitory factor, prolactin inhibitory factor (PIF), now known to be dopamine. Increased prolactin secretion occurs with prolactin-secreting tumours but is a common feature of other pituitary disorders, presumably because they interrupt this inhibitory mechanism. Hence, if dopamine is absent, prolactin secretion is autonomous.

The secretion of prolactin is pulsatile, increases at night, is increased by stress and in women is dependent upon oestrogen status, making it difficult to define a precise upper limit for serum prolactin concentration in normal men and women. Its secretion increases during pregnancy but levels fall to normal within approximately seven days after birth if a woman does not breast feed. With breast feeding,

levels start to decline after about three months, even if breast feeding is continued beyond this time. Excessive levels of prolactin inhibit ovulation and infertility is a common feature in female patients with tumours secreting the hormone. Other features of hyperprolactinaemia are discussed on page 118. Prolactin deficiency is uncommon but does occur, for example, with pituitary infarction; its only manifestation is failure of lactation.

Thyroid-stimulating hormone

Thyroid-stimulating hormone (TSH) is a glycoprotein (molecular weight 28,000 daltons) composed of an α- and a β-subunit; the α-subunit is common to TSH and the gonadotrophins and is almost identical to that of human chorionic gonadotrophin (hCG), but the β-subunit is unique to TSH. TSH binds to specific receptors on thyroid cells and this stimulates the synthesis and secretion of thyroid hormones. The secretion of TSH is stimulated by the hypothalamic tripeptide, thyrotrophin releasing hormone (TRH), and this effect, and probably the release of TRH itself, is inhibited by high circulating levels of thyroid hormones. Dopamine appears to inhibit TSH secretion, but the physiological relevance of this is uncertain.

Thus thyroid hormone synthesis is regulated by a negative feedback system: if serum concentrations of thyroid hormones decrease, TSH secretion increases stimulating thyroid hormone synthesis; if thyroid hormone levels increase, TSH secretion is suppressed. In primary hypothyroidism, TSH secretion is increased; in hyperthyroidism it is decreased. Pituitary tumours secreting TSH occur but are extremely rare. TSH deficiency, however, may occur in pituitary disease and cause secondary hypothyroidism.

Gonadotrophins

Follicle-stimulating hormone (FSH) and luteinizing hormone (LH) are both glycoproteins consisting of two subunits: the β-subunits are unique to each hormone but the α-subunit is the same in each, is present also in TSH and is similar to that in hCG.

The synthesis and release of both hormones are stimulated by the hypothalamic decapeptide, gonadotrophin releasing hormone (GnRH), this effect being modulated by circulating gonadal steroids. GnRH is secreted episodically resulting in pulsatile secretion of gonadotrophins with peaks in serum concentration

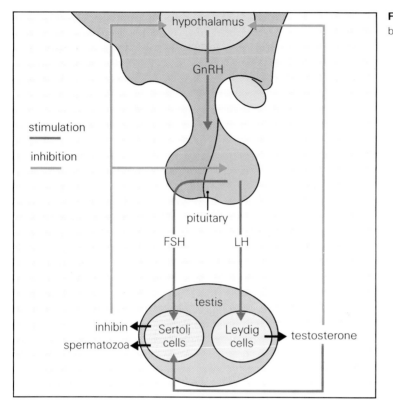

Fig.8.4 Control of testicular function by pituitary gonadotrophins.

occurring at approximately ninety minute intervals. In males, LH stimulates testosterone secretion by Leydig cells and both testosterone and oestradiol, derived from the Leydig cells themselves and from the metabolism of testosterone, feed back to block the action of GnRH on LH secretion. FSH, in concert with high intratesticular testosterone concentrations, stimulates spermatogenesis and its secretion is inhibited by inhibin (Fig.8.4), a hormone produced during spermatogenesis.

In the female, the relationships are more complex. Oestrogen (primarily oestradiol) secretion by the ovary is stimulated primarily by FSH in the first part of the menstrual cycle; both hormones are necessary for the development of the Graafian follicle. As oestrogen concentrations in the blood rise, FSH secretion declines until oestrogens trigger a positive feedback mechanism, causing an explosive release of LH and, to a lesser extent, FSH. The increase in LH stimulates ovulation and development of the corpus luteum but rising levels of oestrogens and progesterone then inhibit FSH and LH secretion; inhibin from the ovaries also appears to inhibit FSH secretion. If conception does not occur, declining levels of oestrogens and progesterone from the regressing corpus luteum trigger menstruation and LH and FSH release, initiating the maturation of further follicles in a new cycle (Fig.8.5). Before puberty, serum levels of LH and FSH are very low and unresponsive to exogenous GnRH. With the approach of puberty, FSH secretion increases first, but as puberty progresses LH secretion increases and the usual adult pattern with the serum levels of LH exceeding those of FSH becomes established.

Increased levels of gonadotrophins are seen in ovarian failure in women, whether pathological or after the menopause. High levels of FSH are seen in azoospermic men and LH is increased if testosterone secretion is decreased. Gonadotrophin-secreting tumours of the pituitary are rare but gonadal failure secondary to failure of gonadotrophin secretion, either as an isolated phenomenon due to hypothalamic dysfunction or as part of generalized pituitary failure, is well recognized. A case of hypogonadotrophic hypogonadism is described in *Case history 11.1*.

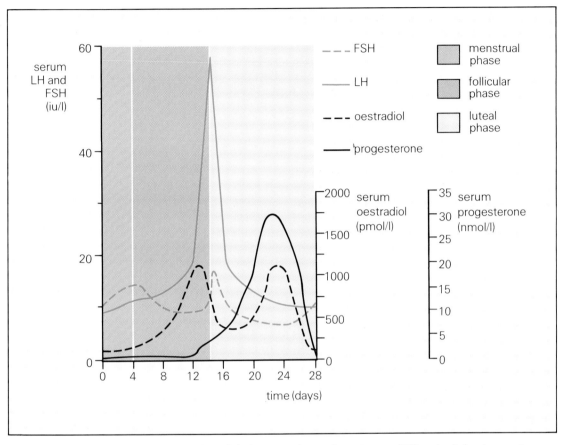

Fig. 8.5 Changes in the serum concentrations of pituitary gonadotrophins during the menstrual cycle. The resultant changes in oestrogens (17β-oestradiol) and progesterone concentration are also shown.

Adrenocorticotrophic hormone

Adrenocorticotrophic hormone (ACTH) is a polypeptide (molecular weight 4500 daltons), comprising a single chain of thirty-nine amino acids. Its biological function, which is to stimulate adrenal glucocorticoid but not mineralocorticoid secretion, is dependent upon the N-terminal twenty-four amino acids. ACTH is a fragment of a much larger precursor pro-opiomelanocortin (molecular weight 31,000 daltons), which is the precursor not only of ACTH but also of β-lipotrophin, itself the precursor of endogenous opioid peptides (endorphins). The control of the release of β-lipotrophin and the endorphins has not been fully elucidated, but ACTH release is controlled by a hypothalamic peptide, corticotrophin releasing hormone (CRH). ACTH secretion is pulsatile and also shows diurnal variation, the serum concentration being highest at approximately 0800h and lowest at midnight. Secretion is greatly increased by stress and is inhibited by cortisol. Thus cortisol secretion by the adrenal cortex is controlled by negative feedback.

Excessive ACTH synthesis is associated with increased pigmentation. This is due to an intrinsic melanocyte-stimulating action of ACTH. Although a separate melanocyte-stimulating hormone (MSH) has been described in other animals it does not occur in man. Increased secretion of ACTH by the pituitary is seen with pituitary tumours (Cushing's disease) and in primary adrenal failure (Addison's disease). The hormone may also be secreted ectopically by non-pituitary tumours. Decreased secretion may be an isolated phenomenon but is more commonly associated with generalized pituitary failure.

MEASUREMENT OF ANTERIOR PITUITARY HORMONES

Hormones produced by the anterior pituitary can be measured in serum by immunoassay, although in some cases the sensitivity of the assays is insufficient to distinguish reliably between normal and reduced concentrations. The pulsatility of the secretion of some of these hormones makes it inappropriate to rely on single measurements for diagnostic purposes. Hence, when measuring trophic hormones, it is often useful to measure both the pituitary hormone and that produced by the target organ. For example, a low serum thyroxine concentration with an elevated serum TSH implies primary hypothyroidism, whereas a low TSH with a low thyroxine suggests decreased pituitary secretion of TSH causing secondary hypothyroidism.

Dynamic tests are important tools in the investigation of pituitary function. There are two types: stimulatory tests, used to investigate suspected pituitary hypofunction, and suppression tests, used to investigate suspected hyperfunction. Tests relating to individual hormones will be described in later sections of this chapter in the context of the clinical conditions in which they are used. In assessing patients with suspected anterior pituitary dysfunction, however, it is often convenient to test the capacity of the gland to secrete GH, prolactin, ACTH, TSH and the gonadotrophins in a single procedure.

The combined pituitary function test (triple bolus test) involves giving a single intravenous bolus of a mixture of TRH, GnRH and insulin. The insulin induces hypoglycaemia, the stress of which stimulates ACTH and GH release. The test is potentially hazardous because of the possible sequelae of hypoglycaemia. A doctor should always be present when the test is performed. It is contra-indicated in patients with a history of fits or ischaemic heart disease and it should not be performed in patients whose 0900h serum cortisol concentration is low. 50ml of 50% dextrose solution must be available for immediate administration should severe hypoglycaemia develop. Giving glucose because of severe symptomatic hypoglycaemia does not invalidate the test. The stress need only be very brief to be effective. It is important that documented hypoglycaemia is obtained, since if it does not occur a failure of ACTH response might be due to the inadequacy of the stimulus rather than to pituitary failure. If hypoglycaemia does not develop, a further dose of insulin must be given. The protocol for this test and the normal responses are shown in Fig.8.6.

It may be preferable to give the insulin by continuous intravenous infusion. The rate can be adjusted until hypoglycaemia develops, whereupon the infusion is stopped. This is a more certain and safer way of producing hypoglycaemia than giving a single bolus of insulin. When the induction of hypoglycaemia is contra-indicated, glucagon may be used to stimulate cortisol and GH secretion instead of insulin.

Because the assay of ACTH is technically difficult, it is usual to measure cortisol instead in these tests. Should an abnormal response be found, it is necessary to demonstrate that the adrenal gland is normally sensitive to ACTH by performing a Synacthen test if this has not been done already. Prolactin secretion is stimulated by both stress and TRH. However, the normal response is variable and since prolactin deficiency is rarely a clinical problem, serum prolactin concentrations are not usually measured in patients undergoing combined tests of pituitary function.

The use of the clomiphene test in the investigation of hypogonadism due to isolated gonadotrophin deficiency is discussed in *Chapter 11*.

DISORDERS OF ANTERIOR PITUITARY FUNCTION

Hypopituitarism

Destructive lesions of the pituitary tend to present with evidence of pituitary hypofunction. Partial hypopituitarism is seen more frequently than complete loss of pituitary function. The presenting features depend on several factors; age is particularly important. Decreased GH secretion is an early feature of pituitary failure (see page 112) but whilst its effects can be dramatic in children, they are minimal in adults. In general, GH and gonadotrophin secretion are affected before ACTH. It is very uncommon for hypothyroidism to be the presenting feature of pituitary failure. Isolated deficiency of some of the anterior pituitary hormones can occur but this is usually congenital. In most such cases, it is apparently due to failure of secretion of a hypothalamic hormone.

In suspected pituitary hypofunction, stimulatory tests are used to assess the ability of the gland to produce hormones. With a suspected pituitary tumour, both excess hormone secretion by the tumour and possible decreases in the production of other

Combined test of pituitary function

Procedure

1. fast patient overnight and weigh

2. insert and heparinize i.v. cannula

3. draw and discard 1ml of blood before collecting each sample and heparinize cannula after each sample is drawn

4. after 30min take basal blood sample and analyze for glucose, cortisol (or ACTH), FSH, LH, TSH and GH

5. give 200µg TRH, 100µg GnRH and 0.15u/kg body weight soluble insulin

6. take blood samples for analysis as follows:

time (minutes)	assay				
	glucose	cortisol	FSH, LH	TSH	GH
0	*	*	*	*	*
15	*				
20			*	*	
30	*	*			*
45	*				
60	*	*	*	*	*
90	*	*			*
120	*	*			*

7. repeat insulin dose at 45min if patient has not become clinically (sweating) or biochemically (blood glucose <2.2mmol/l) hypoglycaemic and extend sampling accordingly

Normal response

cortisol	increment	>200nmol/l
	peak	>500nmol/l (the same criteria apply if glucagon is used)
GH	peak	>20mu/l (after glucagon: 15mu/l in males and 20mu/l in females)
FSH	increment	>1.5 times basal level
LH	increment	>5 times basal level
TSH	increment	≥2mu/l (elderly)
		≥5mu/l (young adults)

Fig.8.6 Combined test (triple bolus test) of pituitary function. In patients thought very likely to be hypopituitary, the insulin dose should be 0.10u/kg body weight; in patients with Cushing's disease or acromegaly, a dose of 0.30u/kg may be used. When glucagon (1mg intramuscularly) is used instead of insulin, blood samples for cortisol and GH should be taken at 30 minute intervals from 90–240 minutes after the injection (GH and cortisol responses occur later than when insulin is used).

hormones must be investigated. The underlying cause must be determined where possible and pituitary tumours should be carefully assessed, by examination of the visual fields, skull radiography and computerized tomography (CT), to define their anatomical extent; this may determine the correct management. Some of the many causes and clinical features of hypopituitarism are indicated in Fig.8.7.

Clinical evidence of posterior pituitary dysfunction (diabetes insipidus) must be sought and, if present, confirmed by appropriate tests (see pages 120 & 121). Diabetes insipidus is uncommon except with large pituitary tumours but can develop, often transiently, after surgery. Even in patients with impaired vasopressin (antidiuretic hormone or ADH) secretion, diabetes insipidus may not be apparent if ACTH secretion is also impaired, since cortisol, the secretion of which is stimulated by ACTH, is necessary for normal water excretion.

CASE HISTORY 8.1

A fifty-year-old man tripped and fell as he was running for a bus. He hit his head against the kerb and was knocked out for a few seconds. An ambulance was called and he was taken to the local hospital.

There was no sign of physical injury on examination but a skull radiograph showed enlargement of the pituitary fossa. The casualty officer questioned the patient further. Over the preceeding twelve months he had lost his libido and found it necessary to shave less frequently than before; he had also noticed some loss of axillary and pubic hair. He frequently felt dizzy when getting out of bed in the morning and despite spending a lot of time in the sun had not acquired his usual summer tan.

Investigations

serum:	cortisol (0900h)	300nmol/l
	GH	<2mu/l
	thyroxine	70nmol/l
	TSH	2mu/l
	testosterone	4nmol/l
	LH	<1.5iu/l

serum:	FSH	<1.0iu/l
	prolactin	<50mu/l

combined glucagon, TRH and GnRH stimulation test:

serum:	
cortisol (maximum)	350nmol/l at 180min
LH, FSH	no increment over basal values
GH	no increment over basal value
TSH	5mu/l at 20min; 3mu/l at 60min

Comment

The clinical features are typical of hypopituitarism (Fig.8.7) and the test results confirm this diagnosis. GH, gonadotrophin and prolactin concentrations are all low; the cortisol concentration is in the lower part of the normal range. These hormones show little or no response to appropriate stimuli. The low testosterone is secondary to the lack of gonadotrophins. In view of the overwhelming evidence for the diagnosis, it was not considered necessary to perform a Synacthen test. The serum thyroxine is near the lower end of the normal range; if this was related to incipient thyroid failure, a higher TSH concentration would be expected. There is a TSH response to TRH, but even this is towards the lower limit of normal.

Cortisol replacement therapy was started immediately and testosterone and thyroxine were also given. Within a few hours, the patient became polyuric and signs of water depletion developed. The serum sodium concentration, which was low on admission (128mmol/l), rose to 149mmol/l. Diabetes insipidus, due to impaired release of vasopressin, can be masked by simultaneous cortisol deficiency and revealed when replacement therapy is started, as in this case.

The patient was given synthetic vasopressin which controlled his polyuria. He subsequently underwent surgery and a chromophobe adenoma was successfully removed. On follow-up, there was no evidence of recovery of pituitary function and he remained on replacement therapy.

Hypopituitarism	
Causes	
Tumours	**Vascular disease**

pituitary tumours:
  adenoma
  craniopharyngioma
cerebral tumours:
  primary
  secondary
	post-partum necrosis (Sheehan's syndrome)
	infarction, especially of tumours
	severe hypotension
	cranial arteritis

	Trauma
Granulomatous disease	**Infection**
sarcoidosis histiocytosis X	meningitis, especially tuberculous syphilis
Hypothalamic disorders	**Iatrogenic**

tumours
functional disturbances, e.g. anorexia
  nervosa and starvation, causing reversible
  hypogonadotrophic hypogonadism
isolated GH and gonadotrophin secretion due
  to impaired secretion of hypothalamic
  releasing hormones
	surgery
	irradiation
	prolonged treatment with glucocorticoids and
	  thyroid hormones causing isolated ACTH
	  and TSH deficiency respectively

Clinical features	
Hormone	**Features of deficiency**
growth hormone	children: growth retardation adults: any tendency to hypoglycaemia may be accentuated
prolactin	failure of lactation

gonadotrophins	children: delayed puberty
	females: oligomenorrhoea, infertility and atrophy of breasts
	            and genitalia
	males: impotence, azoospermia and testicular atrophy
	both sexes: decreased libido, loss of body hair and fine
	            wrinkling of skin

ACTH	weight loss, weakness, hypotension, hypoglycaemia and other
	  features of glucocorticoid deficiency, usually of insidious onset
	  unless stressed; decreased skin pigmentation

Fig.8.7 Main causes and clinical features of hypopituitarism.

Anorexia nervosa

Anorexia nervosa, a disorder characterized by self-imposed starvation and a preoccupation with body size, may clinically resemble hypopituitarism. Amenorrhoea is common to both conditions, due to decreased gonadotrophin secretion. However, pubic and axillary hair, which may be lost in hypopituitarism, is normal in anorexia nervosa and there may even be additional (lanugo) hair on the body. The weight loss of anorexia nervosa is usually severe in comparison with that which can occur in hypopituitarism. Serum cortisol and GH concentrations tend to be elevated in anorexia nervosa.

Growth hormone deficiency

GH deficiency that develops once growth is completed is of no significance. In children, however, it is an uncommon but important cause of growth retardation. Serum GH levels may be very low in normal people which means that, while a random level of greater than 20mu/l excludes significant deficiency, a low concentration in a random blood sample is not diagnostic. GH status can be assessed by various stimulatory tests (Fig.8.8).

CASE HISTORY 8.2

A ten-year-old boy was referred to hospital for investigation of short stature. He had always been small, but his parents became worried when his seven-year-old brother overtook him in height. He had been measured two years before and had only grown 3cm since then. On examination, there was no abnormality apart from his short stature. The history and appropriate tests excluded many of the recognized causes of growth retardation (see page 313).

Investigations

serum GH : 4mu/l (after vigorous exercise)
4mu/l (during documented hypoglycaemia in insulin stress test)

Comment

The diagnosis of GH deficiency depends upon the demonstration of subnormal growth velocity and subnormal serum GH concentrations. Both features are present in this case. If the serum GH concentration is normal (>20mu/l), either after exercise or in a sample obtained while the child is asleep, this obviates the need to perform the more invasive and hazardous insulin stress test. However, the exercise must be sufficient to make the patient breathless, for example, repeated running up and down stairs. In this case, the response was subnormal and to confirm GH deficiency the insulin stress test was performed; the response was again subnormal.

A fall in blood glucose concentration is sufficient to stimulate GH secretion in normal people, but it is advisable to achieve documented hypoglycaemia (serum glucose <2.2mmol/l) to ensure an adequate stimulus. Sex steroids are important in determining the magnitude of the response. Therefore equivocal responses in children with pubertal delay require that the test is repeated after priming with either testosterone in boys or oestrogen in girls.

There was no other evidence of pituitary hypofunction in this boy and no evidence of a destructive pituitary lesion. A diagnosis of idiopathic GH deficiency was made. He was accepted for GH treatment and grew at a normal rate thereafter, although he was always shorter than his peers. Typically, the lost height is not completely restored when GH deficiency is treated.

MANAGEMENT

Until recently, the only source of GH was from pituitary glands of human cadavers and supplies were hence very limited. However, human GH can now be reproduced in bacteria by inserting the appropriate DNA sequence into the bacterial genome causing the organisms to synthesize the hormone. Since most cases of isolated GH deficiency are now known to be due to growth hormone releasing hormone (GHRH) deficiency, GHRH may have a therapeutic role in the future.

Assessment of growth hormone status	
Procedure	**Normal response**
arginine infusion test: give 0.5g/kg body weight (maximum 30g) i.v. over 30min take blood samples at 30min intervals for 2h	peak >15mu/l GH
Bovril (yeast extract) test: give 20g/1.5m^2 body surface area orally in water take blood samples at 30min intervals for 2h	peak >20mu/l GH
exercise test: patient runs up and down stairs as fast as possible for 10min	peak >20mu/l GH
insulin/glucagon test: see Fig.8.6	

Fig.8.8   Procedures for assessing GH status.

Pituitary tumours

Pituitary tumours may be purely destructive but are often functional, producing excessive quantities of a hormone. Prolactin-, GH- and ACTH-secreting tumours are well recognized while tumours secreting gonadotrophins and TSH are very rare. Any pituitary tumour may give rise to clinical features due to the destruction of normal pituitary tissue, that is hypopituitarism, and of intracranial space-occupying lesions such as headache, vomiting and papilloedema. Visual field defects may develop when an upward growing tumour impinges on the optic chiasm and occasionally a patient's sight may be threatened.

Growth hormone excess: acromegaly and gigantism

Acromegaly and gigantism are the result of excessive GH secretion by a pituitary tumour. As a result, there is increased growth of soft tissues and bone. If this occurs before the epiphyses have fused, growth of long bones occurs leading to gigantism. More commonly, GH-secreting tumours occur in adults, producing acromegaly with increased growth of soft tissues, hands, feet, jaw and internal organs. The GH concentration in a random serum sample is usually raised, but the clinical diagnosis should be confirmed biochemically by demonstrating a failure of GH suppression in response to an oral glucose tolerance test. In normal subjects, serum GH concentration falls to less than 2mu/l during this procedure. In acromegaly and gigantism, GH fails to suppress normally and there may even be an increase in concentration. The glucose response may indicate impaired glucose tolerance, or frank diabetes mellitus.

The clinical features of excessive GH secretion are related both to the somatic and to the metabolic effects of the hormone (Fig.8.9). In addition, features due directly to the presence of the pituitary tumour are often present. Hyperprolactinaemia, due either to interference with the normal inhibition of prolactin secretion or to its secretion by the tumour itself, is common in patients with acromegaly but there may be impaired secretion of other pituitary hormones. Acromegaly is occasionally a feature of multiple endocrine adenomatosis (MEA type I).

Clinical features of excessive growth hormone secretion		
Somatic	**Metabolic**	**Local effects of tumour**

increased growth of:
  skin
  subcutaneous tissues
  skull and jaw
  long bones, if before fusion of
    epiphyses
 	
sweating, greasy skin and acne
 	
goitre
 	
cardiomegaly and hypertension
	elevated, non-suppressible serum
	  GH concentration
	
	glucose intolerance
	
	clinical diabetes mellitus
	
	hypercalcaemia and
	  hyperphosphataemia
		headache
		
		visual field defects
		
		hypopituitarism
		
		diabetes insipidus

Fig.8.9 Clinical features of excessive GH secretion.

CASE HISTORY 8.3

A forty-year-old man consulted his general practitioner because he had become impotent. He had also been embarrassed by excessive sweating in the absence of exertion. His wife thought that his facial features had become coarser, while he had recently had to buy a larger pair of shoes than normal because his old ones had become uncomfortable. The practitioner found mild hypertension and a trace of glycosuria and referred him to an endocrine clinic with a presumptive diagnosis of acromegaly.

Investigations

combined pituitary function test (Fig.8.10)

oral glucose tolerance test (Fig.8.11)

serum: prolactin (at 0min)   800mu/l
         testosterone   11nmol/l

visual fields: partial bitemporal hemianopia

skull radiograph: enlarged pituitary fossa with
   erosion of anterior clinoid processes

pituitary CT scan: pituitary tumour with
   suprasellar extension

Comment

The clinical diagnosis is confirmed by the high basal GH level which is not suppressed by glucose. The glucose tolerance test gives a diabetic result; abnormal glucose tolerance is seen in about twenty-five percent of cases of acromegaly, but clinical diabetes mellitus is present in only about ten percent.

The basal prolactin concentration is increased. Gonadotrophins are low and do not increase in response to GnRH. The serum testosterone concentration is at the low end of the normal range due to inadequate testicular stimulation by LH. The cortisol and TSH responses are normal.

The presence of a tumour is confirmed by the radiographic appearances; the optic chiasm lies immediately above the pituitary and compression of it can cause either visual field defects, characteristically a bitemporal hemi- or quadrantanopia, or threaten complete visual failure.

Case history 8.3: combined pituitary function test					
Time (minutes)	Blood glucose (mmol/l)	Serum cortisol (nmol/l)	Serum LH (iu/l)	Serum FSH (iu/l)	Serum TSH (mu/l)
0	4.5	400	2.0	1.5	0.8
15	3.4				
20			2.2	1.7	9.7
30	2.0	700			
45	2.1				
60	3.3	680	2.1	1.5	4.3
90	4.0	600			
120	4.5	550			

Fig.8.10 Results of a combined pituitary function test (see *Case History 8.3*).

Case history 8.3: oral glucose tolerance test		
Time (minutes)	Blood glucose (mmol/l)	Serum GH (mu/l)
0	8.5	22
30	14.5	20
60	14.2	24
90	13.5	25
120	11.5	20

Fig.8.11 Results of an oral glucose tolerance test (see *Case History 8.3*).

MANAGEMENT

Treatment of acromegaly and gigantism is aimed at reducing excessive GH secretion, preventing or treating deficiencies of other pituitary hormones and preventing damage to surrounding structures, particularly the optic nerves, by the tumour. The main modes of treatment are surgery, external irradiation and dopamine agonists, for example, bromocriptine. Surgery is essential if there is suprasellar extension and requires craniotomy. Intrasellar tumours may be resectable from below, using either a trans-sphenoidal or a trans-ethmoidal approach. If there is continuing evidence of excessive GH secretion after surgery, radiotherapy or treatment with bromocriptine can be used. Bromocriptine is a dopamine agonist; it stimulates GH secretion in normal subjects but inhibits it in a proportion of patients with acromegaly.

When there is accompanying hypopituitarism, appropriate replacement treatment with cortisol, gonadal steroids, or gonadotrophins, and thyroxine must be given. All patients with acromegaly and gigantism must be followed up and reassessed regularly for evidence of either recurrence or further loss of normal pituitary function.

Hyperprolactinaemia

Hyperprolactinaemia is a common endocrine abnormality. It is an important cause of infertility in both males and females, impotence in males and menstrual irregularity in females. These effects are thought to be mediated through inhibition of the pulsatility of GnRH by prolactin. The causes and clinical features of hyperprolactinaemia are summarized in Fig.8.12. There may also be features related to the cause of the hyperprolactinaemia. The causes include various drugs, which either block pituitary dopaminergic receptors or deplete the brain of dopamine, in addition to pituitary tumours and destructive pituitary lesions which interfere with the normal inhibition of prolactin secretion. Prolactin-secreting tumours are often very small (<10mm diameter) but larger tumours do occur, which destroy normal pituitary tissue and both erode and expand outside the confines of the pituitary fossa. Prolactin-secreting tumours are at least twice as common as GH-secreting tumours.

Prolactin is secreted in response to both stress and TRH, and normal levels also depend on oestrogen status. It is therefore difficult to define an upper limit of normal for serum prolactin concentration. Less than 400mu/l is probably normal and more than 600mu/l, abnormal. Patients with prolactin-secreting tumours usually have basal levels in excess of 2000mu/l. Dynamic tests are of limited value in the diagnosis of hyperprolactinaemia but, if a tumour is present, patients must be tested for deficient secretion of other anterior pituitary hormones. With small tumours, other pituitary functions remain normal.

MANAGEMENT
The majority of patients with small prolactin-secreting tumours can be cured by either surgery or treatment with bromocriptine; the clinical features regress while hypopituitarism is uncommon. With larger tumours, the chances of effecting a cure are much lower, but shrinkage of the tumour has been reported

Hyperprolactinaemia
Causes
Physiological   stress, sleep, pregnancy and suckling

Drugs
 dopaminergic receptor blockers,
    e.g. phenothiazines and haloperidol
 dopamine-depleting agents,
    e.g methyldopa and reserpine
 others,
    e.g. oestrogens and TRH

Pituitary disorders   prolactin-secreting tumours   tumours blocking dopaminergic inhibition   of prolactin secretion   pituitary stalk section and surgery
Others   hypothyroidism   ectopic secretion   chronic renal failure
Clinical features
Females   oligomenorrhoea and amenorrhoea   infertility   galactorrhoea
Males   impotence   infertility   gynaecomastia

Fig.8.12 Causes and clinical features of hyperprolactinaemia.

in seventy-five percent of cases treated with bromocriptine and can be used as a preliminary to surgery. External irradiation may be helpful in some cases.

Cushing's disease

Cushing's disease, in which increased secretion of cortisol by the adrenal cortex is secondary to increased secretion of ACTH by the anterior pituitary, is discussed in *Chapter 9*. Patients who have been treated for Cushing's disease by adrenalectomy alone may later develop hyperpigmentation and the clinical features of a large pituitary tumour (Nelson's syndrome). The pigmentation is due to the melanocyte-stimulating activity of ACTH and its precursors. Nelson's syndrome is uncommon in patients in whom treatment for Cushing's disease has included pituitary surgery or irradiation in addition to adrenalectomy.

Other conditions related to pituitary tumours

Tumours which secrete TSH and gonadotrophins are very rare. Occasionally, TSH-secreting tumours develop in patients with long-standing, untreated hypothyroidism but these regress when replacement treatment is given. Thirty percent of pituitary tumours, usually chromophobe adenomas, are non-functional. These may present with either features due to the physical pressure of the tumour, hypopituitarism or occasionally be diagnosed incidentally when a skull radiograph is taken for some incidental purpose.

POSTERIOR PITUITARY HORMONES

The posterior pituitary secretes two hormones, vasopressin (antidiuretic hormone or ADH) and oxytocin. These hormones are synthesized in the hypothalamus and pass down nerve axons into the posterior pituitary, from where they are released into the circulation. Oxytocin is involved in the control of uterine contractility and of milk release from the lactating breast. Disorders of its secretion are probably uncommon and are not clinically important. In contrast, vasopressin is essential to life and disorders of its secretion are well recognized.

Vasopressin

Vasopressin has a vital role in the control of the tonicity of the extracellular fluid, and hence indirectly of the intracellular fluid, and of water balance. Excessive secretion results in dilutional hyponatraemia, with a risk of water intoxication, while decreased secretion results in diabetes insipidus, a condition in which there is uncontrolled excretion of water with a tendency to severe dehydration. The syndromes of excessive secretion of vasopressin are discussed on page

Causes of diabetes insipidus

Cranial

tumours:
    craniopharyngioma
    secondary tumours
    pituitary tumours with suprasellar extension
granulomatous disease
meningitis and encephalitis
vascular disorders
trauma
surgery (often transient)
idiopathic

Nephrogenic

congenital
metabolic:
    hypokalaemia
    hypercalcaemia
drugs:
    lithium
    demeclocycline
post-obstructive uropathy
chronic renal disease:
    pyelonephritis
    polycystic disease
    amyloid
    sickle cell disease

Fig.8.13   Causes of diabetes insipidus.

23; they are commonly seen in conditions which do not affect the pituitary directly. Diabetes insipidus, on the other hand, is commonly due to pituitary or hypothalamic disease (Fig.8.13), although it can also be due to a failure of the kidney to respond to the hormone (nephrogenic diabetes insipidus).

In diabetes insipidus, the lack of vasopressin results in polyuria and thirst. Unless the hypothalamic thirst centre is also damaged, thirst leads to increased water intake (polydipsia). The differential diagnosis includes other conditions causing polyuria and polydipsia, for example, diabetes mellitus, chronic renal failure, hypercalcaemia and hypokalaemia. Simple tests will eliminate these possibilities.

A compulsive desire to drink (psychogenic polydipsia) also causes polyuria. However, in this case the polyuria is secondary to increased water intake, while in diabetes insipidus the opposite applies, polydipsia being a response to polyuria. In both conditions, the urine is dilute, but in diabetes insipidus there tends to be an increased serum osmolality (>295mmol/kg) and hypernatraemia, whereas with excessive water intake, a decreased serum osmolality (<280mmol/kg) and hyponatraemia are characteristically present. If a random urine osmolality is greater than 750mmol/kg, diabetes insipidus is excluded.

If there is any doubt about the diagnosis, it can be resolved by performing a water deprivation test (Fig.8.14). This is effectively a biological assay for vasopressin which is difficult to measure in the blood. Patients with diabetes insipidus may become dangerously dehydrated if denied access to water; they may also exercise considerable ingenuity to obtain water. Close supervision is therefore required and the test must always be performed by day.

In essence, a normal subject will concentrate the urine during the eight-hour period and the serum osmolality will not exceed 295mmol/kg. In diabetes insipidus, the urine does not become concentrated and the serum osmolality rises. In patients who are water overloaded before the test is started, the urine may not become concentrated; serum osmolality is usually low and may remain so since vasopressin secretion is only stimulated if it rises above 285mmol/kg. Thus the urine becomes concentrated only if the serum osmolality exceeds this level.

At the end of the eight-hour period, the patient is allowed to drink water and is given 1-desamino-8-D-arginine-vasopressin (desmopressin) a synthetic analogue of vasopressin. In cranial diabetes insipidus, the urine should become concentrated; in patients whose kidneys are insensitive to vasopressin (nephrogenic diabetes insipidus) it remains dilute. If the water deprivation test is to be carried out on a patient with anterior pituitary disease, adequate cortisol replacement must be provided. States of partial diabetes insipidus can be diagnosed by measuring the vasopressin response to infusion of hypertonic saline but the procedure is potentially dangerous and since the assay is not widely available these tests are not yet commonly performed.

CASE HISTORY 8.4

A middle-aged woman, who had undergone mastectomy and local radiotherapy for carcinoma of the breast two years previously, attended for her regular outpatient appointment. There was no sign of recurrence but she complained of increasing thirst over the previous months and that she was passing copious amounts of urine. The thirst became intolerable if she went without water for more than a few hours and her sleep was disturbed by the frequent need to pass urine and have a drink. There was no glycosuria; serum creatinine, potassium and calcium concentrations were all normal. She was admitted for investigation.

Investigations

| random serum: | osmolality | 295mmol/kg |
| | sodium | 144mmol/l |

| urine osmolality | | 90mmol/kg |

Water deprivation test: after six hours water deprivation her weight had fallen from 60kg to 57.6kg; the test was therefore stopped.

| at end of test: | serum osmolality 307mmol/kg |
| | urine osmolality 220mmol/kg |

She was then allowed to drink and was given a dose of desmopressin. Following this her urine osmolality rose to 600mmol/kg.

Comment

The history of intolerable thirst with a slightly raised serum osmolality yet dilute urine is very suggestive of diabetes insipidus and this diagnosis is confirmed by the failure to conserve water and concentrate the urine during water deprivation. She responded to desmopressin, indicating that vasopressin deficiency, rather than renal insensitivity to the hormone, was the cause of her symptoms.

She was treated successfully with regular administration of desmopressin and her symptoms resolved. A skull radiograph was normal but CT scanning revealed a small lesion in the region of the hypothalamus. The patient died one year later, with extensive cerebral metastatic deposits from her breast carcinoma.

Water deprivation test
Procedure
allow fluids overnight before test and give light breakfast with no fluid; no smoking permitted
weigh patient
allow no fluid for 8h; patient must be under constant supervision during this time
every hour thereafter, for 8h: 1. weigh patient; stop test if body weight falls by >3% 2. ask patient to empty bladder; measure volume and osmolality of urine
collect blood for measurement of serum osmolality at 30min and 3.5, 6.5 and 7.5h; stop test if osmolality exceeds 300mmol/kg
after 8h allow patient to drink and give 20µg desmopressin intranasally
collect urine hourly for a further 4h
Results

diabetes insipidus diagnosed if:
    weight loss >3% initial body weight
    serum osmolality exceeds 300mmol/kg
    urine: serum osmolality ratio does not exceed 1.9, provided serum osmolality exceeds 285mmol/kg

nephrogenic diabetes insipidus diagnosed if urine does not become concentrated after 4h of desmopressin

Fig.8.14 Water deprivation test. The first eight hours test for the ability to concentrate the urine and hence differentiate between diabetes insipidus and psychogenic polyuria. The final four hours, after the administration of desmopressin, test for the kidney's ability to respond to vasopressin and therefore differentiate between cranial and nephrogenic diabetes insipidus.

In about one-third of cases of cranial diabetes insipidus there is no obvious underlying cause. Some of these cases have a familial incidence and the onset may then be very sudden. Urine output in diabetes insipidus may exceed 10 l/day although less than this is more usual. Patients with nephrogenic diabetes insipidus are insensitive to vasopressin, which is secreted in either normal or increased amounts in such patients. The congenital form is inherited as an X-linked recessive disorder; the defect is in the adenylate cyclase mechanism responsible for mediating the action of vasopressin after it has become bound to receptors.

Management of diabetes insipidus

Patients must always have access to adequate fluid and whenever possible, the underlying disease should be treated. Cranial diabetes insipidus is usually treated with desmopressin, given as a nasal spray, although mild cases may be successfully treated with chlorpropamide, an oral hypoglycaemic agent which also increases renal sensitivity to vasopressin (hypoglycaemia is a possible side-effect). Patients must learn to monitor their fluid output and input in order to avoid water intoxication. This may be a particular problem if the sensation of thirst is blunted.

Patients with nephrogenic diabetes insipidus, because they do not respond to vasopressin, must maintain an adequate water intake to avoid dehydration. Hydronephrosis and hydroureter secondary to bladder distension may occur and lead to renal impairment. Thiazide diuretics, which induce a state of sodium depletion, increasing renal sodium and water retention, may reduce the polyuria. Potassium supplements or the concomitant use of a potassium-sparing diuretic may be necessary to prevent hypokalaemia.

SUMMARY

The anterior pituitary gland secretes growth hormone and prolactin, and trophic hormones which control the activity of the gonads (luteinizing hormone and follicle-stimulating hormone), thyroid (thyroid-stimulating hormone) and the adrenal cortex (adrenocorticotrophic hormone). The secretion of all these hormones is regulated by hypothalamic hormones, which reach the pituitary through a portal system of blood vessels. The trophic hormones are in addition controlled by feedback mechanisms involving the hormones produced by the respective target organs.

Anterior pituitary hypofunction may result in the inadequate production of one or more hormones and the clinical manifestations depend upon the particular pattern of deficiency. It may either be the result of disease affecting the pituitary itself or be secondary to hypothalamic disease, with failure of production of hypothalamic hormones. Pituitary hypofunction (hypopituitarism) is investigated by tests designed to stimulate the production of pituitary hormones.

Pituitary tumours can cause hypopituitarism by destroying normal pituitary tissue, but may be functional and produce syndromes related to excessive hormone secretion. Pituitary tumours producing prolactin, growth hormone and adrenocorticotrophic hormone are well-recognized but excessive gonadotrophin and thyroid-stimulating hormone secretion is very rare. In addition to their endocrine effects, both functional and non-functional tumours can give rise to clinical features characteristic of intracranial space-occupying lesions.

The posterior pituitary gland secretes oxytocin and vasopressin. Both are synthesized in the hypothalamus and reach the posterior pituitary through nerve axons. Because of this, damage to the posterior pituitary may only cause temporary failure of hormone secretion. Oxytocin stimulates uterine contraction during labour but does not appear to be an essential hormone.

Vasopressin is essential as it controls water excretion by altering the permeability of the renal collecting tubules to water in response to changes in extracellular fluid osmolality. Excessive vasopressin secretion produces water retention with a dilutional hyponatraemia. Defective vasopressin secretion results in diabetes insipidus, with uncontrolled renal water loss. Diabetes insipidus can also be due to renal insensitivity to vasopressin; the two types can be distinguished from each other, and from psychogenic polydipsia, by a water deprivation test.

FURTHER READING

Besser GM & Cudworth AG (1987) *Clinical Endocrinology: An Illustrated Text.* London: Chapman & Hall; Gower Medical Publishing.

Hall R, Anderson J, Smart GA & Besser M (1980) *Fundamentals of Clinical Endocrinology.* 3rd edition. London: Pitman Medical.

Wilson JD & Foster DW (eds) (1985) *Williams – Textbook of Endocrinology.* 7th edition. Philadelphia: WB Saunders Company.

9. The Adrenal Glands

INTRODUCTION

The adrenal glands have two functionally distinct parts, the cortex and the medulla. The adrenal cortex is essential to life; it produces three classes of steroid hormone, the glucocorticoids, mineralocorticoids and androgens. The medulla, however, which is functionally part of the sympathetic nervous system, is not essential to life and its pathological importance is related mainly to the occurrence of rare catecholamine-secreting tumours.

Glucocorticoids, of which the most important is cortisol, are secreted in response to adrenocorticotrophic hormone (ACTH), which is itself secreted by the pituitary in response to the hypothalamic releasing hormone, corticotrophin releasing hormone (CRH). Cortisol, in turn, has a negative feedback effect on ACTH release. The glucocorticoids have many physiological functions (Fig.9.1) and are particularly important in mediating the body's response to stress.

The most important mineralocorticoid is aldosterone. This is secreted in response to the activation of the renin–angiotensin system by a decrease in renal blood flow and other indicators of decreased extracellular fluid volume (Fig.9.2). Secretion of aldosterone is also directly stimulated by hyperkalaemia, despite the fact that potassium loading actually decreases renin release. The main action of aldosterone is to stimulate sodium reabsorption in the distal convoluted tubules in the kidney in exchange for potassium or hydrogen ions; it thus has a central role in the determination of the extracellular fluid (ECF) volume. ACTH has no physiological role in aldosterone release.

The adrenal cortex is also a source of androgens, including dehydroepiandrosterone (DHEA), DHEA-sulphate (DHEAS) and androstenedione. These are the major androgens in the female and are involved with the development of certain secondary sexual characteristics. The clinical effects of excessive adrenal androgens may be a prominent feature of adrenal disorders.

Functions of glucocorticoids
increase protein catabolism
increase hepatic glycogenolysis
increase hepatic gluconeogenesis
inhibit ACTH secretion (negative feedback mechanism)
sensitize arterioles to action of noradrenaline, hence involved in maintenance of blood pressure
permissive effect on water excretion; required for initiation of diuresis in response to water loading

Fig.9.1 Principal physiological functions of glucocorticoids.

ADRENAL STEROID HORMONE BIOSYNTHESIS

The hormones secreted by the adrenal cortex are synthesized from cholesterol by a sequence of enzyme-catalyzed reactions (Fig.9.3). An awareness of these pathways is important for the understanding of congenital adrenal hyperplasia, a group of conditions each characterized by a lack of one of these enzymes.

MEASUREMENT OF ADRENAL STEROID HORMONES

The advent of sensitive and specific immunoassays for the steroid hormones has rendered obsolete the measurement of groups of related hormones by chemical methods, for example, 17-oxosteroids for adrenal androgens and 17-oxogenic steroids which are related to cortisol.

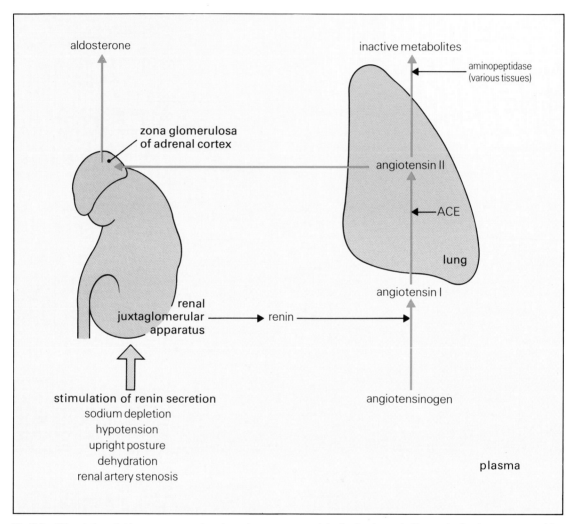

Fig.9.2 Stimulation of aldosterone secretion through activation of the renin–angiotensin system. Renin, released into the plasma from the juxtaglomerular cells of the kidney in response to various stimuli, catalyzes the formation of angiotensin I from angiotensinogen, an α_2-globulin. Angiotensin I is metabolized to an octapeptide, angiotensin II, during its passage through the lungs by angiotensin converting enzyme (ACE). Angiotensin II stimulates the release of aldosterone from the adrenal cortex.

One exception is the continued use of a fluorometric technique for the measurement of 11-hydroxycorticosteroids, the major component of which is cortisol. This is a simple, cheap technique which is adequate for many clinical purposes. Specific immunoassays are available for the measurement of cortisol, aldosterone, 17α-hydroxyprogesterone and the principal adrenal androgens. Serum measurements are most widely used but fluctuations in the serum concentra-

tions of these hormones occur for a number of reasons and the results of single estimations must be interpreted with caution.

Cortisol

Ninety-five percent of cortisol in the circulation is bound to protein, principally to the cortisol-binding

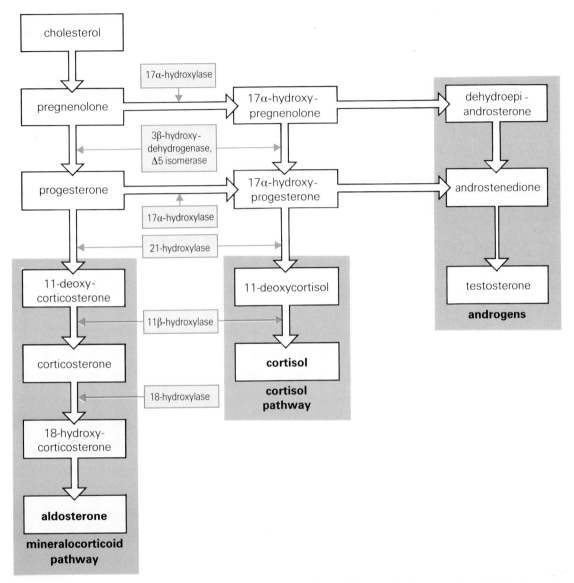

Fig.9.3 Biosynthesis of adrenal steroid hormones. Cortisol and the androgens are synthesized in the zona reticularis and zona fasciculata of the adrenal glands. 18- hydroxylase, required for the synthesis of aldosterone, is present only in the zona glomerulosa.

globulin, transcortin. Thus the amount of free cortisol that can be excreted unchanged in the urine is very small. Transcortin is almost fully saturated at normal cortisol concentrations and it follows that if cortisol production is increased, the concentration of free hormone, and thus the amount filtered at the glomerulus and excreted in the urine, increases at a rate similar to its rate of synthesis. For this reason, measurement of the twenty-four-hour urinary excretion of cortisol, provided that an accurate urine collection can be made, is a sensitive way of detecting an increase in the secretion of the hormone, but not of a decreased secretion.

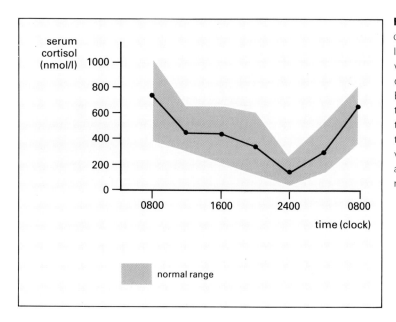

Fig.9.4 Diurnal variation in serum cortisol concentration. Serum cortisol levels are at their highest shortly after waking and then decline throughout the day to reach a nadir in the late evening. Because of this variation, it is important that blood samples are taken at times that coincide either with the peak or the trough, random samples being of little value. The graph shows mean values and the range in a sample of healthy men.

Serum cortisol concentrations show a diurnal variation being highest in the morning and lowest at night (Fig.9.4). Blood for cortisol measurement should usually be drawn between 0800h and 0900h; however, samples are taken at 2300h to detect loss of the diurnal variation, an early feature of adrenal hyperfunction (Cushing's syndrome). Random measurements are rarely of any value in the diagnosis of adrenal disease, except that a high level in a sick patient may reasonably be taken to exclude adrenal failure. However, this should preferably be confirmed by performing a short Synacthen test.

Cortisol is secreted in response to stress, mediated through ACTH, and thus stress must be kept to a minimum if results are to be interpreted correctly. Investigations of adrenal hypo- or hyperfunction often involve measurement of cortisol after attempting to stimulate or suppress its secretion.

When interpreting serum cortisol results, it must be remembered that the synthetic glucocorticoid, prednisolone, can cross-react with cortisol in immunoassays for the hormone. In the presence of this drug, the fluorometric technique for measuring cortisol can be used. Spironolactone, an aldosterone antagonist used as a potassium-sparing diuretic, cross-reacts with cortisol in the fluorometric test, but not the immunoassay. Dexamethasone does not cross-react in either method.

Aldosterone

Serum aldosterone concentration varies with posture and blood should be drawn from patients after they have been recumbent overnight and then, for comparison, after thirty minutes of being upright. Aldosterone secretion is stimulated by renin and therefore it is often appropriate to measure the serum renin activity at the same time as the aldosterone to establish whether aldosterone secretion is autonomous or under normal control. This point is discussed further in connection with Conn's syndrome and hyperaldosteronism (see page 134).

Androgens

Measurements of adrenal androgens are of value in the diagnosis and management of congenital adrenal hyperplasia (see page 135) and in the investigation of virilization in women (see *Chapter 11*).

DISORDERS OF THE ADRENAL CORTEX

Adrenal disorders may present with clinical features related either to hypo- or hyperfunction. In congenital adrenal hyperplasia, a combination of features may be present.

Adrenal hypofunction	
Causes	**Clinical features**
Common autoimmune adrenalitis tuberculosis adrenalectomy	Common tiredness, generalized weakness and lethargy anorexia weight loss dizziness and postural hypotension pigmentation
Less common secondary tumour deposits amyloidosis haemochromatosis histoplasmosis adrenal haemorrhage	Less common hypoglycaemia loss of body hair (female) depression

Fig.9.5 Causes and clinical features of adrenal hypofunction.

Adrenal hypofunction (Addison's disease)

The common causes and clinical features of this disease are listed in Fig.9.5. The cases originally described by Thomas Addison were caused by tuberculosis but autoimmune disease is now the major cause in the United Kingdom. In such cases, adrenal auto-antibodies will usually be present and there may be associated autoimmune disease of other organs, for example, pernicious anaemia.

The majority of the clinical features are due to the lack of glucocorticoids and mineralocorticoids. Increased pigmentation is due to the high serum concentration of ACTH. This hormone has some melanocyte-stimulating activity and its concentration is increased as a result of the loss of negative feedback control by cortisol.

Adrenal failure may develop acutely and is a medical emergency. The clinical features include severe dehydration, shock and hypoglycaemia. Adrenal crisis may be precipitated by stress, such as infection, trauma or surgery, in patients with incipient adrenal failure. Patients on glucocorticoid replacement treatment are also susceptible to adrenal failure in these circumstances if the dosage is not increased. Haemorrhage into the adrenal glands may occur as a complication of anticoagulant treatment and in meningococcal septicaemia, and can result in acute adrenal failure.

The diagnosis of adrenal failure rests on the demonstration of an absent or subnormal cortisol response to an injection of synthetic ACTH (Synacthen test).

CASE HISTORY 9.1

A seventeen-year-old girl presented with a two-month history of tiredness and lethargy. She had noticed that she became dizzy when she stood up. On examination, she had pigmentation of the buccal mucosa and palmar creases and in an old appendicectomy scar. Her blood pressure was 120/80mmHg lying down, but fell to 90/50mmHg when she stood up.

Investigations

serum:	sodium	128mmol/l
	potassium	5.4mmol/l
	urea	8.5mmol/l
blood glucose (fasting)		2.5mmol/l

short Synacthen test:

plasma cortisol:	
0900h	150nmol/l
30min after Synacthen	160nmol/l
60min after Synacthen	160nmol/l
plasma ACTH (0900h)	500ng/l

anti-adrenal antibodies were detectable at a titre of 1 in 20

Comment

On the basis of these results, a diagnosis of primary adrenal failure was made. Her symptoms resolved rapidly after starting glucocorticoid and mineralocorticoid replacement and she remained well thereafter.

Postural hypotension is a common finding in adrenal failure; it is due to a decrease in ECF volume caused by a lack of aldosterone leading to sodium loss. This decrease in ECF volume may also cause a degree of prerenal uraemia as demonstrated in this case. Hyponatraemia is not always present in adrenal failure, particularly in the early stages. Sodium is lost isotonically from the kidneys, but the lack of cortisol may cause water retention and with severe hypovolaemia vasopressin (antidiuretic hormone, ADH) secretion is stimulated. Deficiency of aldosterone is also responsible for potassium retention and thus hyperkalaemia.

The fasting blood glucose is at the low end of normal in this patient; the unopposed action of insulin may cause symptomatic hypoglycaemia.

The 0900h cortisol is low and there is virtually no response to Synacthen. Except in very severe cases, cortisol is measurable in the serum, even though the concentration is low-normal or frankly low. However, this represents the adrenal glands' maximal output since they are already stimulated by the high level of endogenous ACTH.

Patients with adrenal failure require life-long replacement therapy, usually with both hydrocortisone and 9α-fludrocortisone, a synthetic mineralocorticoid. Occasionally patients can be free of symptoms on hydrocortisone alone, but they usually maintain a high salt intake to counter the effect of inadequate aldosterone secretion.

Adrenal failure is often secondary to pituitary failure when, although the adrenal glands are normal, there is decreased stimulation by ACTH. Other features of hypopituitarism may be present (see page 113); such patients, in contrast to those with primary adrenal failure, are not pigmented.

Hypotension may also occur in secondary adrenal failure because the sensitivity of arteriolar smooth muscle to catecholamines is reduced by a lack of cortisol. Hyponatraemia may sometimes be present since the lack of cortisol reduces the ability of the kidneys to excrete a water load, but there is no renal salt wasting since aldosterone secretion is not dependent upon ACTH.

The normal response to Synacthen (tetracosactrin) is shown in Fig.9.6. If the response is in any way abnormal the patient should be assumed to have adrenal failure. In both primary and secondary adrenal failure, the response in the short Synacthen test is absent or blunted (see *Case History 9.1*). This should be regarded as a screening test for adrenal failure; unless the clinical features leave no doubt that primary adrenal disease is responsible, it should be followed by a long Synacthen test (Fig.9.6). In this test, depot Synacthen, which has a longer duration of action, is given daily for three days. Serum cortisol concentration is measured each day and again twenty-four hours after the last dose. In primary adrenal failure, serum cortisol concentrations remain low; in secondary adrenal failure the concentration increases. An alternative version of the long Synacthen test involves only a single dose of depot Synacthen, with serum cortisol estimations made at various times up to twenty-four hours, but this is likely to be less sensitive if secondary adrenal failure is long-standing.

The best differentiation between primary and secondary adrenal failure is provided by measurement of serum ACTH, if this assay is available. In primary adrenal failure, pituitary ACTH production is greatly increased because of the lack of negative feedback control by cortisol, whereas in secondary adrenal failure the serum level of ACTH is low.

Synacthen stimulation tests	
Short test	**Long test**
Procedure	**Procedure**
take blood sample at 0900h for measurement of serum cortisol	day 1: inject 1mg depot Synacthen i.m.
inject 250μg Synacthen i.m.	days 2 and 3: repeat
take further blood samples after 30 and 60min for cortisol measurement	day 4 : measure serum cortisol at 0900h
Results	**Results**

serum cortisol:
  baseline   >190nmol/l
  after Synacthen   increment of 200nmol/l
             with peak of >550nmol/l
	primary adrenal insufficiency: serum cortisol on day 4 <200nmol/l (usually <100nmol/l)
	
	secondary adrenal insufficiency (hypothalamic or hypopituitarism): serum cortisol on day 4 at least 200nmol/l above baseline

Fig.9.6 Synacthen stimulation tests for the diagnosis of adrenal failure. It is important to note that blood should be taken for ACTH assay (if available), before giving Synacthen. It is not necessary to withhold any treatment until after the tests are completed, provided that the drug being used does not cross-react with cortisol, since exogenous steroids do not affect the response of the adrenal gland to ACTH.

Adrenal hyperfunction

In Cushing's syndrome, there is overproduction primarily of glucocorticoids though mineralocorticoid and androgen production may also be excessive. In Conn's syndrome, mineralocorticoids alone are produced in excess.

Cushing's syndrome

The causes and clinical features of Cushing's syndrome are listed in Fig.9.7. Pituitary-dependent adrenal hyperfunction is known specifically as Cushing's disease. The clinical features are due primarily to excessive cortisol but cortisol precursors and indeed cortisol itself have some mineralocorticoid activity. Thus sodium retention, leading to hypertension, and potassium wasting, causing a hypokalaemic alkalosis, are common findings except in iatrogenic disease (synthetic glucocorticoids have no mineralocorticoid activity). Increased production of adrenal androgens may also contribute to the clinical presentation.

There are two diagnostic steps in the investigation of a patient with suspected Cushing's syndrome: the demonstration of high serum cortisol levels and the elucidation of the cause. It is common to see patients who look Cushingoid; however, it is much less common that Cushing's syndrome is the cause. It is therefore often useful to carry out preliminary tests on an outpatient basis, aimed at excluding those patients

Cushing's syndrome
Causes
corticosteroid treatment pituitary hypersecretion of ACTH (Cushing's disease) adrenal adenoma adrenal carcinoma ectopic ACTH secretion by tumours, e.g. carcinoma of bronchus and carcinoid tumours
Clinical features
truncal obesity ('moon face', buffalo hump and protruberant abdomen) thinning of skin purple striae excessive bruising hirsutism, especially in adrenal carcinoma skin pigmentation (only if ACTH elevated) hypertension glucose intolerance muscle weakness and wasting, especially of proximal muscles menstrual irregularities back pain (osteoporosis and vertebral collapse) psychiatric disturbances: euphoria mania depression

Fig.9.7 Causes and clinical features of Cushing's syndrome.

who do not have adrenal disease and identifying those who may, and who thus merit further investigation. Tests used for this purpose (Fig.9.8) are the twenty-four-hour urinary cortisol excretion test and the overnight dexamethasone suppression test. It must, however, be appreciated that outpatient urine collections may be incomplete so that a normal result for the urinary cortisol excretion does not exclude Cushing's syndrome.

In normal subjects, the 0900h cortisol is suppressed by dexamethasone given the night before. Dexamethasone is a synthetic glucocorticoid which binds to cortisol receptors in the pituitary and suppresses ACTH release. A failure of suppression is characteristic of Cushing's syndrome but is not specific since it may also be seen in severe depression, as a result of stress, and in alcoholics; false negative responses vir-

tually never occur. It is important that, if a urine collection is made, the twenty-four-hour period should not include the time when dexamethasone is given.

Loss of diurnal variation of cortisol secretion is an early feature of Cushing's syndrome and the diagnosis is excluded if the serum cortisol concentration at 2300h or 2400h is normal. Since the patient must be resting and not stressed, serum cortisol measurement at night is not a practical outpatient procedure. It necessitates hospital admission, itself a stressful event, with the result that false positive results are common. However, if care is taken to minimize stress (ideally blood is taken from the sleeping patient through a previously inserted cannula after two or three days in hospital), a raised value does indicate pathological over-production of cortisol.

CASE HISTORY 9.2

A thirty-five-year-old male window cleaner presented with muscle weakness. This mainly affected his thighs so that he sometimes had to use his hands to help himself up from a sitting position. He was also finding it difficult to climb ladders at work. He had no other complaints.

On examination, he had a Cushingoid appearance with truncal obesity, proximal muscle wasting, violaceous abdominal striae and a plethoric, 'moon face'. His blood pressure was 180/110mmHg. He admitted that he had noticed the changes in his appearance developing over the past nine months but had been too shy to seek medical advice. It was only when he became concerned that he might not be able to continue working that he consulted his doctor. He was admitted to hospital for further investigation.

Investigations

serum:	sodium	136mmol/l
	potassium	3.2mmol/l
	bicarbonate	33mmol/l
blood glucose (fasting)		7.5mmol/l
serum cortisol:	(0900h)	1150nmol/l
	(2400h)	1100nmol/l
	ACTH (0900h)	180ng/l
urine cortisol excretion		840nmol/24h

Tests for Cushing's syndrome	
Overnight dexamethasone suppression	**24h urine free cortisol**
Procedure	**Procedure**
give 1 mg dexamethasone orally at bedtime	collect urine for 24h
take blood for measurement of cortisol at 0900h	measure total cortisol excretion
Result	**Result**
in normal subject, plasma cortisol falls to <50nmol/l	in normal subject, cortisol excretion <300nmol/ 24h

Fig.9.8 Screening tests for Cushing's syndrome. Note that the exact diagnostic values depend upon the method used to measure cortisol and must be checked with the local laboratory. A normal result in either test excludes Cushing's syndrome.

dexamethasone suppression test:
  0900h serum cortisol after
    0.5mg dexamethasone four times
    daily for two days (low dose)   1000nmol/l

  0900h serum cortisol after
    2.0mg dexamethasone four times
    daily for two days (high dose)   80nmol/l

Comment

The diagnosis is Cushing's disease. The clinical features are typical. The high 0900h cortisol, lack of diurnal variation and high urinary cortisol excretion all suggest adrenal hyperfunction. An overnight dexamethasone suppression test was not performed, because the presentation was so typical that an outpatient screening test was considered unnecessary. However, the formal two-stage test gave a result typical of Cushing's disease, with no cortisol response to the low dose but suppression of secretion on the high dose.

In Cushing's disease, the pituitary usually remains susceptible to feedback by glucocorticoids, but is apparently less sensitive than normal (i.e. a higher concentration of cortisol is necessary to suppress ACTH; Fig.9.9b). In Cushing's syndrome caused by adrenal tumours, whether adenomas or carcinomas, and also in ectopic ACTH secretion, there is usually no response to dexamethasone, even at the higher dose, since pituitary ACTH secretion is already suppressed by the high serum cortisol levels (Fig.9.9c). This patient's serum ACTH is raised; with adrenal tumours, feedback of cortisol to the pituitary suppresses ACTH while with ectopic ACTH secretion, ACTH levels are very high (Fig.9.9d). The results of biochemical tests in the various forms of Cushing's syndrome are summarized in Fig.9.10.

Measurements of serum ACTH concentration are of great value in establishing the cause of Cushing's syndrome. However, the hormone is very labile and serum must be separated rapidly, using a refrigerated centrifuge, and kept deep-frozen until the assay is performed if meaningful results are to be obtained. The assay itself is technically difficult.

This patient has a hypokalaemic alkalosis, a result of renal potassium wasting, and hypertension, a result of sodium retention. The fasting blood glucose is a little above normal; impaired glucose tolerance is common in Cushing's syndrome but clinical features of diabetes are rare, except in ectopic ACTH secretion. If the patient is coincidentally diabetic there may be a marked deterioration in control.

A skull radiograph in this patient showed a normal pituitary fossa; in Cushing's disease, the pituitary tumour secreting ACTH is usually very small, but may be revealed by computerized tomography (CT).

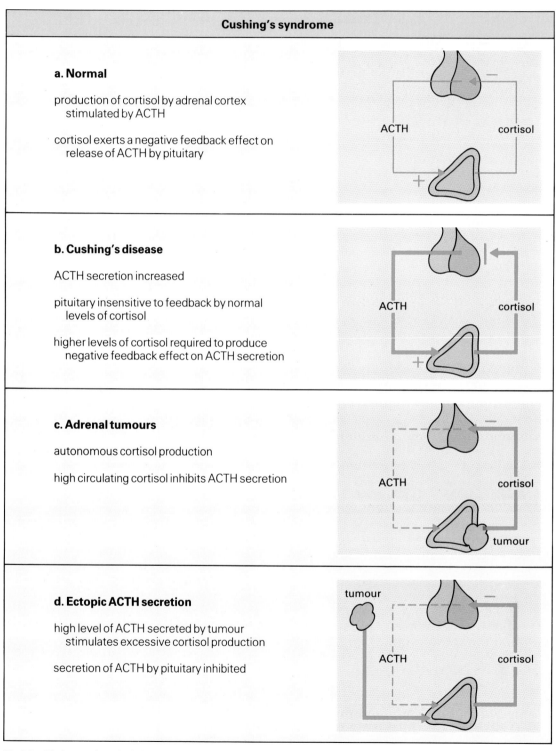

Cushing's syndrome

a. Normal

production of cortisol by adrenal cortex stimulated by ACTH

cortisol exerts a negative feedback effect on release of ACTH by pituitary

b. Cushing's disease

ACTH secretion increased

pituitary insensitive to feedback by normal levels of cortisol

higher levels of cortisol required to produce negative feedback effect on ACTH secretion

c. Adrenal tumours

autonomous cortisol production

high circulating cortisol inhibits ACTH secretion

d. Ectopic ACTH secretion

high level of ACTH secreted by tumour stimulates excessive cortisol production

secretion of ACTH by pituitary inhibited

Fig.9.9 Pituitary–adrenal relationships in Cushing's syndrome.

Results of adrenal function tests in Cushing's syndrome				
Condition	Basal cortisol (nmol/l)	Dexamethasone suppression test		Serum ACTH (ng/l)
		Low dose	High dose	
Cushing's disease	↑ (<1000)	no suppression	suppression	↑ (<200)
adrenal tumour	↑ (variable)	no suppression	no suppression	↓
ectopic ACTH secretion	greatly ↑ (>1000)	no suppression	no suppression	greatly ↑ (>200)

Fig.9.10 Results of adrenal function tests in Cushing's syndrome. With ectopic ACTH secretion by carcinoid tumours, the results of these tests may be identical to those seen in Cushing's disease, as the tumour may have glucocorticoid receptors which will respond to dexamethasone.

The diagnosis of adrenal hyperfunction and the elucidation of the cause may be straightforward. In difficult cases, other investigative techniques are valuable. For example, an adrenal tumour may be demonstrated by CT of the abdomen or a mass be demonstrable on chest radiography in a patient with an ACTH-secreting tumour. However, while the high dose dexamethasone suppression test usually identifies patients with Cushing's disease as opposed to ectopic ACTH-secreting tumours or primary adrenal tumours, both false negative and false positive results can occur. If available, a reliable measurement of the serum ACTH will often resolve the diagnosis.

The avoidance of stress to the patient during investigations for Cushing's syndrome has been emphasized. Obese patients may look Cushingoid and have urine cortisol excretion at the upper limit of the normal range, but usually retain a diurnal variation in serum cortisol concentration. Alcoholics may demonstrate both clinical and biochemical evidence of Cushing's syndrome (pseudo-Cushing's syndrome) but these usually resolve rapidly on withdrawal of alcohol. Occasionally, laboratory tests may suggest a diagnosis of Cushing's syndrome in a patient with severe depression; depression may itself be a feature of Cushing's disease. However, patients with depression have a normal cortisol response to insulin-induced hypoglycaemia; this response is lost even in mild cases of Cushing's disease. The insulin stress test, more commonly used in the investigation of pituitary function (see page 111), may be invaluable in resolving the diagnosis in difficult cases.

The syndrome of ectopic ACTH secretion is discussed in more detail in *Chapter 21*. When the tumour is a carcinoma, patients commonly present with a rapidly progressive illness in which the metabolic features of Cushing's syndrome (hypertension, hypokalaemia, leading to muscle weakness, and hyperglycaemia) predominate and the somatic features may be absent. Ectopic secretion of ACTH may, however, be a feature of carcinoid tumours, particularly bronchial adenomas. These are characteristically slow-growing and may produce a form of Cushing's syndrome that is very difficult to distinguish from true pituitary-dependent Cushing's disease.

MANAGEMENT

The management of Cushing's syndrome depends upon the cause. Adrenal adenomas and, if possible, carcinomas, should be resected. Until recently, Cushing's disease has usually been treated by bilateral adrenalectomy and external pituitary irradiation to prevent continued growth of the pituitary tumour, which may itself eventually give rise to clinical signs and symptoms together with extensive pigmentation (Nelson's syndrome). Trans-sphenoidal hypophysectomy (Cushing's original mode of treatment) has now been reintroduced in some centres but adrenalectomy is also commonly required since the adrenals may have become semi-autonomous. Patients who have undergone hypophysectomy or adrenalectomy will require appropriate steroid replacement therapy for life. When surgery is not possible, and in all cases pending surgery, symptomatic relief may ensue from the use of drugs which block cortisol synthesis, such as metyrapone, which inhibits steroid-11-hydroxylase, and trilostane.

Conn's syndrome

The common causes and clinical features of this condition are listed in Fig.9.11. Conn's syndrome is characterized by excessive production of aldosterone (hyperaldosteronism). In some eighty percent of cases, this is due to an adrenal adenoma, while in the remainder, there is diffuse hypertrophy of the cells of the zona glomerulosa, which produce aldosterone, in both adrenals. The majority of the clinical features are due to hypokalaemia, itself a result of renal potassium wasting. Patients are also hypertensive, a consequence of aldosterone-induced sodium retention. Conn's syndrome is a rare cause of hypertension and accounts for only approximately one percent of all cases, but it is important since it is potentially curable.

Hyperaldosteronism is also seen in patients whose plasma renin activity is increased. This is secondary hyperaldosteronism, since the adrenal glands are responding to their normal trophic stimulus, in contrast to the autonomous secretion of aldosterone in Conn's syndrome, which is termed primary hyperaldosteronism. In primary hyperaldosteronism, plasma renin is low.

Secondary hyperaldosteronism is far more common than the primary form and is associated with a variety of conditions (Fig.9.12) in which renin secretion is stimulated. Patients may or may not be hypertensive, depending on the underlying condition.

When investigating a patient with hypokalaemia and hypertension, many possible causes of secondary hyperaldosteronism can be eliminated either on clinical grounds or on the basis of simple tests. The definitive test to distinguish between primary and secondary hyperaldosteronism involves simultaneous measurement of plasma aldosterone and renin. In primary hyperaldosteronism, plasma renin activity is reduced; in secondary, renin is the cause of the excessive aldosterone secretion and is raised. However, the measurement of renin is technically difficult and should be reserved for the further investigation of those patients in whom there is a high index of suspicion of primary hyperaldosteronism. Such patients are selected on the basis of the results of simpler tests (Fig.9.13).

Primary hyperaldosteronism should be suspected in any hypertensive patient who has a low serum potassium concentration. Diuretics are an important cause

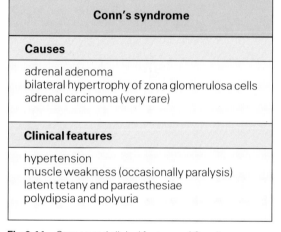

Conn's syndrome
Causes
adrenal adenoma bilateral hypertrophy of zona glomerulosa cells adrenal carcinoma (very rare)
Clinical features
hypertension muscle weakness (occasionally paralysis) latent tetany and paraesthesiae polydipsia and polyuria

Fig.9.11   Causes and clinical features of Conn's syndrome.

of hypokalaemia and indeed the use of diuretics in hypertension is the commonest cause of hypokalaemia and raised blood pressure. It is therefore essential that, if a hypertensive patient has been treated with a diuretic and is found to be hypokalaemic, the patient is given some other antihypertensive treatment for at least two weeks before further investigation is undertaken. Methyldopa and bethanidine are suitable for this purpose. Other causes of hypokalaemia (see Fig.3.3) should be sought and eliminated.

In Conn's syndrome, the hypokalaemia is caused by renal potassium wasting. Potassium should be absorbed maximally in the presence of hypokalaemia and in a hypokalaemic patient who is not on diuretics, renal potassium excretion of more than 30mmol/24h is very suggestive of primary hyperaldosteronism. In early cases, or if patients have a low salt intake, frank hypokalaemia may be intermittent or the serum potassium concentration only marginally abnormal. Under these circumstances, the patients should be given a high sodium intake (200mmol/24h) and repeated measurements of serum potassium made. Sodium loading increases the amount of sodium reaching the distal renal tubules. In a normal subject, the loading inhibits aldosterone secretion but if there is autonomous secretion this sodium is reabsorbed and there is a reciprocal increase in the amount of potassium excreted.

Conditions associated with secondary hyperaldosteronism

Common

congestive cardiac failure
cirrhosis of liver with ascites
nephrotic syndrome

Less common

renal artery stenosis
sodium-losing nephritis
Bartter's syndrome
renin-secreting tumours

Fig.9.12 Conditions associated with secondary hyperaldosteronism.

Screening tests for Conn's syndrome

serum potassium (off diuretics) <3.5mmol/l
urine potassium >30mmol/24h
serum and urine potassium as above with
    sodium loading (200mmol/24h)
exclusion of other causes of hypokalaemia
exclusion of secondary hyperaldosteronism

Fig.9.13 Screening tests for Conn's syndrome.

CASE HISTORY 9.3

A thirty-five-year-old woman was found to have a blood pressure of 190/110mmHg by her general practitioner at a routine health check. He prescribed a thiazide diuretic but a week later she returned to the surgery complaining of severe muscle weakness and constipation. The doctor arranged an urgent consultation at the local hospital where her serum potassium was found to be 2.6mmol/l. The diuretic was stopped, her blood pressure was controlled with methyldopa and she was given oral potassium supplements. After three weeks, her serum potassium concentration was only 3.0mmol/l. A twenty-four-hour urine collection contained 70mmol potassium. Conn's syndrome was suspected and she was admitted to the hospital for further investigation.

Investigations

	recumbent	upright for 30min
plasma:		
aldosterone	1320pmol/l	1340pmol/l
renin activity	<0.5pmol/h/ml	<0.5pmol/h/ml

Comment

A CT scan of the abdomen showed a small mass arising from the left adrenal gland. This was removed surgically. She made a rapid recovery after the operation and repeated checks showed her to be normokalaemic and normotensive.

Giving a diuretic may provoke symptomatic hypokalaemia in mild hyperaldosteronism. The hypokalaemia in this condition is characteristically resistant to potassium supplementation. The diagnosis of Conn's syndrome is indicated by the high aldosterone and low renin while the normal response to standing is blunted. In normal subjects, the decrease in renal blood flow caused by standing stimulates renin release and hence aldosterone secretion. In hyperaldosteronism, renin is suppressed and there is little or no increase in the serum aldosterone concentration when the patient stands up.

MANAGEMENT
If Conn's syndrome is shown to be due to a tumour, this should be removed surgically. In patients with bilateral adrenal hyperplasia, treatment with spirono-lactone, a diuretic which antagonizes the action of aldosterone, may be sufficient to control the blood pressure. Spironolactone is also used in patients with tumours while they await surgery.

Congenital adrenal hyperplasia

This syndrome encompasses a group of inherited metabolic disorders of adrenal steroid hormone biosynthesis. The clinical features of each depend upon the position of the defective enzyme in the synthetic pathway, which ultimately determines the pattern of hormones and precursors that is produced (see Fig.9.3).

21-hydroxylase deficiency, with an incidence of 1 in 12,000 of live births in the United Kingdom, accounts for more than ninety-five percent of all cases of congenital adrenal hyperplasia. The majority of the remaining five percent are due to deficiency of 11β-hydroxylase. 21-hydroxylase deficiency is often incomplete and adequate cortisol synthesis can thus be maintained by increased secretion of ACTH by the pituitary. It is this that causes hyperplasia of the glands. Because of the enzyme block, the substrate of the enzyme (17α-hydroxyprogesterone) accumulates and there is increased formation of adrenal androgens (see page 125).

Female infants affected by congenital adrenal hyperplasia may be born with ambiguous genitalia but when the enzyme block is only partial the condition may not present until early adulthood with hirsutism or amenorrhoea. Males may present with precocious pseudopuberty. In about one-third of neonates with 21-hydroxylase deficiency the enzyme deficiency is complete; these present shortly after birth with a life-threatening salt-losing state in which both cortisol and aldosterone production are insufficient to maintain normal homoeostasis. The partial and complete forms of 21-hydroxylase deficiency appear to be two separate entities, the manifestations of the condition running true-to-type within affected families.

Diagnosis is made by demonstrating an elevated concentration of either 17α-hydroxyprogesterone in the serum or of its metabolite, pregnanetriol, in the urine. Treatment involves replacement of cortisol, and mineralocorticoid if necessary, which should suppress the excessive ACTH production and hence the excessive androgen synthesis. Treatment is monitored by measurement of either serum 17α-hydroxyprogesterone or DHEA.

Partial 11β-hydroxylase deficiency is also more common than complete deficiency of the enzyme. It is characterized by hypertension, due to the accumulation of 11-deoxycorticosterone, a substrate of the defective enzyme which has salt-retaining properties. There is excessive androgen production. The diagnosis rests upon the demonstration of an increased serum concentration of either 11-deoxycortisol or of its urinary metabolite. The management is similar to that of the more common 21-hydroxylase deficiency, that is replacement of the missing hormone.

Other forms of congenital adrenal hyperplasia involving, for example, 17-hydroxylase, 18-hydroxylase (thus affecting aldosterone secretion only) and steroid 3β-hydroxydehydrogenase, Δ5 isomerase, are very rare. Some indication of their consequences, in terms of adrenal steroid metabolism, can be seen by studying Fig.9.3.

DISORDERS OF THE ADRENAL MEDULLA

The main interest of the adrenal medulla in clinical chemistry relates to phaeochromocytomas. These are tumours which secrete catecholamines, the normal secretory product of the organ, and which are a rare, but treatable, cause of hypertension. Approximately ten percent of phaeochromocytomas are found in extramedullary tissue that shares the same embryological origin, that is chromaffin tissue derived from neuro-ectoderm. Catecholamines can also be produced by tumours of embryologically related tissue, for example, the carotid bodies, and by neuroblastomas, rare tumours occuring only in infants and young children. These tumours form part of a group known as APUD (amine precursor uptake and decarboxylation) tumours (see pages 273 & 274).

Patients with phaeochromocytomas usually present with hypertension; although this may be episodic, it is usually sustained. Other features include palpitation, flushing and abdominal discomfort. These tumours are a rare cause of hypertension (approximately 0.5% of all cases). While hypertension itself is common, it is important to have available a screening test that identifies those patients likely to have a phaeochromocytoma, and who should be subjected to more definitive investigation, and that screens out those in whom this probability is very low.

The metabolism of catecholamines is outlined in Fig.9.14. Adrenaline (epinephrine) and noradrenaline (norepinephrine) are metabolized by catechol-O-methyltransferase (COMT) to metadrenaline and normetadrenaline, respectively. They are also converted by the consecutive action of monoamine oxidase and COMT to 4-hydroxy-3-methoxymandelic acid (HMMA), also known as vanillylmandelic acid (VMA).

Measurement of urinary HMMA is the most reliable screening test although the measurement of total urinary metanephrines (i.e. metadrenaline and normetadrenaline) is also used. Various foodstuffs, including tea, coffee, chocolate, bananas and vanilla, may react in the screening test for HMMA and should be avoided. Some drugs also interfere in the test, depend-

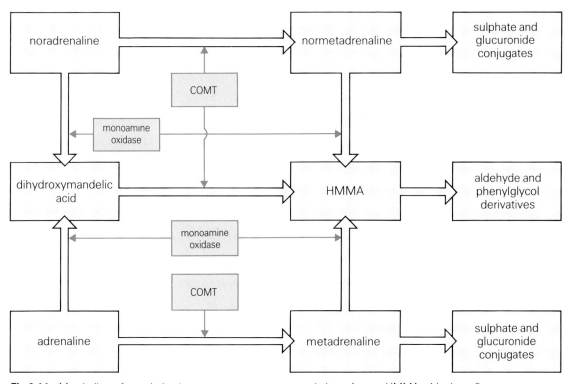

Fig.9.14 Metabolism of catecholamines. Normetadrenaline, HMMA, and noradrenaline are measured in the urine. COMT=catechol-O- methyltransferase; HMMA=4-hydroxy-3-methoxymandelic acid.

ing on the method used, and advice should be sought from the laboratory before either collecting samples or starting treatment.

If the screening test is negative, the diagnosis can be excluded. False negative responses occur only very rarely; they are more likely in a patient whose hypertension is intermittent and it is important that in such cases a urine collection is made at a time when the blood pressure is raised. If the screening test is strongly positive, serum adrenaline and noradrenaline should be measured. If these levels are raised, localization procedures, such as selective venous cannulation with catecholamine measurements and isotope and CT scanning should be carried out, guided by the fact that extra-adrenal phaeochromocytomas tend to secrete noradrenaline in excess of adrenaline, the reverse being true of tumours of the adrenal medulla itself.

When the screening test is not clearly diagnostic, the pentolinium test may be used as an aid to diagnosis. Pentolinium is a sympathetic ganglion-blocking drug that reduces catecholamine secretion in normal sub-

jects but not in patients with phaeochromocytomas; in such patients, secretion is autonomous. Blood is taken for catecholamine content before and fifteen minutes after giving 2.5mg pentolinium by intravenous injection. Although still potentially dangerous, this procedure is more reliable and less hazardous than the phentolamine test which has been used in the past. This α-adrenergic blocking drug causes dramatic hypotension in patients with phaeochromocytomas.

Although phaeochromocytomas are benign in ninety percent of cases, all tumours should be removed surgically. However, this is a potentially hazardous operation since large quantities of catecholamines may be released into the circulation during the procedure. It should be noted that ten percent of patients with phaeochromocytomas have multiple tumours. The tumours may be a component of the Sipple syndrome (multiple endocrine adenomatosis type IIa, see pages 279 & 280) and thus evidence of other relevant endocrine disorders should be sought in affected patients.

SUMMARY

The adrenal cortex secretes three classes of steroid hormones, glucocorticoids, androgens and mineralocorticoids. The secretion of glucocorticoids, of which cortisol is the most important, is controlled by adrenocorticotrophic hormone (ACTH), while ACTH secretion is subject to feedback inhibition by cortisol. Control is also exerted from the higher centres through the hypothalamus. Cortisol secretion shows a diurnal variation, with peak serum levels in the morning and a trough in the late evening. Cortisol is essential to life; it is involved in the response to stress and with other hormones regulates many pathways of intermediary metabolism. Its metabolic action is largely catabolic. Androgen secretion is also stimulated by ACTH; these hormones have a role in determining secondary sexual characteristics in the female but do not appear to have a specific role in the male.

Aldosterone stimulates sodium reabsorption in the distal tubule of the kidneys. It is an important determinant of the extracellular fluid volume. Its secretion is controlled by the renin–angiotensin system, in response to changes in blood pressure and blood volume.

Adrenal failure is most frequently due to organ-specific autoimmune destruction of the glands although there are many other causes. It can present acutely as a medical emergency with hypoglycaemia and circulatory collapse due to renal salt wasting. In more chronic cases, lassitude, weight loss and postural hypotension are frequent clinical features. Diagnosis depends upon demonstrating a failure of the adrenal to produce cortisol in response to ACTH (Synacthen test). Pituitary disease can cause secondary adrenal failure by interfering with normal ACTH secretion.

Over-production of adrenal cortical hormones can affect predominantly cortisol (producing Cushing's syndrome), or aldosterone (Conn's syndrome). In addition, Cushing's syndrome can be secondary to excess ACTH production by either a pituitary or a non-endocrine tumour (ectopic ACTH production), or be iatrogenic, due to treatment with high doses of corticosteroids or ACTH. Clinical features of Cushing's syndrome include characteristic somatic changes, muscle weakness, glucose intolerance, hypokalaemia and hypertension. Patients with Conn's syndrome develop hypertension and hypokalaemia. The diagnosis of these conditions involves first demonstrating high, non-suppressible concentrations of the hormones and then determining the cause.

The various syndromes of congenital adrenal hyperplasia are inherited metabolic disorders of adrenal steroid hormone biosynthesis. The clinical features derive from a mixture of under-production of either cortisol or aldosterone, or both, and increased production of androgens. The most common type is steroid 21-hydroxylase deficiency.

The adrenal medulla produces catecholamines but is not essential to life. There appear to be no clinical sequelae from decreased adrenal medullary activity but tumours of the glands (neuroblastomas and phaeochromocytomas) can produce excessive quantities of catecholamines. These cause hypertension and other clinical features related to increased sympathetic activity.

FURTHER READING

Besser GM & Cudworth AG (1987) *Clinical Endocrinology: An Illustrated Text.* London: Chapman & Hall; Gower Medical Publishing.

Hall R, Anderson J, Smart GA & Besser M (1980) *Fundamentals of Clinical Endocrinology.* 3rd edition. London: Pitman Medical.

Wilson JD & Foster DW (eds) (1985) *Williams – Textbook of Endocrinology.* 7th edition. Philadelphia: WB Saunders Company.

10. The Thyroid Gland

INTRODUCTION

The thyroid gland secretes three hormones: thyroxine (T_4) and triiodothyronine (T_3), both of which are iodinated derivatives of tyrosine (Fig.10.1), and calcitonin, a polypeptide hormone. T_4 and T_3 are produced by the follicular cells but calcitonin is secreted by the C cells which are of separate embryological origin. Calcitonin is functionally unrelated to the other thyroid hormones, possibly being involved in calcium homoeostasis, and disorders of its secretion are rare (see *Chapter 14*). Thyroid disorders in which there is either over- or under-secretion of T_4 and T_3 are, however, common.

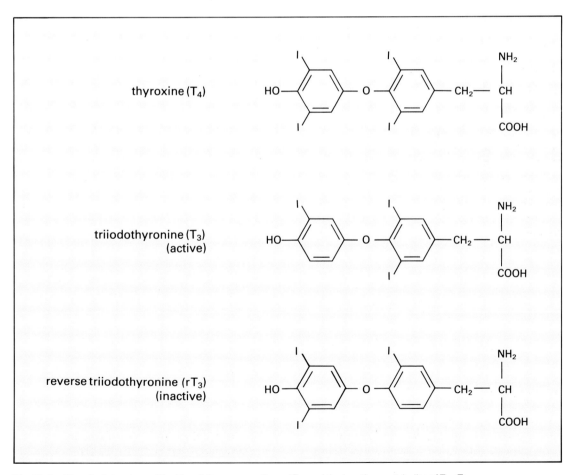

thyroxine (T_4)

triiodothyronine (T_3)
(active)

reverse triiodothyronine (rT_3)
(inactive)

Fig.10.1   Chemical structure of the thyroid hormones, T_4 and T_3, and the inactive metabolite of T_4, rT_3.

Thyroxine synthesis and release are stimulated by the pituitary trophic hormone, thyroid-stimulating hormone (TSH). The secretion of the trophic hormone itself is controlled by negative feedback by the thyroid hormones (see page 107) which modulate the response of the pituitary to the hypothalamic hormone, thyrotrophin releasing hormone (TRH; Fig.10.2).

The major product of the thyroid gland is T_4. Some T_3 is secreted directly (in thyroid disease the proportion may be greater), but most is derived from T_4 by deiodination in peripheral tissues, particularly the liver and kidney. T_3 is 3–4 times more potent than T_4. Deiodination can also produce reverse triiodothyronine (rT_3; see Fig.10.1) which is physiologically inactive. It is produced instead of T_3 in various stressful states, such as starvation, and the formation of either the active or inactive metabolite of T_4 appears to play an important part in the control of energy metabolism. The anterior pituitary is also active in converting T_4 to T_3. It is thought that the pituitary senses thyroid hormone status through a change in the concentration of T_3 due to deiodination within the pituitary cell nuclei.

THYROID HORMONES

Functions

Thyroid hormones are essential for normal growth and development and have many effects on metabolic processes. They act by entering cells and binding to specific receptors in the nuclei, where they stimulate the synthesis of a variety of species of mRNA, thus stimulating the synthesis of polypeptides including hormones and enzymes. Their most obvious overall effect on metabolism is to stimulate the basal metabolic rate but the precise molecular basis of this action is unknown. Thyroid hormones also increase the sensitivity of the cardiovascular and nervous systems to catecholamines.

Synthesis

Thyroid hormone synthesis involves a number of specific enzyme-catalyzed reactions, beginning with the uptake of iodine by the gland and culminating in the iodination of tyrosine residues in the protein,

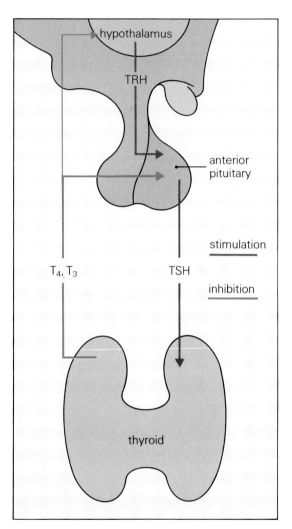

Fig.10.2 Control of TSH secretion. TSH is released from the pituitary in response to the hypothalamic hormone, TRH. TSH release is inhibited by high circulating levels of thyroid hormones which either feed back on the hypothalamus to reduce TRH release or, more importantly, act directly on the pituitary gland by modulating its sensitivity to TRH.

thyroglobulin (Fig.10.3); these reactions are all stimulated by TSH. Rare congenital forms of hypothyroidism due to inherited deficiencies of each of the various enzymes concerned have been described.

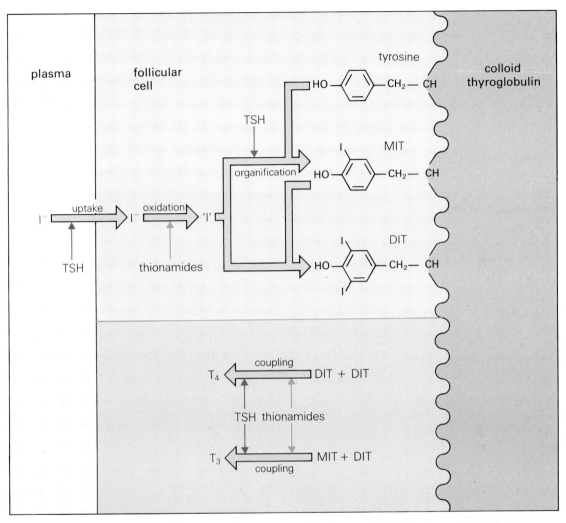

Fig.10.3 Biosynthesis of the thyroid hormones. The iodination and condensation reactions involve tyrosine residues that are an integral part of the thyroglobulin polypeptide. The thyroid hormones remain protein-bound until they are released from the cell. The precise nature of the active iodine moiety ('I') is unknown. Once formed, it is rapidly incorporated into tyrosine residues to form monoiodotyrosine (MIT) and diiodotyrosine (DIT). Antithyroid thionamide drugs, such as carbimazole, act by inhibiting the formation of the active moiety or by preventing the coupling of DIT to form T_4.

Thyroglobulin is stored within the thyroid gland in colloid follicles. These are accumulations of thyroglobulin-containing colloid surrounded by thyroid follicular cells. Release of thyroid hormones (stimulated by TSH) involves pinocytosis of colloid by follicular cells, fusion with lysosomes to form phagocytic vacuoles and proteolysis (Fig.10.4). Thyroid hormones are thence released into the blood stream. Proteolysis also results in the liberation of mono- and diiodotyrosines (MIT and DIT); these are usually degraded within thyroid follicular cells and their iodine retained and re-utilized. A small amount of thyroglobulin also reaches the blood stream.

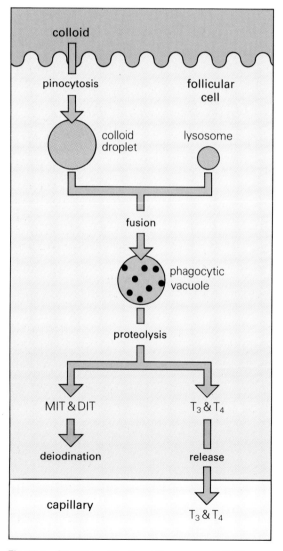

Fig.10.4 Secretion of the thyroid hormones. Colloid is taken up into follicular cells by pinocytosis and undergoes lysosomal proteolysis resulting in the release of thyroid hormones.

Thyroid hormones in blood

The normal serum concentrations of T_4 and T_3 are 60–150nmol/l and 1.0–2.9nmol/l, respectively. Both hormones are extensively protein bound, some 99.95% of T_4 and 99.5% of T_3 being bound, principally to a specific thyroxine-binding globulin (TBG) and to a lesser extent to prealbumin and albumin. TBG is approximately one-third saturated at normal concentrations of thyroid hormones (Fig.10.5). It is generally accepted that it is only the free, or non protein-bound, thyroid hormones that are physiologically active. Although the total T_4 concentration is normally fifty times that of T_3, the different extents to which these hormones are bound to protein mean that the free T_4 concentration is only 2–3 times that of free T_3. In the tissues, most of the effects of T_4 probably result from its conversion to T_3, which means that T_4 itself is essentially a prohormone.

The precise physiological function of TBG is unknown; subjects who have a genetically determined deficiency of the protein show no clinical abnormality. It has, however, been suggested that the extensive binding of thyroid hormones to TBG provides a buffer which maintains the free hormone levels constant in the face of any tendency to change. The binding may also reduce the amount of thyroid hormones lost through the kidneys and it has been suggested that the binding protein may have a specific role in facilitating hormone uptake by cells. However, for the clinician and pathologist, protein binding provides a major obstacle to the laboratory assessment of thyroid status.

The most widely used assays for thyroid hormones measure their total concentration (not the free, physiologically active fraction) which is obviously very dependent upon the concentration of binding protein present in the blood. If this increases (Fig.10.6) the temporary fall in free hormone concentration caused by increased protein binding would stimulate TSH release and restore the free hormone concentration to normal. Conversely, if the protein concentration were to fall, the reverse would occur. In either situation, there is a change in the concentrations of total hormones, but the free hormone concentrations remain normal. Thus measurement of total hormone concentrations can give misleading information.

	Serum concentration		Extent of protein binding (%)	Half-life (days)
	total (nmol/l)	free (pmol/l)		
T_4	60–150	9.0–26.0	99.95	6–7
T_3	1.0–2.9	3.0–9.0	99.5	1–1.5

Fig.10.5 Thyroid hormones in blood. Each laboratory should determine its own normal range for serum concentrations.

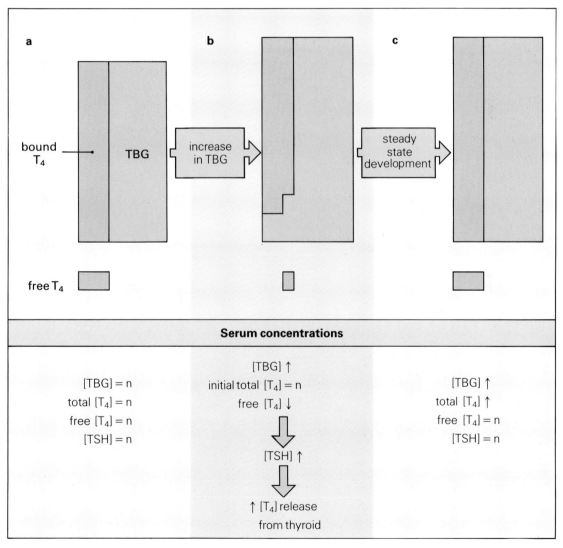

Fig.10.6 Effect of an increase in TBG concentration on serum T_4 levels. (a) in the initial steady state, TBG is one-third saturated with T_4. (b) TBG levels increase causing more T_4 to be bound, thus reducing the free T_4 concentration. This stimulates TSH secretion which leads to an increase in the release of T_4 from the thyroid. (c) the new T_4 is re-distributed between the bound and the free states leading to a new steady state with the same free T_4 level but an increased total T_4.

This is a matter of considerable practical importance since changes in the concentrations of the binding proteins occur in many circumstances (Fig.10.7). Further, certain drugs, for example, salicylates and phenytoin, will displace thyroid hormones from their binding proteins, thus reducing the total, but not the free, hormone concentrations once a new steady state is attained. If an attempt is made to assess thyroid status in a patient who is not in a steady state, the results may be bizarre and totally misleading.

Only small amounts of T_4 and T_3 are excreted by the kidneys due to their extensive protein binding. The major route of thyroid hormone degradation is by deiodination and metabolism in tissues, but they are also conjugated in the liver and excreted in bile.

Causes of abnormal serum TBG levels
Increase
genetic
pregnancy
oestrogens, including
oestrogen-containing oral
contraceptives
Decrease
genetic
protein-losing states,
e.g. nephrotic syndrome
malnutrition
malabsorption
acromegaly
Cushing's disease
corticosteroids (high dose)
severe illness
androgens

Fig.10.7 Causes of abnormal serum concentrations of TBG.

TESTS OF THYROID FUNCTION

Laboratory tests of thyroid function are required to assist in the diagnosis and monitoring of thyroid disease. No single test is suitable for these purposes and it is usual for laboratories to have a repertoire of several thyroid function tests, and to adopt a strategy whereby one of a number of logical sequences or combinations of these tests is performed according to the clinical indications.

Total thyroxine

Measurement of the total T_4 concentration in serum is carried out using an immunoassay. The assay reflects the activity of the thyroid gland and is in widespread use as the first-line test of function. However, its use for this purpose is complicated by the fact that T_4 is extensively protein bound, as discussed. Further, the range of concentrations seen in the euthyroid population is wide and there is overlap between values at the lower end of the range and those characteristic of hypothyroidism and, to a lesser extent, the upper end and those seen in hyperthyroidism. Thus, further tests are frequently required to precisely define a patient's thyroid status.

Total triiodothyronine

A raised total T_3 is a reliable indicator of hyperthyroidism. It may be used to support this diagnosis when the total T_4 is elevated and to confirm or refute it when the T_4 is towards the upper limit of the normal range, and hence is compatible with, but not diagnostic of, hyperthyroidism.

Serum T_4 and T_3 levels generally change in the same direction, but this is not invariable. In approximately five per cent of cases of thyrotoxicosis, the serum T_4 is normal (T_3-toxicosis). Very rarely in thyrotoxicosis the T_4 is unequivocally high but the T_3 is not (T_4-toxicosis). This is usually seen in patients with thyrotoxicosis who develop an intercurrent illness and in whom there is a decrease in peripheral conversion of T_4 to T_3 (see below); the T_3 becomes elevated once they have recovered. Additionally, in some patients treated for hyperthyroidism, whether by drugs, radioactive iodine or surgery, the total T_4 may be found to be low and the TSH slightly raised, but the T_3 is normal and the patients are clinically euthyroid.

This has been termed 'compensated hypothyroidism' and may represent a transition to true hypothyroidism.

In non-thyroidal illness, there is commonly a decrease in the peripheral conversion of T_4 to T_3; this occurs, for example, in starvation, following surgery and in many acute and chronic illnesses. Some β-adrenergic blocking drugs, such as propranolol, also inhibit the conversion of T_4 to T_3. In such patients, the total T_3 is low in relation to the T_4; in very sick patients, both the T_4 and T_3 may be in the hypothyroid range yet the serum TSH is usually normal and the patient is, as far as can be assessed, clinically euthyroid. Although the concentration of binding proteins is frequently reduced in these patients, this is not the only reason for the decreased hormone levels, since free hormone levels are also decreased. This syndrome of 'sick euthyroidism' can present considerable diagnostic confusion. The results of thyroid function tests are rarely helpful in the very sick patient and such investigations are best avoided. A similar pattern of results is commonly seen in patients in chronic renal failure.

Free hormone concentrations

The measurement of free hormone concentrations poses major technical problems since the binding of free hormones in an assay, by an antibody, for example, will disturb the equilibrium between bound and free hormone and cause release of hormone from binding proteins. Various techniques have been developed which allow the estimation of free T_4 and T_3 concentrations in the serum. Such measurements, in theory, circumvent the problems associated with protein binding and should render obsolete the techniques for the indirect assessment of free hormone concentrations, such as the resin uptake test, calculation of the free thyroxine index or measurement of the T_4 to TBG ratio. However, with gross abnormalities of binding proteins, the results of measurements of the free hormones may be misleading due to technical problems in the test methods. Further, in non-thyroidal illnesses, free thyroid hormone concentrations may be low yet the patient is clinically euthyroid ('sick euthyroidism', see above).

Free thyroid hormone assays are now used in some laboratories in preference to total hormone assays but results must be interpreted with caution in patients with non-thyroidal illness. In pregnancy, for reasons that are not understood, the normal range for free T_4 in euthyroid women decreases as the pregnancy progresses. There is an increase in TBG, due to the increased oestrogen levels, and in total T_4, but these are disproportionate, causing the level of free T_4 to fall. Thyroid status, as assessed clinically, does not change.

Thyroid-stimulating hormone

Since the release of TSH from the pituitary is controlled through negative feedback by thyroid hormones, it should, in theory, be possible to assess thyroid status by examining the state of the feedback loop, that is, by measuring serum TSH concentration. The measurement of serum TSH is a powerful test in the diagnosis of primary hypothyroidism. In overt disease, serum TSH levels are unequivocally raised, often to more than one hundred times the normal value. If the TSH is normal or low, it may reliably be inferred that the patient is not hypothyroid. The only exception will be patients with hypothyroidism secondary to pituitary hypofunction, but this is rarely a cause of diagnostic confusion since such patients are uncommon and usually have other clinical features of hypopituitarism.

Conversely, in cases of thyrotoxicosis, the TSH concentration should be very low, since its secretion should be suppressed by the high circulating levels of thyroid hormones. (Cases of thyrotoxicosis secondary to pituitary tumours secreting TSH have been described but are exceptionally rare.) Until recently, radioimmunoassays for TSH were insufficiently sensitive to be able to differentiate reliably between normal and low levels of TSH. However, highly sensitive assays are now becoming available which are capable of discriminating between hyperthyroid and euthyroid levels of TSH.

Such assays are now being introduced as first-line tests of thyroid function instead of total or free T_4. They should avoid the problems associated with binding proteins and the difficulties in interpreting T_3 and T_4 concentrations (total or free) in non-thyroidal illness. In some very sick patients, the hypothalamo–pituitary–thyroid axis as a whole is affected and patients may appear clinically euthyroid but have a low T_4 and low TSH. In practice, the results are usually so bizarre that, when viewed in the clinical context, they do not cause diagnostic confusion.

Thyrotrophin releasing hormone (TRH) test

In this test, serum TSH is measured immediately before, and twenty and sixty minutes after giving the patient 200μg of TRH intravenously (Fig.10.8). The normal response is an increase in TSH concentration of 1–20mu/l in twenty minutes, with a reversion towards the basal value at sixty minutes.

In hyperthyroidism, there is either little (<1mu/l) or no TSH response to injected TRH. This may also be seen in ophthalmic Graves' disease and in euthyroid patients with small, functioning thyroid adenomas or multinodular colloid goitres. This situation has been called 'borderline thyrotoxicosis'. In such patients, the thyroid is probably functioning autonomously, that is, it is not under the control of the pituitary, so that TSH production is suppressed but the thyroidal over-activity is not sufficient to render the patient clinically thyrotoxic. A similar lack of response to TRH is seen in patients treated for thyrotoxicosis who have been rendered euthyroid. This may reflect continuing sub-clinical hyperthyroidism, but it can also be due to a delay in the return of pituitary responsiveness to circulating thyroid hormone levels after the pituitary has been suppressed during the time the patient was thyrotoxic.

The TRH test is best used as a test of exclusion; a normal response excludes hyperthyroidism but a 'flat' response, although not diagnostic, is characteristic of hyperthyroidism. In addition to the circumstances mentioned above, it may also be seen, for instance, in some cases of acromegaly. However, since highly sensitive TSH assays will distinguish between normal and suppressed levels of TSH, and a normal level excludes hyperthyroidism, it will become unnecessary to carry out TRH tests for this purpose in laboratories where this new TSH assay has been introduced.

In primary hypothyroidism, there is an exaggerated response to TRH; if the basal TSH is unequivocally raised, the result of the TRH test affords no additional diagnostic information but can be valuable in patients whose results are borderline. However, recent work suggests that the response to TRH can be predicted from the basal TSH. Subclinical or biochemical hypothyroidism, in which patients appear euthyroid but have a thyroid function test result suggestive of hypothyroidism, is common. As in the investigation of suspected hyperthyroidism, a normal TSH response to TRH excludes hypothyroidism but an abnormal response, though characteristic, is not diagnostic.

Patients with hypopituitarism rarely lose TSH production entirely. Thus, in secondary hypothyroidism the TSH response to TRH may be normal or low-normal, but is rarely completely absent. A delayed response to TRH is characteristic of hypothalamic disorders, with the TSH concentration at sixty minutes exceeding that at twenty minutes (see Fig.10.8).

Other tests of thyroid function

Other biochemical disturbances, not involving the thyroid hormones, occur in thyroid disease but are of no value diagnostically; examples are hypercalcaemia and hyperphosphataemia, in some cases of thyrotoxicosis, and hypercholesterolaemia and hyponatraemia in hypothyroidism.

Techniques involving the use of radioactive isotopes for the investigation of thyroid disease are of two types. Tests involving the quantification of radioactive iodine uptake were introduced before specific tests for thyroid hormones were available, but they are now virtually obsolete. Thyroid scinti-scanning, however, is in common use. In this technique, a dose of an isotope, usually ^{99m}Tc, is given and its distribution within the thyroid is determined using a gamma camera. This technique allows the identification of 'hot' (active) or 'cold' (inactive and potentially malignant) nodules in patients with lumps in the thyroid. It can also distinguish between Graves' disease (uniformly increased uptake), multinodular goitre (patchy uptake) or an adenoma (single 'hot' spot) in patients with thyrotoxicosis, and detects aberrant or ectopic thyroid tissue.

A number of auto-antibodies to thyroid antigens have been detected in the serum of patients with thyroid disease. For example, the thyroid-stimulating immunoglobulins are thought to be pathogenic in Graves' disease. Assays for them are not yet generally available, but there are indications that their measurement may be valuable in the diagnosis of Graves' disease as a cause of thyrotoxicosis in the absence of the characteristic eye signs, and in the prediction of relapse of Graves' disease after a course of antithyroid drugs. In contrast, there is no direct evidence that the anti-microsomal and anti-thyro-globulin antibodies, present in high titre in most patients with Hashimoto's thyroiditis, and in some with Graves' disease, are pathogenic. Anti-thyro-globulin antibodies may bind thyroid hormones and give rise to abnormalities in the results of thyroid function tests. Other antibodies have been detected which stimulate or block the growth of thyroid cells

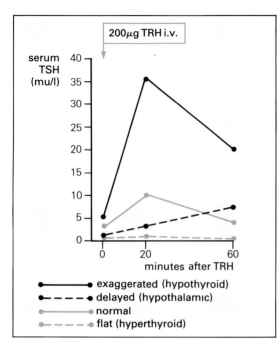

Fig.10.8   TRH test. 200µg TRH is given intravenously and serum TSH is measured at 0, 20 and 60 minutes. Typical responses are shown.

and block the TSH receptor. The role of these antibodies in the pathogenesis of thyroid disease remains unknown.

DISORDERS OF THE THYROID

The metabolic manifestations of thyroid disease relate either to excessive or inadequate production of thyroid hormones (hyperthyroidism and hypothyroidism, respectively). The clinical syndrome that results from hyperthyroidism is thyrotoxicosis. Myxoedema is often used to describe the clinical sequelae of hypothyroidism, but strictly refers to the dryness of the skin, coarsening of the features and subcutaneous swelling characteristic of severe hypothyroidism. Patients with thyroid disease may present with a thyroid swelling or goitre. Thyroid function should be assessed in patients with goitres although many will be euthyroid. A goitre is more commonly present in patients with thyrotoxicosis than in those with myxoedema. However, it is not invariably present in either of these conditions.

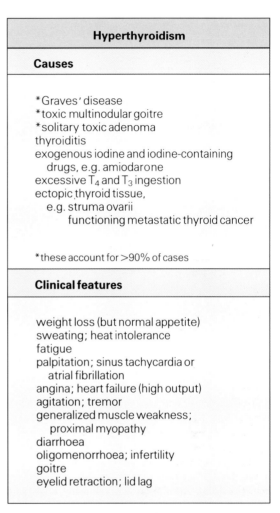

Fig.10.9   Causes and clinical features of hyperthyroidism. Periorbital oedema, proptosis, diplopia, ophthalmoplegia, corneal ulceration, loss of visual acuity, pretibial myxoedema and thyroid acropachy are features of Graves' disease only.

Hyperthyroidism

The major causes and clinical features of hyperthyroidism are shown in Fig.10.9. The commonest single cause is Graves' disease, an autoimmune disease characterized by the presence of thyroid-stimulating antibodies in the blood. These auto-antibodies bind to TSH receptors in the thyroid and stimulate them in the same way as TSH, through activation of adenylate cyclase and the formation of cyclic AMP.

CASE HISTORY 10.1

A twenty-six-year-old bank clerk presented with a three-month history of weight loss, despite increased appetite. She had noticed that her hands were shaky and this was making it difficult for her to write. Although it was a cold day when she consulted her general practitioner, she was wearing only light clothing and said that she did not feel the cold. Her doctor observed that her eyes were rather prominent and that her thyroid gland was slightly enlarged, and made a clinical diagnosis of thyrotoxicosis.

Investigations

serum: T_4 >200nmol/l
       T_3 8nmol/l
      free T_4 40pmol/l

An isotope thyroid scan showed diffusely increased uptake throughout the thyroid.
    Thyroid auto-antibodies were present in the serum in high titre.

Comment

The clinical diagnosis of thyrotoxicosis is supported by the results of the thyroid function tests. The clinical features here are typical (see Fig.10.9) but they are often less obvious in milder cases and may closely mimic an anxiety state (see *Case History 10.2*). Some clinical features are specific to Graves' disease and have a clinical course independent of the hyperthyroidism. Thus patients may present with the ocular manifestations of Graves' disease (ophthalmic Graves' disease) yet be clinically euthyroid. However, such patients usually have a suppressed TSH, show no TSH response to TRH and eventually become clinically thyrotoxic.

There are three options for the treatment of thyrotoxicosis: anti-thyroid drugs, radioactive iodine and surgery (sub-total thyroidectomy). Treatment with β-adrenergic blocking drugs may provide temporary symptomatic relief but has no effect on the underlying disease process. Very rarely, patients with thyrotoxicosis present with, or develop, thyroid storm or crisis, a medical emergency whose features include hyperpyrexia, dehydration and cardiac failure. The diagnosis of thyroid storm is made on clinical grounds and although thyroid function tests should be performed to confirm the diagnosis, treatment must not be delayed pending the results of the tests.

Unless they have a very large goitre, it is usual to treat younger patients with Graves' disease with anti-thyroid drugs. These suppress thyroid hormone synthesis, though not the release of pre-formed hormone so that there is usually a delay before any response is seen. They may affect the underlying disease process through an immunosuppressive action. Untreated Graves' disease has a natural history of remission and relapse. Some patients (30–40%) only have a single episode of hyperthyroidism. It is usual to give anti-thyroid drugs for a period of approximately one to two years. Thyroid status is monitored during this time and thereafter. If the patient relapses, a further course of drug treatment or another mode of treatment is indicated.

Once a patient has been rendered euthyroid with anti-thyroid drugs, a reduction in dose is usually possible and the maintenance dose is determined on the basis of regular clinical and laboratory assessments. The total serum T_4 (if binding proteins are normal), T_3 or free T_4 may be used for this purpose. The TSH level may remain suppressed for a considerable period after the patient has become euthyroid. If the concentration is in the normal range, it may be inferred that the patient is euthyroid, but a suppressed TSH level does not necessarily mean that the patient is still hyperthyroid.

Long-term follow-up of patients treated for Graves' disease is essential. Patients treated with anti-thyroid drugs may relapse, occasionally after many years; some, on the other hand, become hypothyroid. Recurrences of the disease also occur in patients treated surgically or with radioactive iodine. Up to thirty-five percent of patients treated surgically and the majority of patients treated with radioactive iodine will eventually become hypothyroid, sometimes as long as ten years or more after treatment. Hypothyroidism that develops within six months of either surgery or radioactive iodine treatment may be temporary, but abnormalities of thyroid function tests, in particular a normal serum T_4 concentration but a slightly raised TSH level, may persist though the patient remains clinically euthyroid. This is illustrated in Fig.10.10.

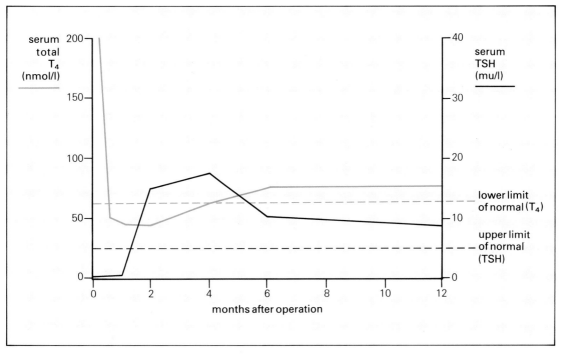

Fig.10.10 Changes in serum T_4 and TSH levels in a patient with Graves' disease treated by partial thyroidectomy. Total T_4 falls immediately after the operation; TSH initially remains suppressed but eventually responds to the low T_4, due to lack of negative feedback, and stimulates the thyroid. A normal T_4 is thence maintained by the remaining thyroid tissue, but this requires a greater than normal TSH drive.

CASE HISTORY 10.2

Shortly before her final examinations, a medical student experienced sleep disturbances, tachycardia with palpitation and noticed that her hands were warm and sweaty. Her doctor was sure that her symptoms were due to anxiety, but was persuaded to take a blood sample for thyroid function tests.

Investigations

serum: T_4 165nmol/l
        T_3 2.9nmol/l
        free T_4 24pmol/l

These results were considered to be equivocal and a TRH test was performed.

serum TSH: 0min   1.2mu/l
             20min   5.4mu/l
             60min   3.1mu/l

Comment

The total T_4 is elevated and the T_3 is at the upper limit of normal; the free T_4 is near the upper limit of normal. Mild thyrotoxicosis could not be excluded on the basis of these results. However, the TSH response to TRH is normal, indicating that the patient is euthyroid. It transpired that she was taking an oestrogen-containing oral contraceptive pill and the resulting increase in serum TBG would explain the abnormal total hormone concentrations.

Two algorithms for the investigation of possible hyperthyroidism are shown in Fig.10.11; one uses T_4 or free T_4 as the initial test, the other, TSH, assuming that a highly sensitive method is available.

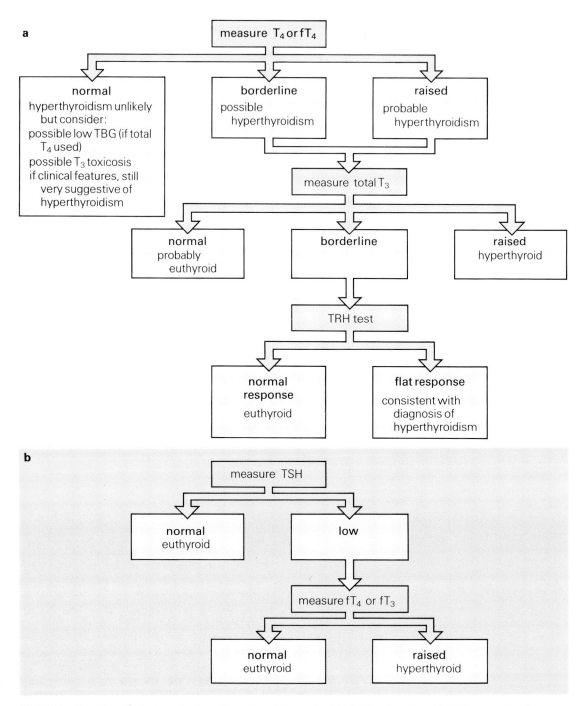

Fig.10.11 Algorithms for the investigation of hyperthyroidism, using (a) total T_4 or free T_4 and (b) TSH as the first-line test.

Hypothyroidism	
Causes	**Clinical features**
*atrophic hypothyroidism	lethargy; tiredness
*autoimmune hypothyroidism (Hashimoto's thyroiditis)	cold intolerance
	dryness and coarsening of skin and hair
*post-surgery, radioactive iodine, anti-thyroid drugs (e.g. carbimazole) and other agents (e.g. lithium)	hoarseness
	weight gain
	slow relaxation of muscles and tendon reflexes
congenital	many others including:
dyshormonogenic	anaemia, typically macrocytic, non-megaloblastic
secondary (pituitary or hypothalamic disease)	but pernicious in 10% of cases
iodine deficiency	dementia; psychosis
	constipation
	bradycardia; angina; pericardial effusion
	muscle stiffness
*these account for >90% of cases	carpal tunnel syndrome
	infertility; menorrhagia, galactorrhoea

Fig.10.12 Causes and clinical features of hypothyroidism. Children with hypothyroidism may present with growth failure, delayed pubertal development or a deterioration in academic performance.

Hypothyroidism

There are many possible causes of primary hypothyroidism (Fig.10.12) but hypothyroidism may also occur secondary to either decreased trophic stimulation in hypopituitarism or in hypothalamic disease. It is, however, very rare for patients with pituitary failure to present with clinical features of hypothyroidism alone. The commonest cause of hypothyroidism is atrophic myxoedema, the end result of autoimmune destruction of the gland. The clinical manifestations (Fig.10.12) are variable and may result in the patient being referred to almost any specialist department in a hospital.

CASE HISTORY 10.3

A senior civil servant was persuaded to seek medical advice because of his increasingly bizarre behaviour. He had slowed-up mentally, becoming indecisive and had given up his daily game of squash, claiming that he no longer had the energy to play. Whereas in the past he had annoyed his colleagues by opening windows even on cold days, he now no longer objected to them remaining closed. His appearance had changed, his skin appearing sallow and his hair coarse and lacking lustre.

His general practitioner suspected hypothyroidism and elicited a history of recent constipation in addition to the other typical features. Bradycardia was present and when the peripheral reflexes were tested, slow quadriceps relaxation was noted. There was no goitre.

Investigations

serum: T_4 4nmol/l
TSH > 100mu/l

Comment

The clinical diagnosis of hypothyroidism is confirmed by the low serum T_4, reflecting decreased thyroid activity, and greatly elevated TSH, produced in response to the decreased feedback by thyroid hormones.

Hypothyroidism is treated by replacement of thyroid hormones, usually T_4. T_3 has a short plasma half-life and is preferred when treating hypothyroid patients with ischaemic cardiac disease. The increase in metabolic rate and demand for oxygen prompted by hormone replacement may precipitate angina or myocardial infarction; a much more rapid response to a decrease in dosage is seen if T_3 rather than T_4 is used. T_3 is also preferable in the initial treatment of patients in myxoedema coma (see below).

In the laboratory, treatment is monitored by measuring serum TSH. If this is raised, the dosage is insufficient to render the patient euthyroid. If normal or low, the circulating thyroid hormone level must be sufficient to inhibit TSH release. Possible over-treatment cannot be reliably diagnosed in the laboratory. In patients treated with T_4, the serum concentrations associated with a clinically euthyroid state are somewhat higher than the normal euthyroid range, presumably because there is no contribution to endogenous hormone activity by secreted T_3. In patients treated with T_3, this rather than T_4 should be measured if insufficient information is provided by the TSH alone.

Occasionally, patients with hypothyroidism present as an emergency with stupor and hypothermia. This myxoedema coma has a high mortality. In addition to thyroid hormone re-

placement, usually with T_3, possible coexistent adrenal insufficiency must be treated with hydrocortisone and appropriate measures taken to treat any infection, heart failure, electrolyte imbalance and to restore body temperature to normal.

CASE HISTORY 10.4

Thyroid function tests were performed on an elderly widower with bronchopneumonia, because he had been slightly hypothermic on admission (core temperature 35.5°C).

Investigations

serum: T_4 50nmol/l
free T_4 8pmol/l
TSH 2.4mu/l

Comment

The total T_4 is low. This could possibly be due to decreased levels of binding proteins, a consequence of poor dietary protein intake, which commonly occurs in the elderly. If this were the only cause the free T_4 would be normal. However, the TSH is normal and in primary hypothyroidism it should be raised. There are two possible explanations: secondary (pituitary or hypothalamic) hypothyroidism or sick euthyroidism. Hypothyroidism is very rarely the sole feature of pituitary hypofunction. If in doubt, it is prudent to measure the 0900h cortisol as a rough guide to the integrity of the pituitary–adrenal axis. *Case history 8.1* gives an example of hypopituitarism which includes partial secondary hypothyroidism.

This patient's tests were repeated immediately before his discharge. Total and free T_4 were normal, confirming that he indeed had the sick euthyroid syndrome.

An algorithm for the investigation of suspected hypothyroidism is given in Fig.10.13. Either T_4 or TSH can be used as the first-line test. A highly sensitive TSH assay is not required, since conventional techniques have sufficient sensitivity to discriminate between raised and normal values, but in practice it is

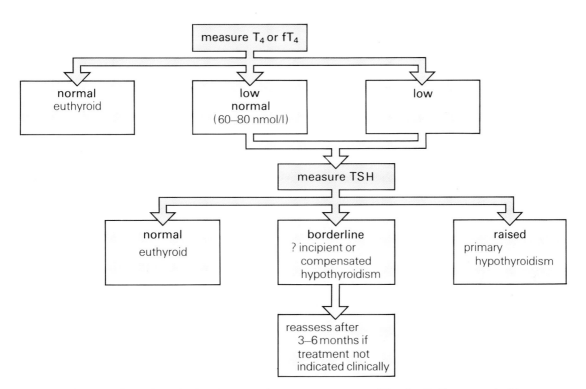

Fig.10.13 Algorithm for the investigation of suspected hypothyroidism. Either T₄ or TSH can be used as the first-line test. In cases of incipient or compensated hypothyroidism, reassessment could involve measurement of T_4 or free T_4. However, a decision on whether to treat the patient is better made on clinical grounds.

likely to be used if the TSH is also the first-line test for hyperthyroidism.

Thyroiditis

Inflammation of the thyroid, or thyroiditis, may be due to infection (usually viral) or autoimmune disease. In viral thyroiditis, associated with coxsackie, mumps and adenovirus, the inflammation results in a release of pre-formed colloid and there is an increase in the level of thyroid hormones in the blood stream. Patients may become transiently, and usually only mildly, thyrotoxic. This phase persists for up to six weeks and is followed by a similar period in which thyroid hormone output may be decreased, although not sufficiently to cause symptoms. Thereafter, normal function is regained.

Hashimoto's thyroiditis, an autoimmune condition, has been mentioned as a cause of hypothyroidism. Auto-antibodies are present in high titre and the disease is associated with the presence of other organ-specific autoimmune diseases. Very occasionally, transient hyperthyroidism may occur early in the course of the disease due, as in viral thyroiditis, to increased release of pre-formed colloid.

Goitre

Patients with goitres may be eu-, hyper- or hypothyroid. The thyroid hormone status of such patients should be investigated and may give a clue to the cause of the goitre. The laboratory has no part to play in the diagnosis of thyroid cancer, except calcitonin-secreting medullary carcinoma. When patients with thyroid cancer are treated by ablative doses of radioactive iodine and put on replacement thyroxine, the efficacy of treatment can be assessed by measuring serum thyroglobulin levels. Since small amounts of thyroglobulin are normally released from the gland together with thyroid hormones, persistent thyroid activity can be inferred if thyroglobulin is present in the serum.

SCREENING FOR THYROID DISEASE

Congenital hypothyroidism is sufficiently serious and common (1 in 4000 live births in the United Kingdom but considerably higher in some other countries) for it to be worthwhile to screen for the condition. Untreated, affected children become cretins, with very low intelligence and impaired growth and motor function. Treatment by replacement of T_4 is simple and effective. The screening method requires measurement of serum TSH in a capillary blood sample, taken at 6–8 days after birth. It is convenient to take the sample at the same time that the baby is being screened for phenylketonuria.

Hypo- and hyperthyroidism are both common in the elderly (more so the former) with a combined prevalence of five percent; there are many clinical manifestations of hypothyroidism, and hyperthyroidism may present atypically in the elderly (see *Chapter 24*). The serum total T_4 alone is an inadequate screening test; if it is used, further tests are required in up to fifty percent of cases to investigate equivocal results. Elderly people often have low binding proteins, due, for example, to poor nutrition or coexistent disease. Even if the free T_4 is used as the screening test, the sick euthyroid syndrome may cause a patient's thyroid status to be incorrectly assigned. Early results with highly sensitive TSH assays suggest that this is the most efficient screening test. Although other tests, such as a free T_4, are required when the TSH result is borderline, such tests are needed in fewer than twenty percent of cases.

SUMMARY

The thyroid gland secretes two iodine-containing hormones, thyroxine (T_4) and triiodothyronine (T_3). T_4 is secreted in the greater amount and is in part metabolized to T_3 in peripheral tissues; T_3 is the more active hormone. The synthesis and secretion of thyroid hormones is stimulated by the pituitary hormone, thyroid-stimulating hormone (TSH). The release of TSH is in turn controlled by thyrotrophin releasing hormone from the hypothalamus. T_4 and T_3 exert negative feedback inhibition of TSH release.

The thyroid hormones are essential for normal growth and development, and also control basal metabolic rate and stimulate many metabolic processes.

T_4 and T_3 are extensively protein-bound in the blood (T_4 to an even greater extent than T_3), to thyroxine-binding globulin, albumin and prealbumin, the free, and physiologically active, fractions being less than one percent of the total. Factors which affect the concentration of the binding proteins can alter total hormone concentrations without affecting the free fraction, and thus erroneously suggest the presence of an abnormality of thyroid function.

Thyroid status can be assessed biochemically by measurement of thyroid hormones in the serum. The concentration of T_4 is greater but is more affected by changes in the levels of binding proteins than T_3. Measurement of the free fractions of the hormones is feasible but technically more difficult than measurement of the total. Alternatively, thyroid status can be assessed indirectly by measuring TSH: in hyperthyroidism, negative feedback results in suppression of TSH release and the hormone is undetectable in the serum; in hypothyroidism, unless it is a result of pituitary disease, TSH release is stimulated and serum levels are elevated.

Patients with thyroid disease may present due to overactivity of the gland (hyperthyroidism, leading to thyrotoxicosis) or underactivity (hypothyroidism, leading to myxoedema). Patients in either category may have enlargement of the gland (goitre) but patients with goitres can be euthyroid. Both hyper- and hypothyroidism are commonly the result of auto-immune disease although there are many other causes. The measurement of specific auto-antibodies can provide useful diagnostic information in thyroid disease. Options for the treatment of hyperthyroidism include anti-thyroid drugs, radioactive iodine and surgery; patients with hypothyroidism require hormone replacement.

The thyroid also secretes calcitonin, a polypeptide hormone which may be involved in calcium homoeostasis.

FURTHER READING

Hall R, Anderson J, Smart GA & Besser M (1980) *Fundamentals of Clinical Endocrinology*. 3rd edition. London: Pitman Medical.

Wilson JD & Foster DW (eds) (1985) *Williams–Textbook of Endocrinology*. 7th edition. Philadelphia: WB Saunders Company.

11. The Gonads

INTRODUCTION

Androgens and testicular function

The testes are responsible for the synthesis of the male sex hormones, or androgens, and for the production of spermatozoa. The most important androgen, both in terms of potency and the amount secreted, is testosterone. Other testicular androgens include androstenedione and dehydroepiandrosterone (DHEA). These weaker androgens are also secreted by the adrenal glands but adrenal androgen secretion does not appear to be physiologically important in the male. In the female, however, it contributes to the development of certain secondary sexual characteristics, in particular the growth of pubic and axillary hair. The pathological consequences of increased adrenal androgen secretion are discussed in *Chapter 9* and on page 161.

Testosterone is a powerful anabolic hormone. It is vital to the development of secondary sexual characteristics in the male and is essential for spermatogenesis. It is secreted by the Leydig cells of the testis under the influence of luteinizing hormone (LH). Spermatogenesis is not only dependent upon testosterone but also upon the Sertoli cells of the seminiferous tubules in the testes (see page 108). The development of the Sertoli cells is stimulated by follicle-stimulating hormone (FSH).

Testosterone levels in the serum are very low before puberty but then rise rapidly to reach normal adult levels. A slight decline in concentration may be seen in the elderly. Testosterone is also present in females, at a much lower concentration, about one-third being derived from the ovaries and the remainder from the metabolism of adrenal androgens. In the circulation, approximately ninety-seven per cent of testosterone is protein-bound, principally to sex hormone-binding globulin (SHBG) and to a lesser extent to albumin and other proteins. Only the free testosterone is available to tissues. However, the biological activity of testosterone is mainly due to dihydrotestosterone (DHT). This is formed from testosterone in target tissues in a reaction catalyzed by the enzyme, 5α-reductase. In a rare condition in which there is

deficiency of this enzyme, DHT cannot be formed; male internal genitalia develop normally (Wolffian duct development in the fetus is testosterone-dependent) but masculinization, which requires DHT, is incomplete. In states of androgen insensitivity, defects of the receptors for either testosterone or DHT, or both, can cause a spectrum of clinical abnormalities ranging from gynaecomastia to pseudohermaphroditism.

Specific assays are available for testosterone, DHT and other androgens, and SHBG. There is now no indication for the measurement of urinary 17-oxosteroids in the assessment of gonadal function. These are metabolites of androstenedione to which some testosterone is metabolized, but two-thirds of urinary 17-oxosteroids are derived from adrenal androgens.

Oestrogens and ovarian function

The cyclical control of ovarian function during the reproductive years is discussed in *Chapter 8*. The principal ovarian hormone is 17β-oestradiol, but some oestrone is also produced. Oestrogens are also secreted by the corpus luteum and the placenta.

Oestrogens are responsible for the development of many female secondary sexual characteristics. They also stimulate the growth of ovarian follicles and the proliferation of uterine endometrium during the first part of the menstrual cycle. They have important effects on cervical mucus and vaginal epithelium, and on other functions associated with reproduction.

Serum levels of oestrogens are low before puberty. During puberty, oestrogen synthesis increases and cyclical changes in concentration occur thereafter until the menopause, unless pregnancy occurs. After the menopause, serum oestrogen concentrations fall to very low levels. In the serum, oestrogens are transported bound to protein, sixty percent to albumin and the remainder to SHBG. Only 2–3% remains unbound. Oestrogens stimulate the synthesis of SHBG and also that of other transport proteins, notably thyroxine-binding globulin (TBG) and transcortin, and thus increase total thyroxine and total cortisol concentrations in the serum. Oestradiol is present in low concentrations in the serum of normal men. Approximately one-third is secreted by the testis, the remainder being derived from the metabolism of testosterone in the liver and in adipose tissue. The uses of serum and urinary measurements of oestrogens are considered on page 162. Slowly rising or sustained high levels of oestrogens together with progesterone inhibit pituitary gonadotrophin secretion by negative feedback, but the rapid rise in oestrogen concentration which occurs prior to ovulation stimulates LH secretion (positive feedback).

Progesterone

Progesterone is an important intermediate in steroid hormone biosynthesis but is secreted in appreciable quantities only by the corpus luteum and the placenta. Its concentration in the serum rises during the second half of the menstrual cycle but then falls if conception does not take place. In the plasma, it is extensively bound to albumin and transcortin; only 1–2% is free. Progesterone has many important effects on the uterus, including preparation of the endometrium for implantation of the conceptus, and also on the cervix, vagina and breasts. It is pyrogenic and mediates the increase in basal body temperature that occurs with ovulation. Progesterone can be measured in serum and this assay is used in the investigation of infertility in women (see page 160).

Sex hormone-binding globulin

SHBG binds both testosterone and oestradiol in the plasma, though it has greater affinity for testosterone. The serum concentration of SHBG in males is about half that in females. Factors which alter SHBG concentration (Fig.11.1) alter the ratio of free testosterone to free oestradiol. If SHBG concentration decreases, the ratio of free testosterone to free oestradiol is increased, though there is an absolute increase in the concentration of both hormones. If SHBG concentration increases, the ratio decreases. Thus in either sex, the effect of an increase in SHBG is to increase oestrogen-dependent effects while a decrease in SHBG increases androgen-dependent effects (Fig.11.2).

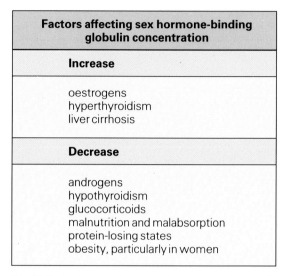

Factors affecting sex hormone-binding globulin concentration
Increase
oestrogens hyperthyroidism liver cirrhosis
Decrease
androgens hypothyroidism glucocorticoids malnutrition and malabsorption protein-losing states obesity, particularly in women

Fig.11.1 Factors which cause an increase or a decrease in the concentration of sex hormone-binding globulin (SHBG).

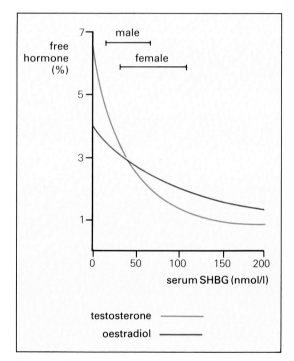

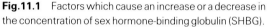

Fig.11.2 Effect of a change in serum SHBG concentration on free oestradiol and testosterone concentrations. A decrease in SHBG increases free testosterone concentration more than free oestradiol and thus is androgenic; an increase in the concentration of SHBG is anti-androgenic. The normal ranges of SHBG in males and females are shown.

DISORDERS OF MALE GONADAL FUNCTION

Hypogonadism

Hypogonadism may be primary or occur secondarily to pituitary or hypothalamic disease. Some of the causes are indicated in Fig. 11.3. Primary hypogonadism may occur with defective seminiferous tubule function or defective Leydig cell function. The former leads to infertility through decreased production of spermatozoa, but masculinization is usually normal. Defective Leydig cell function, on the other hand, results in a failure of testosterone-dependent functions, including spermatogenesis. The effects of decreased testosterone secretion depend upon the time of onset of the disorder. Secondary sexual characteristics are partially preserved if secretion is lost after puberty.

The basic biochemical characteristics that distinguish between primary and secondary hypogonadism are not always clear-cut. This is partly because most currently available assays for gonadotrophins are insufficiently sensitive to distinguish between low and normal concentrations. Provocative tests, such as the gonadotrophin releasing hormone (GnRH) test and the clomiphene test, may be necessary to distinguish between primary and secondary hypogonadism.

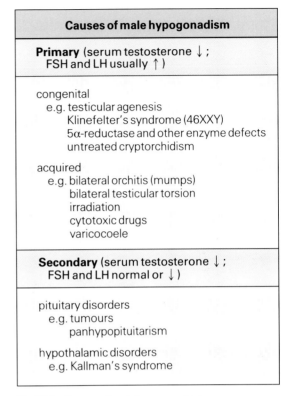

Causes of male hypogonadism

Primary (serum testosterone ↓ ; FSH and LH usually ↑)

congenital
  e.g. testicular agenesis
       Klinefelter's syndrome (46XXY)
       5α-reductase and other enzyme defects
       untreated cryptorchidism

acquired
  e.g. bilateral orchitis (mumps)
       bilateral testicular torsion
       irradiation
       cytotoxic drugs
       varicocoele

Secondary (serum testosterone ↓ ; FSH and LH normal or ↓)

pituitary disorders
  e.g. tumours
       panhypopituitarism

hypothalamic disorders
  e.g. Kallman's syndrome

Fig.11.3 Causes of male hypogonadism.

CASE HISTORY 11.1

A twenty-year-old man presented with impotence. On examination, he was eunuchoid; there was only sparse pubic and axillary hair, the genitalia were infantile, muscular development was poor and his span exceeded his height with a sole–pubic symphysis distance greater than symphysis to crown.

Investigations

serum: testosterone   3nmol/l
         LH   <1·5iu/l
         FSH   <1·5iu/l

clomiphene test (3mg/kg body weight clomiphene citrate daily for seven days):

serum: LH   <1·5iu/l
         FSH   <1·5iu/l

gonadotrophin releasing hormone (GnRH) test (100µg GnRH i.v.):

time (mins)	FSH (iu/l)	LH (iu/l)
0	<1·5	<1·5
20	2·0	2·0
60	2·5	3·0

(after 100µg GnRH subcutaneously daily for two weeks)

0	3·5	4·5
20	8·4	21·5
60	4·5	8·0

Comment

The age of onset of normal puberty may sometimes be delayed until eighteen years of age and hypogonadism should be diagnosed with caution in patients who are younger than this. The low testosterone and gonadotrophins in this case suggest a lesion at the level of either the pituitary or hypothalamus. This is confirmed by the failure of response to clomiphene. This drug competes with gonadal steroids for hypothalamic receptors and in normal men results in an increase in gonadotrophin secretion and thus testosterone secretion. Patients with pituitary lesions may have clinical or biochemical evidence of other pituitary abnormalities (none was present in this case).

Further differentiation between pituitary and hypothalamic causes of reduced gonadotrophin secretion may be provided by the results of a GnRH test (see page 111). A diminished or absent response is seen with pituitary lesions, while with a hypothalamic defect, the response is characteristically delayed (cf. TRH test, page 146). In this case, there is little response to GnRH. In hypothalamic GnRH deficiency, the pituitary may be insensitive to exogenous GnRH, but this can be corrected by 'priming' the pituitary with repeated injections of the hormone. When the GnRH test was repeated after GnRH priming, this patient's response was normal, indicating a hypothalamic, rather than a pituitary defect. He was later found to be anosmic (lacking a sense of smell). The association between anosmia and hypogonadotrophic hypogonadism is called Kallman's syndrome. The eunuchoid habitus is a direct consequence of testosterone deficiency; this hormone promotes epiphyseal fusion and when its secretion is inadequate, there is continued growth of long bones which become disproportionate to the axial skeleton.

Although biochemical tests are important in establishing that a patient has primary, rather than secondary, gonadal failure they are less useful in distinguishing between the various causes of primary hypogonadism. In general, seminiferous tubule defects are associated with a raised serum FSH concentration; Leydig cell defects are associated with a raised serum LH concentration. Human chorionic gonadotrophin (hCG), which has an action similar to LH, can be used to test Leydig cell function (Fig.11.4). Semen analysis will provide an indication of seminiferous tubule function and testicular biopsy is valuable in patients with low sperm counts if the cause is not obvious clinically. Careful clinical examination is essential in all cases of gonadal failure.

The treatment of hypogonadism in males should be directed towards the underlying cause wherever possible. Testosterone is given in testosterone deficiency syndromes, but if fertility is required, treatment must be with gonadotrophin replacement or, in hypothalamic disorders, pulsatile GnRH administration.

Gynaecomastia

Breast development in males is usually related to a disturbance of the balance of oestrogens to androgens. It may occur physiologically in neonates as a result of exposure to maternal oestrogens. During puberty, approximately fifty percent of normal boys develop gynaecomastia due to temporarily increased secretion of oestrogens relative to androgens. In both instances, the gynaecomastia resolves spontaneously. Mild gynaecomastia may also occur in the elderly, as a result of a decrease in testosterone secretion.

Gynaecomastia occurring at other times should be regarded as pathological. The principal causes are shown in Fig.11.5. The cause may be obvious from either the history or clinical examination. Measurement of serum testosterone, gonadotrophins, SHBG and prolactin, and assessment of liver and possibly thyroid function will help to distinguish between them. Karyotyping is required to diagnose Klinefelter's syndrome in which an additional X-chromosome is present (46XXY); chest and skull radiographs, and tests of pituitary and adrenal function may be of use.

DISORDERS OF FEMALE GONADAL FUNCTION

The climacteric

During the climacteric, progressive ovarian failure causes a decline in ovarian oestrogen secretion and eventually menstruation ceases; the menopause is the last menstrual period. The only oestrogen secreted after the menopause is the small amount produced

Human chorionic gonadotrophin (hCG) test	
Procedure	**Results**
day 0: 0900h; take blood for testosterone give 2000iu hCG i.m.	normal response: serum testosterone level increases to above upper limit of normal range
day 3: 0900h; give 2000iu hCG i.m.	primary testicular failure: little or no response
day 5: 0900h; take blood for testosterone	secondary testicular failure: response may be normal

Fig.11.4   Human chorionic gonadotrophin test for primary testicular failure.

from metabolism of adrenal androstenedione in adipose tissue. The serum concentrations of pituitary gonadotrophins become greatly elevated; this change is a more reliable indication of ovarian failure than serum oestrogen concentrations, which show considerable variability. Metabolic changes which occur after the menopause include an increase in serum low density lipoprotein concentration and in serum urate concentration. Oestrogen deficiency is one of the major factors which contribute to the development of post-menopausal osteoporosis.

Infertility

Infertility may be due to many causes, including both male and female hypogonadism. The investigation of a couple complaining of infertility involves thorough clinical and laboratory assessment. Semen must be examined to determine that an adequate number of normal, motile sperm is present. A regular ovarian cycle is likely to be ovulatory but this should be confirmed; the rise in basal body temperature which follows ovulation is a useful indicator. Serum progesterone concentration should be measured on day twenty-one of the menstrual cycle; during the luteal phase progesterone secretion from the corpus luteum increases to a peak if ovulation has occurred. Hyperprolactinaemia is a common cause of infertility, particularly in women, and measurement of serum prolactin concentration is essential in the investigation of infertility.

Causes of gynaecomastia

Physiological (see text)

Decreased androgen activity
hypogonadism:
  primary
  secondary
anti-androgens, e.g. cyproterone,
  cimetidine and spironolactone
increased SHBG (liver cirrhosis and
  hyperthyroidism)
androgen insensitivity

Increased oestrogen activity
oestrogen-secreting tumours,
  e.g. Leydig cell and adrenal carcinoma
gonadotrophin-secreting tumours,
  e.g. bronchial carcinoma (hCG)
increased peripheral oestrogen production
  from androgens (liver cirrhosis and
  hyperthyroidism)
increased gonadotrophin secretion in
  re-feeding after starvation
drugs with oestrogenic activity,
  e.g. oestrogens, digoxin and
  tetrahydrocannibinol

Others
hyperprolactinaemia
idiopathic
other drugs, e.g. phenothiazines and
  methyl dopa

Fig.11.5   Causes of gynaecomastia.

If there is no evidence of ovulation, or the woman has oligomenorrhoea or amenorrhoea, tests of ovarian and pituitary function are required. The investigation of defective male gonadal function has been discussed on page 157. If an endocrine cause for infertility can be found, hormone replacement may restore fertility.

Amenorrhoea and oligomenorrhoea

The commonest cause of amenorrhoea in a woman of child-bearing age is pregnancy and this possibility, however unlikely, should always be excluded. Amenorrhoea may be primary (menstruation has never occurred) or secondary. The endocrine causes are shown in Fig.11.6. Hyperprolactinaemia is responsible for about twenty-five per cent of cases. Weight loss is another common cause; the pulsatile secretion of GnRH, and thus of gonadotrophins, is lost when body weight falls and menstruation almost always ceases if the weight falls below seventy-five percent of the ideal, but may cease with smaller losses. Regular menstruation returns if weight is regained.

The indirect assessment of ovarian function is provided by the progestogen challenge test. In this test, a progestogen is given for five days; subsequent vaginal bleeding indicates that sufficient oestrogen was present to produce endometrial proliferation. If no bleeding occurs, the test may be repeated after giving an oestrogen, to distinguish between oestrogen deficiency and local, uterine disease.

If a failure of oestrogen secretion is suspected, serum gonadotrophins should be measured (the normal fluctuation in oestrogen secretion makes oestradiol measurements less useful for this purpose). Raised gonadotrophins (FSH is a more sensitive indicator than LH) indicate ovarian failure; low or normal levels are seen with a pituitary or hypothalamic lesion. This possibility should be investigated further by anatomical studies and dynamic testing of the hypothalamus–pituitary axis in a manner analogous to that described for male hypogonadism. The investigation of patients with oligo- or amenorrhoea who are virilized is outlined in the next section.

The management of amenorrhoea depends upon the cause. Patients with hyperprolactinaemia often respond to bromocriptine (see page 118). In hypo-pituitarism, oestrogen replacement is appropriate, but if fertility is required treatment must be with gonadotrophins. Patients with hypothalamic disease are treated with either GnRH or clomiphene (which

Endocrine causes of amenorrhoea and oligomenorrhoea
Primary ovarian failure
gonadal dysgenesis, e.g. Turner's syndrome
premature menopause, e.g. autoimmune disease
Pituitary disorders
isolated gonadotrophin deficiency
tumours:
causing decreased gonadotrophin secretion
causing hyperprolactinaemia
panhypopituitarism, e.g. post-partum necrosis
Hypothalamic disorders
weight loss
intensive exercise
Others
thyrotoxicosis; severe hypothyroidism
congenital adrenal hyperplasia
polycystic ovary syndrome

Fig.11.6 Endocrine causes of amenorrhoea and oligomenorrhoea. Severe systemic disease of any nature can cause amenorrhoea.

stimulates gonadotrophin secretion). With gonadotrophin and clomiphene treatment, it may be necessary to give human chorionic gonadotrophin (hCG) to stimulate the mid-cycle LH peak and thus stimulate ovulation. Careful monitoring of the serum oestradiol concentration is necessary with this form of treatment, to detect excessive ovarian stimulation which carries a risk of multiple pregnancy. hCG can also be used to induce ovulation in women with anovulatory cycles.

Patients with ovarian failure cannot become pregnant, but cyclical oestrogen and progestogen therapy will induce regular menstruation if the uterus is normal.

Hirsutism and virilism

Hirsutism is an increase in body hair in an androgen-related distribution. It may be accompanied by other features of virilism such as clitoromegaly, temporal balding and by menstrual irregularity. There is considerable racial variation in the amount of body hair in women and what may be regarded as normal in some races may be thought excessive by others.

The cause is usually excessive exposure of tissues to androgens. This may be due either to increased androgen secretion or a low level of SHBG, which increases the free testosterone fraction. In some cases, there appears to be an increased sensitivity to androgens. The causes of hirsutism and virilism are indicated in Fig.11.7.

The commonest causes are so-called 'idiopathic' hirsutism (cause unknown) and the polycystic ovary syndrome. The most important step in management is to detect any specific cause, for which specific treatment may be available, for example, congenital adrenal hyperplasia.

A menstrual history is essential to make a diagnosis. Menstruation is normal in idiopathic hirsutism. However, patients with polycystic ovary syndrome may have oligomenorrhoea or secondary amenorrhoea. Virilizing features other than hirsutism are not seen in either condition but do occur in conditions associated with marked increases in androgen secretion, for example, adrenal tumours.

The investigations that should be performed will depend upon the clinical features, since these may suggest a specific diagnosis, such as Cushing's syndrome. The first test should be measurement of the serum testosterone. A concentration greater than 7nmol/l (normal <2.5nmol/l) strongly suggests a serious cause, for example, an adrenal tumour. (Ideally, because of the pulsatility of testosterone secretion, the concentration should be measured in pooled serum from three blood samples drawn at 20min intervals.) In idiopathic hirsutism and polycystic ovary syndrome, the serum testosterone secretion may be normal or slightly raised. SHBG concentration may be reduced but this information is not usually helpful in management. In general, the management of hirsutism in these cases is not dependent upon the cause.

In patients with severe hirsutism or other features of virilization, measurement of adrenal androgens is useful. Greatly elevated levels of either dehydroepiandrosterone sulphate or androstenedione suggest an adrenal tumour. A high 17α-hydroxyprogesterone concentration is characteristic of congenital adrenal hyperplasia, which occasionally presents in adolescents and young adults.

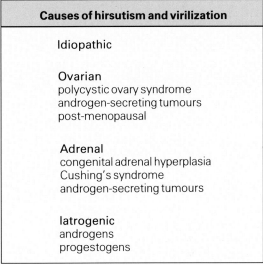

Causes of hirsutism and virilization

Idiopathic

Ovarian
polycystic ovary syndrome
androgen-secreting tumours
post-menopausal

Adrenal
congenital adrenal hyperplasia
Cushing's syndrome
androgen-secreting tumours

Iatrogenic
androgens
progestogens

Fig.11.7 Causes of hirsutism and virilization. Idiopathic causes and the polycystic ovary syndrome account for the great majority of cases.

CASE HISTORY 11.2

A young woman consulted her doctor because she was embarrassed by excessive hair on her upper lip, lower abdomen and thighs. She was moderately obese. Her periods had always been irregular.

Investigations

serum: testosterone   3.5nmol/l
        LH (early follicular)   14iu/l
        FSH (early follicular)   3iu/l

ultrasound examination of the ovaries:
  multiple cysts present, bilaterally

Comment

These findings are characteristic of the polycystic ovary syndrome. The testosterone concentration is slightly raised; FSH concentration is low and LH increased (hence increased LH:FSH ratio). Further, the ovaries are polycystic. Clinical features of this syndrome include obesity, hirsutism, menstrual irregularity and infertility, and any of these may occur in any combination with the hormonal abnormalities. A high LH:FSH ratio($\geqslant 3:1$) during the early follicular phase of the menstrual cycle is particularly characteristic of this syndrome and serum androstenedione concentration, if measured, is usually found to be increased. The syndrome shows wide variation in severity; in some cases in which the characteristic endocrine abnormalities are present, the ovaries are found to be normal. The cause of the polycystic ovary syndrome is unkown.

Conditions such as adrenal tumours and congenital adrenal hyperplasia require specific treatment. In idiopathic hirsutism and the polycystic ovary syndrome, the management depends upon the severity of the hirsutism and whether fertility is required. In mild cases, with no menstrual disturbance, the excessive hair may be acceptably treated by cosmetic means. In more severe cases, various endocrine manipulations may be useful. These include the use of a synthetic glucocorticoid to suppress adrenal androgen excretion and, if this fails, the anti-androgen drug, cyproterone. If fertility is required, cyproterone must not be used; it may be possible to restore fertility with gonadotrophins or clomiphene.

PREGNANCY

Many physiological and metabolic changes take place in the body during pregnancy. These include changes in the concentrations of hormones directly related to pregnancy and resulting secondary metabolic changes.

Specific hormonal changes

Human chorionic gonadotrophin

Fertilization of the ovum prevents the regression of the corpus luteum. Instead, the corpus luteum enlarges, stimulated by the glycoprotein hormone, human chorionic gonadotrophin (hCG), produced by the trophoblast (the developing placenta). This hormone (assays usually measure the β-subunit, see page 282) can be detected in the blood as early as six days after conception and may be detectable in the urine by fourteen days. Its detection in the urine provides a highly sensitive and specific test for the diagnosis of pregnancy. The secretion of β-hCG begins to fall by 10–12 weeks, although it remains detectable in the urine throughout pregnancy. hCG is also produced by some tumours; its use as a tumour marker is discussed in *Chapter 21*.

Oestrogens

The stimulated corpus luteum secretes large amounts of oestrogens and progesterone, but after six weeks the placenta becomes the major source of these hormones. Typical changes in the excretion of pregnanediol (the main urinary metabolite of progesterone), oestriol and hCG are shown in Fig.11.8.

The major route for the synthesis of oestriol involves both placental and fetal enzymes. Measurements of maternal serum oestriol or urinary oestriol excretion can be used to monitor fetoplacental function during pregnancy (Fig.11.9). These measurements correlate well with indices of fetal development. There is, however, considerable diurnal variation in both serum oestriol concentration and oestriol excretion. Because of this, serial measurements give a more reliable indication of an individual patient's oestriol status. They may also show an upward or downward trend, but in practice rapid falls in oestriol are rare and the cause is often clinically obvious.

Progesterone

A low serum progesterone early in pregnancy suggests a poor luteal function and may thus indicate a cause of recurrent abortion. Progesterone measurements are not useful as a test of fetoplacental function later in pregnancy.

Placental polypeptides

Various placental products are detectable in maternal blood during pregnancy. A specific, heat-stable isoenzyme of alkaline phosphatase is produced by the placenta but its level in the serum is unreliable as an indicator of placental function. Human placental lactogen (hPL), also known as human chorionic somatomammotrophin (hCS), has an important

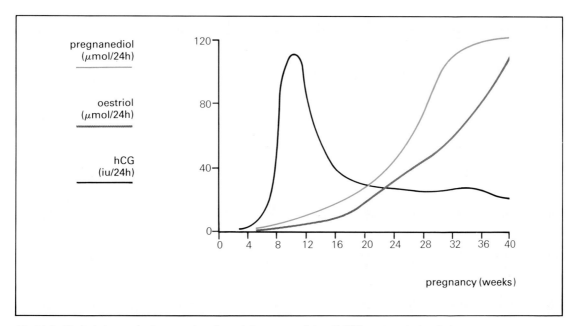

Fig.11.8 Typical changes in the excretion of oestriol, pregnanediol and hCG in maternal urine during pregnancy.

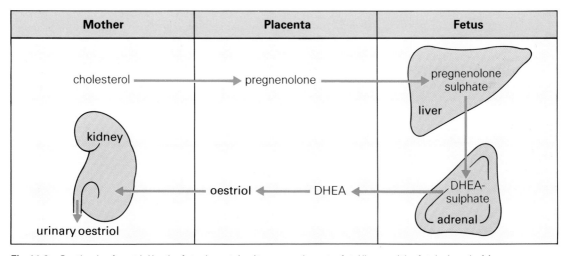

Fig.11.9 Synthesis of oestriol by the fetoplacental unit. This is the major source of oestriol during pregnancy; essential enzyme-catalyzed reactions take place in the placenta, fetal liver and the fetal adrenals. Many intermediate steps and minor pathways have been omitted for clarity.

physiological role in pregnancy, acting like growth hormone to stimulate protein synthesis and calcium absorption. Serial measurements of its serum concentration or urinary excretion in maternal urine have been used to monitor placental function. Its synthesis is not dependent upon fetal enzymes so that theoretically it should not be used as an indicator of fetal well-being. In practice, however, impairment of fetal growth is usually due to placental insufficiency. In comparative studies, the difference in performance between oestriol and hPL measurements as an index of fetal well-being has been shown to be marginal.

163

A protein called pregnancy-specific β_1-glycoprotein is synthesized by the placenta and is detectable in maternal blood. Its usefulness, if any, in the monitoring of pregnancy, has yet to be determined.

Biochemical tests of placental function

Biochemical tests of placental function are of use primarily in the identification of the fetus at risk of placental insufficiency, with consequent growth retardation and other sequelae. They have little or no role in the detection of the two other categories of fetal risk: congenital abnormalities and premature labour. Advanced techniques of fetal ultrasound are more efficient at detecting fetal growth retardation. As a result, biochemical tests of placental function are now performed much less frequently, though they are still useful if advanced fetal ultrasound scanning is not readily available.

Secondary metabolic changes

Many of the metabolic changes that occur in pregnancy are discussed elsewhere in this book. Those that suggest the presence of a pathological process are summarized in Fig.11.10. For example, the oestrogen-mediated increase in thyroxine-binding globulin (TBG) increases the total thyroxine concentration in plasma and may suggest hyperthyroidism. Similarly, an increase in transcortin increases serum cortisol concentration although, in addition, there is a slight increase in cortisol secretion during pregnancy.

Maternal monitoring

Patients with medical conditions may require close monitoring during pregnancy. For example, strict control of diabetes mellitus is vital and entails frequent monitoring of glycosylated haemoglobin and blood glucose. The close cooperation of the laboratory is also required for the monitoring of patients with thyroid disease during pregnancy.

Urine should be tested for proteinuria and glycosuria at clinic attendances; the presence of proteinuria may be an early sign of pre-eclampsia. The renal threshold for glucose is decreased during pregnancy but if more than a trace of glycosuria is detected, it is advisable to perform an oral glucose tolerance test to exclude hitherto undiagnosed diabetes.

Pre-eclampsia is a condition, peculiar to pregnancy, characterized by hypertension, proteinuria and oedema. If left untreated, it can lead to severe hypertension and renal failure. An increase in serum urate concentration can be a sensitive indicator of deteriorating renal function in this condition. Rapid analysis of samples is necessary as pre-eclampsia can progress very quickly.

Fetal monitoring

The antenatal diagnosis of inherited metabolic disease in early pregnancy and the use of maternal serum α-fetoprotein measurements to screen for neural tube defects are considered in *Chapter 18*. Fetoscopy, a technique which allows the collection of pure samples of fetal blood, is becoming more widely available and is a considerable advance in the management of fetal disease.

Fetal lung maturity can be assessed by measurement of phospholipids in amniotic fluid. Phospholipids, principally lecithin and sphingomyelin, are components of surfactant, which lowers the surface tension of alveoli and facilitates expansion and aeration of the lungs at birth. Lack of surfactant predisposes the infant to respiratory distress after birth. Lecithin and sphingomyelin are synthesized by the fetal lungs from twenty-four weeks of gestation and are initially present in the amniotic fluid in equal concentrations. From 32–34 weeks, the concentration of lecithin increases disproportionately, corresponding to increasing fetal lung maturity and a decreasing risk of respiratory distress. A ratio of lecithin to sphingomyelin (L:S ratio) in amniotic fluid of greater than 2.0 implies adequate fetal lung maturity. Such measurements are useful when considering premature delivery of a fetus. The measurement of phosphatidyl glycerol in amniotic fluid, although technically more demanding, provides a more specific test of fetal lung maturity.

During labour, once the cervix is sufficiently dilated, fetal blood hydrogen ion can be measured in capillary samples obtained from the scalp. A level of more than 60nmol/l (pH <7.22) suggests potentially dangerous fetal hypoxaemia. A continuous, direct measurement of fetal P_{O_2} can be obtained using a transcutaneous oxygen electrode.

Metabolic changes during pregnancy and use of oral contraceptives			
Change	**Cause**	**Pregnancy**	**Oral contra-ceptive use**
↓ urea	↑ GFR; ↑ plasma volume	*	
↓ albumin	↑ plasma volume	*	
↓ total protein	↑ plasma volume	*	
↑ total thyroxine	↑ TBG	*	*
↑ cortisol	↑ transcortin	*	*
↑ copper	↑ caeruloplasmin	*	*
glycosuria	↓ renal threshold	*	
↓ glucose tolerance (but normal fasting levels)		*	
↑ triglyceride (VLDL)	↑ oestrogens (antagonism of actions of insulin)	*	*
↓ LDL cholesterol		*	variable
↑ HDL cholesterol		*	variable
↑ alkaline phosphatase	placental isoenzyme	*	

Fig.11.10 Metabolic changes which occur during pregnancy and the use of oral contraceptives. The changes refer to serum concentrations except where indicated. Oestrogens tend to decrease low density lipoprotein (LDL) cholesterol and increase high density lipoprotein (HDL) cholesterol; progestogens have the opposite effect.

Metabolic effects of oral contraceptives

Oral contraceptives contain either a combination of an oestrogen and a progestogen or a progestogen alone. In addition to suppressing ovulation, these contraceptives have a number of metabolic effects similar to some of those that occur in normal pregnancy (Fig.11.10).

SUMMARY

The principal female sex hormone, or oestrogen, is 17β-oestradiol, secreted by the ovaries. The principal male sex hormone, or androgen, is testosterone, secreted by the testes. The secretion of both these hormones is stimulated by pituitary luteinizing hormone (LH). Spermatogenesis and the maturation of ovarian follicles is dependent upon testosterone and oestradiol, respectively, and pituitary follicle-stimulating hormone (FSH). The secretion of LH and FSH is in turn controlled by gonadotrophin releasing hormone, released from the hypothalamus, and subject to feedback control by the gonadal hormones. Androgens are also produced by the adrenals and in males there is some production of oestrogens by metabolism from androgens.

Both testosterone and oestradiol are transported in the plasma bound to sex hormone-binding globulin (SHBG), with only about three percent of each hormone being in free solution. Because of the greater avidity of testosterone for SHBG, factors that increase the concentration of SHBG tend to increase oestrogen-dependent effects while those that decrease it increase androgen-dependent effects.

The secretion of all these hormones is pulsatile. The secretion of testosterone in men is maintained throughout life but in women, after the menopause, oestrogen secretion declines.

Both male and female hypogonadism can be either primary or secondary to either pituitary or hypothalamic dysfunction. Measurement of the appropriate gonadal hormone and the gonadotrophins, often after attempted stimulation of their secretion, will usually indicate the correct diagnosis and permit rational treatment.

Hormone measurements are also valuable in the investigation of gynaecomastia in males and virilism in females. The commonest feature of excessive androgenization is hirsutism which is frequently idiopathic. Another common cause is the polycystic ovary syndrome, where it is associated with menstrual irregularity and infertility. However, the presence of severe hirsutism and virilism should suggest the possibility of an androgen-secreting tumour of the adrenals or ovaries.

The laboratory investigation of infertility also depends heavily upon hormone measurements, though many non-endocrine factors must also be considered. A prime consideration is to establish whether ovulation is taking place; this can be inferred from the finding of an increase in serum progesterone concentration on day twenty-one of the menstrual cycle.

Pregnancy can be diagnosed by the detection of the β-subunit of human chorionic gonadotrophin in the urine. Biochemical tests, for example, urinary oestriol and serum placental lactogen concentration, may be used to monitor fetal well-being during pregnancy, although in many centres this is now done by ultrasound examination. Pregnancy causes a number of physiological changes in biochemical variables, including an increase in the serum concentrations of hormone-binding proteins. As a result, total levels of, for example, thyroxine and cortisol, are increased during pregnancy although the free hormone concentrations are normal.

FURTHER READING

Besser GM & Cudworth AG (1987) *Clinical Endocrinology: An Illustrated Text*. London: Chapman & Hall; Gower Medical Publishing.

Hall R, Anderson J, Smart GA & Besser M (1980) *Fundamentals of Clinical Endocrinology*. 3rd edition. London: Pitman Medical.

Wilson JD & Foster DW (eds) (1985) *Williams – Textbook of Endocrinology*. 7th edition. Philadelphia: WB Saunders Company.

12. Diabetes Mellitus

INTRODUCTION

The concentration of glucose in the blood is normally subject to rigorous control, rarely falling below 2.5mmol/l or exceeding 8.0mmol/l in healthy subjects whether fasting or in a post-prandial state. This control is achieved through the concerted action of various hormones of which the most important are insulin and glucagon.

Insulin is a polypeptide, secreted by the pancreatic β-cells of the islets of Langerhans in response to a rise in blood glucose concentration. It is synthesized as a prohormone, proinsulin. This molecule undergoes cleavage prior to secretion to form insulin and C-peptide (Fig.12.1). Insulin secretion is also stimulated by various gut hormones, including glucagon and gastric inhibitory peptide, GIP (glucose-dependent insulinotrophic peptide). Insulin promotes the removal of glucose from the blood and its storage in the form of glycogen. It also stimulates the synthesis of fat from glucose and its storage in adipose tissue as triglyceride. If the blood glucose level falls, insulin secretion is inhibited and stored glucose mobilized. The actions of insulin are summarized in Fig.12.2.

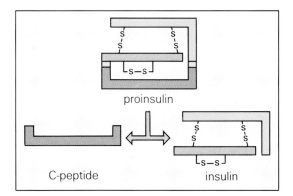

Fig.12.1 Biosynthesis of insulin. The cleavage of proinsulin produces insulin, consisting of two polypeptide chains linked by disulphide bridges, and C-peptide.

Principal actions of insulin and glucagon			
	Liver	Adipose tissue	Muscle
Insulin	Increased protein synthesis fatty acid synthesis glycogen synthesis Decreased ketogenesis gluconeogenesis	Increased glucose uptake fatty acid synthesis triglyceride synthesis Decreased lipolysis	Increased glucose uptake glycogen synthesis protein synthesis Decreased proteolysis
Glucagon	Increased glycogenolysis gluconeogenesis ketogenesis	Increased lipolysis	

Fig.12.2 Principal actions of insulin and glucagon. The actions of insulin on the liver decrease glucose output while those of glucagon increase its output.

Glucagon is a polypeptide secreted by the α-cells of the pancreatic islets; its secretion is decreased by a rise in the blood glucose concentration. In general, its actions oppose those of insulin: it stimulates glycogenolysis and gluconeogenesis and promotes lipolysis and ketogenesis (Fig.12.2). The combined effects of insulin and glucagon are shown diagrammatically in Fig.12.3.

Disturbances of glucose homoeostasis may result in hypoglycaemia or hyperglycaemia. Hypoglycaemia is discussed in *Chapter 13*.

This chapter is devoted primarily to diabetes mellitus, a condition characterized by abnormal glucose tolerance and a tendency to hyperglycaemia.

MEASUREMENT OF GLUCOSE CONCENTRATION

Serum glucose concentration tends to be 10–15% higher than that of whole blood because a given volume of red cells contains less water than the same volume of serum. The difference is of little significance at normal concentrations except in the interpretation of the results of glucose tolerance tests. However, when the glucose concentration is changing rapidly there may be a considerable discrepancy because of delayed equilibration of glucose across the red cell membranes.

Red blood cells *in vitro* continue to utilize glucose, so that unless a blood sample can be analyzed immediately, it is essential to collect it into a tube containing sodium fluoride to inhibit glycolysis. Potassium oxalate is used as an anticoagulant in such 'fluoride-oxalate' tubes, and plasma obtained from this blood is thus unsuitable for the measurement of potassium concentration.

DIABETES MELLITUS

Aetiology and pathogenesis

Diabetes mellitus is a common condition, with a prevalence of approximately 1-2% in the Western world. Diabetes can occur secondarily to other diseases, for example, chronic pancreatitis, following pancreatic surgery and in conditions where there is increased secretion of hormones antagonistic to insulin, as in Cushing's syndrome and acromegaly. Secondary diabetes is, however, uncommon. Most cases of diabetes mellitus are primary, that is, they are

not associated with other conditions, and fall into two distinct types. In Type I (insulin-dependent diabetes mellitus — IDDM) there is effectively no insulin secretion. In Type II (non insulin-dependent diabetes mellitus — NIDDM) either insulin is secreted in amounts insufficient to prevent hyperglycaemia or there is insensitivity to its actions. Overall, some twenty-five percent of patients are insulin dependent; most with NIDDM can be treated by diet with or without oral hypoglycaemic drugs, for example, sulphonylureas and biguanides.

IDDM usually presents acutely in younger people, with symptoms developing over a period of days or only a few weeks; it was formerly called juvenile-onset diabetes. However, there is evidence that the appearance of symptoms is preceded by a 'prediabetic' period of several months during which growth failure, a fall in insulin response to glucose and various immunological abnormalities can be detected. NIDDM tends to present more chronically in the middle-aged and elderly with symptoms developing over months or even longer; it was formerly called maturity-onset diabetes. The old nomenclature is inaccurate since some young diabetic patients are not insulin dependent while IDDM may occasionally present in older people. Some of the characteristics of IDDM and NIDDM are shown in Fig. 12.4.

NIDDM often shows a familial incidence but inheritance is not predictable and environmental factors are also involved. For example, obesity is present in approximately forty percent of patients with NIDDM and the incidence in women is greater in the multiparous than the nulliparous. The pancreatic islet cells are usually histologically normal in NIDDM, but in many cases they appear to be insensitive to glucose and insulin secretion is decreased. Other patients with NIDDM have normal or increased serum insulin concentrations but are insensitive to its effects because of a decreased number of insulin receptors. In some patients, both factors are important.

IDDM shows no familial incidence but there is a strong association with certain histocompatibility antigens, for example, HLA-DR3 and DR4. An individual's HLA antigens are genetically determined and many autoimmune diseases are associated with particular HLA specificities, suggesting that IDDM may also be an autoimmune disease and, further, that the susceptibility to it is in part governed by genetic factors. There is also evidence that viruses, for instance, Coxsackie B4, are aetiological agents in

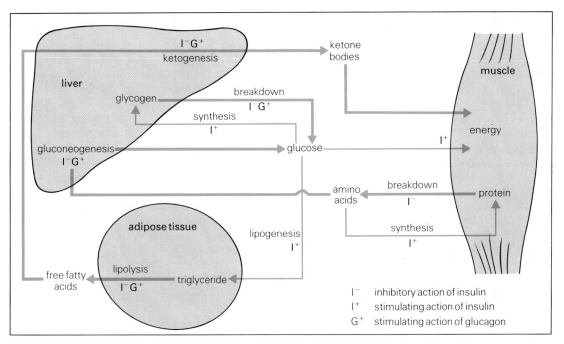

Fig.12.3 Combined effects of insulin and glucagon on substrate flows between liver, adipose tissue and muscle. When the ratio of the concentrations of insulin to glucagon falls (e.g. during starvation), there is increased hepatic glucose and ketone production and decreased tissue glucose utilization. When the ratio is high (e.g. after a meal), glucose is stored as glycogen and converted into fat.

Major characteristics of IDDM and NIDDM		
Feature	**IDDM**	**NIDDM**
typical age of onset	childhood, young adult	middle-age, elderly
onset	acute	gradual
habitus	lean	often obese
weight loss	usual	uncommon
ketosis-prone	usually	usually not
serum insulin concentration	low or absent	often normal
family history of diabetes	uncommon	common
HLA association	DR3, DR4	none

Fig.12.4 Major characteristics of insulin-dependent (IDDM, Type I) and non insulin-dependent (NIDDM, Type II) diabetes mellitus.

IDDM. Inflammation of pancreatic islets (insulitis) leads to islet cell destruction and thus a lack of insulin. Islet cell antibodies can be detected in the serum. It is suggested that activated T cells directed against viral antigens may also react with islet cell antigens in susceptible individuals and lead to cell destruction. However, it remains possible that the immunological changes seen in patients with IDDM are a result of the islet cell destruction, rather than its cause.

Pathophysiology and clinical features

There are two aspects to the clinical manifestations of diabetes mellitus: those related directly to the metabolic disturbance and those related to the long-term complications of the condition. The prevalence of the long-term complications (nephropathy, neuropathy, retinopathy and arteriopathy) increases with duration of the disease, but bears no clear relationship to the severity as inferred from measures of metabolic control. Some patients remain free of complications even after having the condition for many years. There is evidence that the development of nephropathy, neuropathy and retinopathy are related to the glycosylation of proteins, for example, those in the glomerular basement membrane, but this is not conclusive. Associations have also been reported between these complications and certain abnormalities of the immune system. The long-term complications of diabetes are a significant source of morbidity and mortality, but with the exception of nephropathy their diagnosis is largely clinical. In contrast, the management of the acute metabolic disturbances seen in diabetes mellitus requires close collaboration between the physician and the laboratory staff.

The hyperglycaemia of diabetes mellitus is mainly a result of increased production of glucose by the liver and, to a lesser extent, of decreased removal of glucose from the blood stream. In the kidneys, filtered glucose is normally completely reabsorbed in the proximal tubules, but at blood glucose concentrations much above 10mmol/l (the renal threshold), reabsorption becomes saturated and glucose appears in the urine. There is some variation in the threshold between individuals and it is higher in the elderly; it is lower during pregnancy. Glycosuria results in an osmotic diuresis, increasing water excretion and raising the serum osmolality, which in turn stimulates the thirst centre. Osmotic diuresis and thirst cause the classical symptoms of polyuria and polydipsia.

Untreated, the metabolic disturbances may become profound, with the development of life-threatening ketoacidosis, non-ketotic hyperglycaemia or lactic acidosis.

Diagnosis

The diagnosis of diabetes mellitus depends upon the demonstration of hyperglycaemia. In a patient with classical symptoms and signs, this may be inferred from the presence of glycosuria. Under these circumstances, a fasting plasma glucose concentration exceeding 7.8mmol/l, or a random value exceeding 11.1mmol/l, will confirm the diagnosis of diabetes mellitus. In patients who are asymptomatic, either of these limits must be exceeded on more than one occasion for the diagnosis to be made. In doubtful cases, it may be necessary to measure the serum glucose two hours after an oral glucose load, or perform a formal oral glucose tolerance test (OGTT). The World Health Organization (WHO) criteria for the diagnosis of diabetes mellitus are shown in Fig.12.5. People whose blood glucose concentration is raised, but not sufficiently to meet these criteria, are classified as having 'impaired glucose tolerance'. Such people should be reviewed annually; some will eventually become overtly diabetic.

A formal OGTT needs to be carried out only when results are equivocal. It should be emphasized that if the diagnosis is clear from the clinical features and confirmed by a fasting or post-prandial blood glucose measurement, an OGTT is superfluous. The indications for this test, and the test protocol, are given in Fig.12.6. Intravenous glucose tolerance tests are only used in research.

Management

There are many aspects to the management of diabetes mellitus. The education of the patient is vital as he will have the disease for the rest of his life and must, to a considerable extent, be responsible for his own treatment, albeit with guidance from a physician. Regular follow-up is essential: to monitor treatment and to detect early signs of complications, particularly retinopathy which can in many cases be treated successfully.

The aims of treatment are twofold: to alleviate symptoms and prevent the acute metabolic complications of diabetes, and to prevent the long-term complications. The first of these objectives is usually attainable with dietary control with or without oral hypoglycaemic agents, principally sulphonylureas, in patients with NIDDM, and with dietary control and insulin in patients with IDDM.

Diagnostic blood glucose concentrations (mmol/l)				
Diagnosis	Time of sample	Venous whole blood	Venous plasma	Capillary whole blood
diabetes mellitus	fasting 2h post-glucose load	≥6.7 ≥10.0	≥7.8 ≥11.1	≥6.7 ≥11.1
impaired glucose tolerance	fasting 2h post-glucose load	<6.7 6.7–10.0	<7.8 7.8–11.1	<6.7 7.8–11.1

Fig.12.5 Diagnostic blood glucose concentrations.

The oral glucose tolerance test		
Indications	**Procedure**	**Results**
equivocal fasting/random blood glucose concentrations unexplained glycosuria, particularly in pregnancy clinical features of diabetes mellitus or its complications with normal blood glucose concentrations diagnosis of acromegaly, see page 115	patient should eat normal diet, containing at least 250g carbo-hydrate per day for three days fast patient overnight take basal blood sample for glucose determination give 75g glucose in water orally take further blood samples at 30, 60, 90 and 120 min for glucose determination patient should rest throughout test; smoking not permitted; drinks of water are allowed	diabetes mellitus can be diagnosed if venous whole blood glucose concentration exceeds 10.0mmol/l (capillary whole blood, 11.1mmol/l) at 120 min and at one other time during test

Fig.12.6 The oral glucose tolerance test.

There is considerable circumstantial evidence that good control of the blood glucose concentration may delay or even prevent the development of long-term complications of diabetes mellitus. To this end, it is appropriate, particularly in the younger patient, to attempt to maintain the blood glucose concentration within the physiological range, although in practice this may be difficult to achieve. Excessive insulin dosage, for example, may cause hypoglycaemia. In an elderly patient, on the other hand, it is sufficient to alleviate the acute symptoms even if the blood glucose concentration remains high. Many such patients with NIDDM remain symptom free even with blood glucose concentrations as high as 15mmol/l and are in no danger of developing ketoacidosis. Attempting to improve control, perhaps by introducing a harsher dietary regime or by giving oral hypoglyaemic drugs in addition to diet is inappropriate when the patient is unlikely to live long enough for complications to develop. It is important to point out that, despite treatment, the fluctuations in blood glucose concentration that occur in most diabetic patients are still greater than those which occur in normal subjects.

Monitoring treatment

The efficacy of treatment in diabetes is monitored clinically, by ensuring that the patient's symptoms are controlled, and in the laboratory, by semi-quantitative measurement of urine sugar concentration, for example, by using tablets or reagent strips, measurement of the blood glucose concentration and measurement of the concentration of glycosylated proteins in the blood. Many diabetic patients test their urine for glucose at home. Home blood glucose monitoring, using capillary blood and a reagent strip, is increasingly practised. Capillary blood may also be collected on to filter paper, dried and sent to the laboratory for measurement of glucose concentration. However, the blood glucose level at any one time, even if standardized, for example, after an overnight fast or two hours post-prandial, may be entirely unrepresentative of overall diabetic control. Furthermore, although the urine will contain glucose if the blood concentration has exceeded the renal threshold at any time since the bladder was last emptied, the amount of glucose in the urine is only an approximate guide to the severity of any hyperglycaemia and is dependent upon the individual patient's renal threshold. If this is low, there may be considerable glycosuria with only slightly elevated blood glucose concentrations. In addition, testing the urine for glucose is of no value for detecting hypoglycaemia.

The discovery that haemoglobin undergoes non-enzymatic glycosylation *in vivo* has facilitated assessment of diabetic control over a longer term. The rate of formation of glycosylated haemoglobin is proportional to the blood glucose concentration; the reaction proceeds through a reversible stage but once the stable product (HbA_1) is formed, it remains in the red cell for the lifetime of that cell. The proportion of haemoglobin in the glycosylated form (normal <7%; up to 20% in poorly controlled diabetes) effectively 'integrates' the blood glucose concentration over the previous 6–8 weeks. Results are not, however, reliable in patients with decreased red cell life spans, for example, haemolytic anaemia, and in some of the methods used to measure glycosylated haemoglobin, haemoglobin variants may cross-react and give false high results. The measurement of glycosylated serum proteins may prove to be of value as a shorter-term index of control. Serum fructosamine concentration is proportional to the amount of glycosylated albumin present and is used for this purpose.

CASE HISTORY 12.1

A young insulin-dependent diabetic patient attended the outpatient department for his regular follow-up and reported that he had been symptom-free since his last clinic attendance. He had not bothered to test his urine at home and did not like pricking his finger to obtain capillary blood for testing.

Investigations

blood glucose (two hours after breakfast)	18 mmol/l
urine glucose (early morning)	2%
glycosylated haemoglobin	9%

Comment

The glycosylated haemoglobin suggests that diabetic control is good, despite the patient's apparent lack of interest, a high blood glucose concentration and glycosuria. It transpired that he had been to a party the night before and had eaten considerably more than usual. He did not want to admit this, since he was, to use his own words, 'fed up with being lectured at'.

Since the proportion of glycosylated haemoglobin reflects the mean blood glucose concentration over the previous few weeks, its measurement is particularly useful whenever there is a discrepancy between the patient's history and blood or urine glucose measurements. It will, for example, be high in patients who are generally poorly controlled but who make a special effort to comply with their treatment before attending a clinic, in order to please their doctor by having a normal blood glucose concentration and no glycosuria.

METABOLIC COMPLICATIONS OF DIABETES

Ketoacidosis

Ketoacidosis may be the presenting feature of IDDM, or may develop in a patient known to be diabetic who omits to take his insulin or whose insulin dosage becomes inadequate because of an increased requirement, for example, as a result of infection, any acute illness such as myocardial infarction, trauma or emotional disturbance.

CASE HISTORY 12.2

An eighteen-year-old girl consulted her family practitioner because of tiredness and weight loss. On questioning, she admitted to feeling thirsty and had noticed that she had been passing more urine than normal. The doctor tested her urine and found glycosuria. He arranged for her to be seen at the hospital's diabetic clinic the next day. By then, however, she felt too ill to get out of bed, developed vomiting and became drowsy. Her doctor visited her at home and arranged for immediate admission to hospital. On examination she was found to have a blood pressure of 95/60 mmHg with a pulse rate of 112/min and cold extremities. She had deep, sighing respiration (Kussmaul's respiration) and her breath smelt of acetone.

Investigations

serum:
	sodium	130 mmol/l
	potassium	5.8 mmol/l
	bicarbonate	5 mmol/l
	urea	18 mmol/l
	creatinine	140 μmol/l
	glucose	32 mmol/l

arterial blood hydrogen ion   89 nmol/l   (pH 7.05)

P_{CO_2} 2.0 kPa   (15 mmHg)

Comment

The clinical and biochemical features are typical of diabetic ketoacidosis (Fig.12.7). She has hypotension, tachycardia and cold extremities, suggesting marked extracellular fluid depletion (sodium depletion). The low bicarbonate and high hydrogen ion concentrations with hyperventilation and thus a decreased P_{CO_2} indicate a non-respiratory acidosis with partial respiratory compensation. There is renal impairment (raised urea and creatinine) and the disproportionate increase in urea in comparison with creatinine is typical of dehydration compounded by increased urea production due to amino acid breakdown (see below).

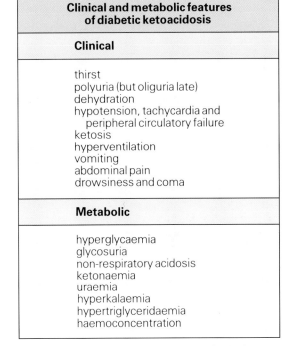

Clinical and metabolic features of diabetic ketoacidosis
Clinical
thirst polyuria (but oliguria late) dehydration hypotension, tachycardia and   peripheral circulatory failure ketosis hyperventilation vomiting abdominal pain drowsiness and coma
Metabolic
hyperglycaemia glycosuria non-respiratory acidosis ketonaemia uraemia hyperkalaemia hypertriglyceridaemia haemoconcentration

Fig.12.7   Clinical and metabolic features of diabetic ketoacidosis.

Hyperkalaemia is commonly present and is a result of the combined effects of a shift of intracellular potassium (due to insulin lack, since insulin promotes cellular potassium uptake, acidosis and tissue catabolism) and decreased renal excretion. However, in spite of the hyperkalaemia, there is always considerable potassium depletion. The serum sodium concentration is usually decreased, because of sodium depletion and replacement of ionized extracellular solutes by glucose. Although this patient is markedly hyperglycaemic, clinically severe ketoacidosis can sometimes occur with only a moderate increase in the blood glucose concentration (10–15 mmol/l).

The pathogenesis of the acidosis and hyperglycaemia are discussed below. Although the term 'diabetic coma' is often used synonymously with diabetic ketoacidosis, many patients have a normal level of consciousness when they present. Drowsiness is common, but only ten percent of patients are actually comatose.

Pathogenesis

The sequence of events which leads to hyperglycaemia and the consequences of this are illustrated in Figs 12.3 and 12.8. An increase in the ratio of the concentration of glucagon to that of insulin in the portal blood decreases the hepatic concentration of fructose 2,6-bisphosphate, a key regulatory intermediate. This results in inhibition of phosphofructokinase, and thus of glycolysis, and activation of fructose 1,6-bisphosphatase, thus stimulating gluconeogenesis. At the same time, glycogen breakdown is promoted and glycogen synthesis is inhibited. Decreased peripheral utilization of glucose, resulting from insulin lack and preferential metabolism of free fatty acids and ketones as energy substrates, contributes to the hyperglycaemia but is less important than the increased rate of glucose production.

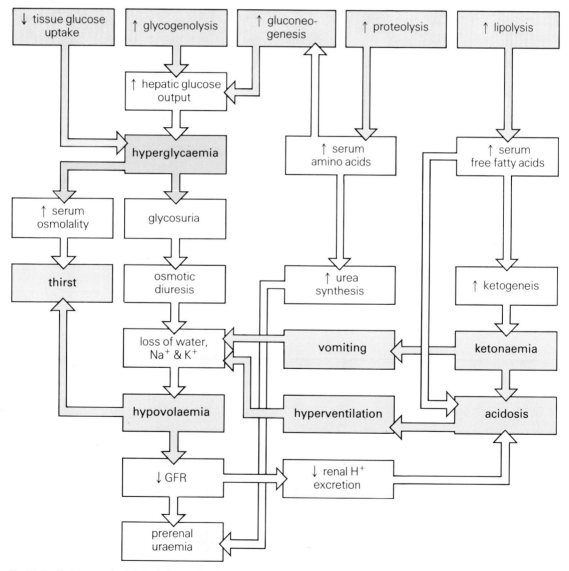

Fig.12.8 Pathogenesis of diabetic ketoacidosis, indicating the consequences of a decreased insulin:glucagon ratio. Hyperkalaemia is invariably present, in spite of total body potassium depletion, as a result of loss of potassium from the tissues to the ECF and, as the glomerular filtration rate falls, to decreased renal excretion.

Glycosuria causes an osmotic diuresis and hence fluid depletion which is exacerbated by the hyperventilation and vomiting. The decrease in plasma volume leads to renal hypoperfusion and prerenal uraemia. As the glomerular filtration rate falls, so does the rate of urine production and the patient, initially polyuric, becomes oliguric. Established renal failure is an uncommon but recognized consequence of diabetic ketoacidosis.

Insulin lack causes increased lipolysis, with increased release of free fatty acids into the blood from adipose tissue, and decreased lipogenesis. In the liver,

fatty acids normally undergo complete oxidation, are re-esterified to triglycerides or are converted to acetoacetic and β-hydroxybutyric acids (ketogenesis). Ketogenesis is promoted in uncontrolled diabetes by the high ratio of glucagon to insulin. The mechanisms involved are shown in Fig.12.9. Some acetoacetate is spontaneously decarboxylated to acetone. Ketones stimulate the chemoreceptor trigger zone, causing vomiting. Acetoacetic and β-hydroxybutyric acids are the major acids responsible for the acidosis but free fatty acids and lactic acid also contribute.

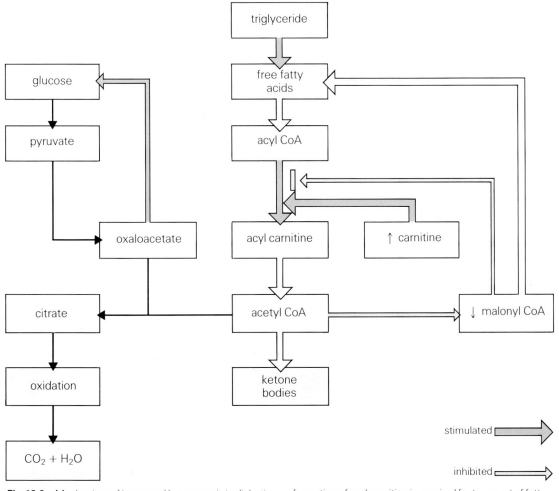

Fig.12.9 Mechanism of increased ketogenesis in diabetic ketoacidosis. The formation of free fatty acids increases as a result of increased lipolysis. Glucagon increases carnitine formation (mechanism unknown) and inhibits the synthesis of malonyl CoA, an intermediate in fatty acid synthesis, which normally inhibits acyl-carnitine synthesis. The formation of acyl carnitine is required for transport of fatty acids into mitochondria, where ketogenesis takes place. Supplies of oxaloacetate necessary for the oxidation of acetyl CoA in the citric acid cycle are diverted instead to gluconeogenesis.

Management

Diabetic ketoacidosis is a medical emergency. The treatment entails replacement of lost fluid and minerals, and reversal of the metabolic disturbance with insulin. Any identifiable precipitating event, for example, infection, must also be treated.

Isotonic saline is given intravenously to replace lost fluid. The rate at which it is given will depend upon the precise circumstances, but it should usually be given rapidly, at least initially, to restore the extracellular fluid volume to normal. Careful monitoring of fluid input and output is essential, and it may be necessary to place a central intravenous catheter to monitor the central venous pressure. If there is gastric stasis, the gastric contents must be aspirated. Catheterization of the bladder may be necessary.

Potassium supplements are required; insulin causes rapid potassium uptake into cells so that although patients are usually hyperkalaemic at presentation, hypokalaemia will develop during treatment if potassium is not replaced. Regular monitoring of the serum potassium concentration is essential.

Insulin is best given by constant intravenous infusion at a rate of 6–10units/h. The blood glucose concentration must be monitored and, once it has fallen to near normal levels, the intravenous fluid is changed to five percent dextose and the rate of insulin infusion decreased to maintain euglycaemia until it is possible to establish oral food and water intake, and a conventional regimen of subcutaneous insulin injections.

It is seldom necessary to give bicarbonate, except in the severest cases, since restoration of normal renal perfusion allows excretion of the hydrogen ion load and regeneration of bicarbonate, while restoration of normal metabolism reduces the production rate. If bicarbonate is used, only small quantities should be given at one time and the effect monitored by measurement of the arterial hydrogen ion concentration. Rapid correction of an acidosis may impair oxygen delivery to the tissues, through an effect on the affinity of haemoglobin for oxygen, result in over-compensation with production of an alkalosis and paradoxically increase the cerebro-spinal fluid (CSF) hydrogen ion concentration because of delayed equilibration of the bicarbonate between the serum and the CSF. The response to treatment of a typical patient with diabetic ketoacidosis is shown in Fig.12.10.

The deficits present in a patient with ketoacidosis are considerable and may exceed five litres of water and 500mmol each of sodium and potassium. Considerable depletion of other ions, in particular phosphate, may occur in ketoacidosis. The serum phosphate concentration should be monitored, but specific replacement therapy is not usually required. Other biochemical abnormalities that may be seen include an increase in serum amylase activity, hyper-triglyceridaemia and, in the severely shocked patient, an increase in transaminases.

Non-ketotic hyperglycaemia

Not all patients with uncontrolled diabetes develop ketoacidosis. In NIDDM, severe hyperglycaemia can develop (blood glucose concentration >50mmol/l) with extreme dehydration and a very high serum osmolality, but with no ketosis and minimal acidosis. This complication is often referred to as hyperosmolar non-ketotic hyperglycaemia, but patients with ketoacidosis usually also have increased serum osmolality, although not to the same extent.

CASE HISTORY 12.3

A middle-aged widow, who lived alone, was admitted to hospital after her son found her semi-conscious at home. He had not seen her for a week but she had seemed well then. On examination, she was extremely dehydrated but not ketotic. Her respiration was normal.

Investigations

serum:		
	sodium	149mmol/l
	potassium	4.7mmol/l
	bicarbonate	18mmol/l
	urea	35mmol/l
	creatinine	180μmol/l
	glucose	54mmol/l
	total protein	90g/l
	osmolality	370mmol/kg

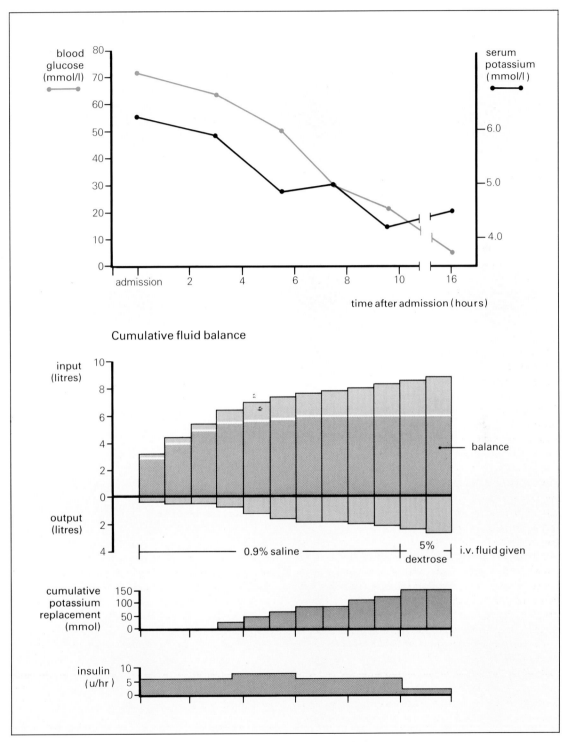

Fig.12.10   Response to treatment in a patient with severe diabetic ketoacidosis, indicating a fluid deficit of six litres on admission.

Comment

The serum osmolality is very high, reflecting the severe hyperglycaemia. This has caused an osmotic diuresis, resulting in a decrease in glomerular filtration with retention of urea and creatinine and an increase in serum protein concentration due to the loss of water from the serum. The serum bicarbonate concentration is a little below normal because of the decreased renal hydrogen ion excretion. The serum sodium concentration is often raised in this condition, reflecting loss of water in excess of sodium as a result of the sustained osmotic diuresis.

Non-ketotic hyperglycaemia occurs only in NIDDM. There is sufficient insulin secretion to prevent the excessive lipolysis and to oppose the ketogenic action of glucagon that are essential for the generation of ketoacidosis (the concentrations of insulin required to do this are lower than those needed to prevent increased glucose production). Blood glucose concentrations are in general higher than in ketoacidosis. Perhaps because vomiting is not a feature, patients do not become acutely ill so quickly. A history of thirst and polyuria was subsequently obtained from this patient, who had not been previously diagnosed as having diabetes. She had also been assuaging her thirst with copious amounts of sweetened carbonated drinks.

Management

Rehydration and the administration of insulin are the most essential aspects of treatment. Insulin is given by constant intravenous infusion but a satisfactory decrease in glucose concentration is often attained using a lower rate of infusion than that employed in ketoacidosis. In view of the hypertonicity, hypotonic ('half-normal') saline is given although it may be wise to give a litre of normal saline initially, specifically to expand the extracellular compartment and also to prevent too rapid a fall in osmolality (see page 16). Potassium supplements are required, but less is needed than in ketoacidosis. Careful monitoring of the glucose concentration and fluid balance are essential. Heparin is sometimes given prophylactically in view of the hyperviscosity and attendant danger of

thrombosis. In contrast to ketoacidosis, continued treatment with insulin is seldom required once the acute illness is over; patients can usually be managed either on a diet or on a diet with oral hypoglycaemic drugs.

Lactic acidosis

Lactic acidosis is an uncommon complication of diabetes. It was formerly chiefly seen in patients treated with phenformin, a biguanide oral hypoglycaemic drug, but is now more usually associated with severe systemic illness, for example, severe shock and pancreatitis. It is discussed in more detail on pages 42 and 43.

Hypoglycaemia in diabetic patients

Hypoglycaemia may occur in patients with NIDDM who are treated with long-acting oral hypoglycaemic agents (particularly chlorpropamide, a sulphonyl-urea), and most patients with IDDM have occasional episodes of hypoglycaemia. The presence of glycosuria does not exclude hypoglycaemia, as the renal threshold may have been exceeded since the bladder was last emptied. The diagnosis must rest on the blood glucose concentration, but if there is any doubt, it is always safe to give glucose to an unconscious or confused diabetic patient pending the results becoming available.

Diabetic nephropathy

The renal disease that may complicate diabetes tends to progress towards end-stage renal failure, although there is considerable variation in its time course. Proteinuria is the main feature of diabetic nephropathy and microalbuminuria (the presence of <20mg/l of albumin in the urine which is not detectable by reagent strips) is the earliest biochemical abnormality. There is encouraging evidence to suggest that excellent control of the blood glucose concentration and treatment of any hypertension may abolish early microalbuminuria and thus prevent or delay the onset of renal impairment.

GLYCOSURIA

Although diabetes mellitus is the commonest cause of glycosuria, it is also seen in patients with a low renal threshold for glucose. This may occur as an isolated and harmless abnormality (renal glycosuria), can develop during pregnancy, and is a feature of congenital and acquired generalized disorders of proximal renal tubular function (the Fanconi syndrome, see page 70).

A positive test for reducing substances in the urine when using, for example, testing tablets, is given by a number of substances other than glucose, some of which are indicated in Fig.12.11. Reagent strips containing glucose oxidase are specific for glucose.

GLUCOSE IN CEREBROSPINAL FLUID

Glucose concentration in the CSF is commonly measured in patients suspected of having bacterial meningitis, since it is usually decreased as a result of bacterial metabolism. If the CSF is frankly purulent, the measurement of CSF glucose provides no useful additional information. The CSF glucose concentration is approximately sixty-five percent of the blood glucose concentration and CSF glucose should always been interpreted in the light of the glucose concentration of a blood sample obtained at the same time.

Substances giving a positive reducing test in urine
glucose
lactose (during lactation and the last trimester of pregnancy)
galactose (in galactosaemia and galactokinase deficiency)
fructose (in hereditary fructose intolerance and essential fructosuria)
pentoses (after eating certain fruits and in essential pentosuria)
homogentisic acid (in alkaptonuria)
glucuronides of drugs
salicylic acid (in aspirin overdose)
ascorbic acid (with high vitamin C intake)
creatinine (only in high concentration)

Fig.12.11 Substances giving a positive reducing test in urine.

SUMMARY

Diabetes mellitus is a condition characterized by abnormal glucose tolerance with a tendency to hyperglycaemia and is due to a relative or absolute deficiency of insulin. It may occur secondarily to obvious pancreatic disease but the majority of cases are idiopathic. Type I, or insulin-dependent diabetes (IDDM), typically affects younger patients. It usually has an acute onset; there is strong evidence of an auto-immune pathogenesis. Type II, or non insulin-dependent diabetes (NIDDM), typically affects middle-aged and elderly people and has a more gradual onset. Genetic and environmental factors are important in its pathogenesis.

Hyperglycaemia leads to glycosuria and causes an osmotic diuresis, producing the classical clinical features of polyuria and thirst. If inadequately treated, patients with IDDM may develop diabetic ketoacidosis. In this condition, hyperglycaemia, together with increased lipolysis, proteolysis and ketogenesis, leads to severe dehydration, mineral loss, prerenal uraemia and a profound non-respiratory acidosis. Patients with NIDDM appear to have sufficient insulin secretion to prevent the excessive lipolysis and ketogenesis which are essential to the production of ketoacidosis. Instead, inadequate treatment may lead to the development of very severe hyperglycaemia and dehydration, producing a non-ketotic, hyperosmolar state.

Both ketoacidosis and non-ketotic hyperosmolar coma are medical emergencies; their management involves rehydration and insulin replacement with general supportive measures and treatment of any specific preexisting or complicating factors. Patients treated with insulin, and in some circumstances those treated with oral hypoglycaemic drugs, are prone to develop hypoglycaemia if the dose is inappropriate to the body's requirements.

In the longer term, patients with diabetes are at risk of developing neuropathy, retinopathy, nephropathy and arterial disease. The evidence that these complications are related to persistent hyperglycaemia is strong but is not proven.

The treatment of diabetes is aimed at relieving symptoms and preventing both the short- and long-term complications. The efficacy of treatment, whether with insulin, oral hypoglycaemic drugs or dietary modification alone, can be assessed clinically and by measurements of blood and urine glucose concentrations. However, the blood glucose gives an indication of control only at a single moment and urine glucose reflects blood glucose concentration over at most only a few hours. For the assessment of control over longer periods, the extent to which haemoglobin is glycosylated appears to reflect the overall blood glucose concentration over a period of several weeks, while the measurement of other glycosylated proteins may provide useful information about control over intermediate periods.

FURTHER READING

Nattrass M (ed.) (1986) *Recent Advances in Diabetes – 2*. London / Edinburgh: Churchill Livingstone.

Nattrass M & Santiago JV (eds) (1984) *Recent Advances in Diabetes – 1*. London / Edinburgh: Churchill Livingstone.

Wilson JD & Foster DW (eds) (1985) *Williams – Textbook of Endocrinology*. 7th edition. Philadelphia: WB Saunders Company.

13. Hypoglycaemia

INTRODUCTION

Glucose homoeostasis is described in *Chapter 12*. Hypoglycaemia, conventionally (although arbitrarily) defined as a blood glucose concentration of 2.2mmol/l or less, may be caused by a decrease in the rate of delivery of glucose to the blood or by an increase in the rate of its removal. The sources of this glucose are dietary carbohydrate, gluconeogenesis and glycogenolysis. The last two processes are capable of maintaining the blood glucose concentration within normal limits even when no glucose is being absorbed from the gut. Thus a lack of dietary carbohydrate alone is not a sufficient cause for hypoglycaemia, although it may be contributory when gluconeogenesis or glycogenolysis are defective. The hormones involved in glucose homoeostasis are listed in Fig.13.1.

As would be expected, hypoglycaemia is a feature of a number of endocrine diseases; however, it may also develop in the absence of any overt hormonal disturbance if the normal homoeostatic mechanisms are overwhelmed as, for example, in tumour associated hypoglycaemia (see page 186), or are ineffective, such as in severe liver disease (see page 184).

HYPOGLYCAEMIA

Causes

It is conventional, and convenient, to divide the causes of hypoglycaemia into those causing a low blood glucose concentration during *fasting* and those in which it follows a stimulus (*reactive* hypoglycaemia), including the stimulus of a meal (post-prandial hypoglycaemia). It is usually possible to distinguish

Hormones involved in glucose homoeostasis		
Hormone	**Effect on blood glucose**	**Principal mechanism of action**
insulin	↓	inhibition of gluconeogenesis (L) stimulation of glycogen synthesis (L and M) stimulation of cellular uptake of glucose (M)
adrenaline	↑	stimulation of glycogenolysis (L and M)
glucagon	↑	stimulation of glycogenolysis (L) stimulation of gluconeogenesis (L)
cortisol	↑	stimulation of gluconeogenesis (L) permissive effect on action of glucagon (L)
growth hormone	↑	mobilization of triglyceride (glucose sparing)
L = effect on metabolism in liver M = effect on metabolism in muscle		

Fig.13.1   Hormones involved in glucose homoeostasis.

between these categories from the patient's history.

Episodes of reactive hypoglycaemia can occur in patients with fasting hypoglycaemia although, in contrast, fasting hypoglycaemia is virtually never a feature of those conditions associated with reactive hypoglycaemia. The causes of hypoglycaemia are summarized in Fig.13.2 and the pathogenesis indicated in each case. These conditions are discussed further in the following sections.

Clinical features

Glucose is an essential energy substrate for the nervous system, at least in the short-term; during starvation, adaptation occurs and ketone bodies can be utilized. The clinical features of hypoglycaemia are the result of dysfunction of the nervous system (neuroglycopenia) and the effects of catecholamines which are released in response to the stimulus provided by the low blood glucose.

In acute hypoglycaemic episodes, the symptoms and signs are characteristic (Fig.13.2), but in chronic hypoglycaemia they may be atypical and bizarre. Typical clinical features are more likely to occur if the blood glucose falls rapidly and if hypoglycaemic episodes are separated by periods of normoglycaemia. If the blood glucose concentration falls rapidly, symptoms may develop at concentrations somewhat higher than 2.2mmol/l. The clinical features of hypoglycaemia are likely to be enhanced if cerebral blood flow is impaired, while they may be attenuated in patients taking β-adrenergic blocking drugs, such as propranolol.

Diagnosis

The two stages in the diagnosis of hypoglycaemia are: confirmation of the low blood glucose concentration and elucidation of the cause. Allusion has been made to the considerable variation in the blood glucose concentration at which symptoms of hypoglycaemia begin to appear. In children and young adults, symptoms will usually be present only with a concentration less than 2.2mmol/l. The elderly tend to be more sensitive to a low blood glucose, perhaps because of impaired homoeostatic responses or decreased cerebral perfusion resulting from atheroma. Neonates, however, usually develop symptoms only when the blood glucose is less than 1.5mmol/l. Blood for glucose analysis must be collected in a tube containing fluoride, to inhibit glycolysis, and an anticoagulant.

Clinical features caused by hypoglycaemia should be confirmed by giving glucose either by mouth or parenterally, as appropriate. The symptoms due to acute neuroglycopenia and catecholamine release should subside immediately, but those attributable to chronic hypoglycaemia do not.

The cause of the hypoglycaemia may be obvious from the patient's history, particularly in reactive hypoglycaemia. With fasting hypoglycaemia, many possible causes can be eliminated by simple tests; investigations are therefore often directed towards the detection of a possible insulin-secreting tumour.

HYPOGLYCAEMIC SYNDROMES

Reactive hypoglycaemia

Drug-induced hypoglycaemia

INSULIN
Occasional episodes of hypoglycaemia are not uncommon in insulin-dependent (IDDM) diabetic patients; occasionally a diabetic patient will deliberately administer excessive insulin to gain admission to hospital. More commonly, hypoglycaemia is related to a missed meal or some other factor.

CASE HISTORY 13.1

A young male jogger collapsed during a ten-mile fun run. He was conscious but disorientated and his speech was incoherent. He was taken to hospital where a finger-prick stick test for blood glucose concentration showed this to be very low. Blood was sent to the laboratory for confirmation of the diagnosis. He was given 25g glucose and recovered rapidly. He then admitted that he was insulin-dependent; he had injected his normal dose of insulin that morning and eaten his usual breakfast. The laboratory reported a blood glucose concentration on admission of 1.6mmol/l. He was given further carbohydrate by mouth and was discharged that evening with a normal blood glucose level to be reviewed in the diabetic clinic the next day.

Hypoglycaemia	
Causes	**Clinical features**

Reactive hypoglycaemia
drug-induced:
    insulin
    sulphonylureas
    others
post-prandial:
    gastric surgery
    essential (idiopathic) reactive hypoglycaemia
alcohol-induced
inherited metabolic disorders:
    galactosaemia
    hereditary fructose intolerance
 	
Fasting hypoglycaemia
hepatic and renal disease (rare)
endocrine disease:
    adrenal failure
    pituitary failure
    isolated ACTH deficiency
inherited metabolic disorders:
    glycogen storage disease Type I
hyperinsulinism:
    insulinoma
    nesidioblastosis
non-pancreatic neoplasms
alcohol-induced fasting hypoglycaemia
various forms of neonatal hypoglycaemia
	Acute
	due to neuroglycopenia:
	    anxiety
	    detachment
	    hunger
	    ataxia
	    dizziness
	    diplopia
	    paraesthesiae
	    hemiparesis
	    convulsions
	    coma
	due to sympathetic stimulation:
	    palpitation and tachycardia
	    profuse sweating
	    facial flushing
	    tremor
	
	Chronic neuroglycopenia
	personality changes
	memory loss
	psychosis
	dementia

Fig.13.2 Major causes and clinical features of hypoglycaemia. Chronic neuroglycopenia is seen mainly in patients with insulin-secreting tumours; the features of acute neuroglycopenia are classically seen in diabetic patients who have taken too much insulin, but may occur with other forms of reactive hypoglycaemia.

Comment

Insulin requirements in insulin-dependent diabetic patients are reduced by exercise. It is an important part of the education of diabetic patients that they are aware of this and can thus reduce their insulin dose or increase their carbohydrate intake accordingly. Diabetic patients should always carry sugar and a means of identification to facilitate treatment in an emergency.

OTHER DRUGS

The sulphonylureas (hypoglycaemic drugs used in the treatment of non insulin-dependent diabetes – NIDDM) can also cause hypoglycaemia. Chlorpropamide, the most commonly implicated, has a long plasma half-life and, since it is eliminated only by the kidneys, tends to accumulate in patients with impaired renal function.

Hypoglycaemia may occasionally occur in patients treated with β-adrenergic blocking drugs but only when other contributory factors, such as starvation or exercise, are involved. Children, but not adults, poisoned with salicylates may develop severe hypoglycaemia. Hypoglycaemia has also been reported in patients who have taken overdoses of paracetamol and, in these cases, it is probably related to the severe liver damage that this drug can cause.

Post-prandial hypoglycaemia

In patients who have undergone gastric surgery, either with a gastrointestinal anastomosis or a pyloroplasty, hypoglycaemia developing 90–150min after a meal, particularly a meal rich in sugar, is common. There is rapid passage of glucose into the small intestine and release of hormones which stimulate insulin secretion. The insulin response is excessive and hypoglycaemia ensues as glucose absorption from the gut falls off rapidly, rather than slowly as it does when gastric emptying is normal.

Symptoms suggestive of hypoglycaemia following meals may be described by people who have not undergone surgery (essential or idiopathic post-prandial hypoglycaemia). Although transient hypo-glycaemia is common from 90–150min after taking 75g glucose orally in a glucose tolerance test, it is often asymptomatic and the relevance of hypoglycaemia after this artificial stimulus is questionable. Further, low blood glucose concentrations are not always demonstrable in such subjects, nor is there objective evidence of disordered glucose homoeostasis. Nevertheless, dietary manipulations, of which the addition of guar or other vegetable fibre to the diet is the most effective, may give considerable sympto-matic relief.

Post-prandial hypoglycaemia may be an early feature of NIDDM. It is probably due to delayed but excessive secretion of insulin in response to a glucose load, with secretion persisting for longer than is appropriate.

Alcohol and reactive hypoglycaemia

Insulin- and drug-induced reactive hypoglycaemia are potentiated by alcohol. Alcohol also increases insulin release in response to an oral glucose load and this may enhance any tendency to post-prandial reactive hypoglycaemia. Alcohol-induced fasting hypogly-caemia is considered in *Case history 13.3*.

Other causes of reactive hypoglycaemia

Various inherited metabolic diseases have reactive hypoglycaemia as a feature. Since these are usually first recognized in children they are considered in the discussion of neonatal and childhood hypoglycaemia (see pages 186–188).

Sudden cessation of hypertonic dextrose infusion, being given as part of a parenteral feeding regimen, can precipitate hypoglycaemia, especially when insulin has been given concomitantly. Hypoglycaemia may also occur after dialysis against a glucose-rich dialysate.

Fasting hypoglycaemia

Hepatic and renal disease

Although the liver is central to glucose homoeostasis, its functional reserve is so great that hypoglycaemia is a rare feature of hepatic disease. It may occur, however, with the rapid, massive hepatocellular destruction that can follow poisoning with paracetamol and other toxins. The kidneys are the only organs other than the liver capable of gluconeo-genesis; they are also responsible for insulin degradation. These facts may in part explain the severe hypoglycaemia that is occasionally a feature of terminal renal disease.

Endocrine disease

Deficiency of hormones antagonistic to insulin is a recognized but uncommon cause of hypoglycaemia. Lack of cortisol may be either due to primary adrenal failure or secondary to panhypopituitarism; hypo-glycaemia may be a feature of both. Mild hypo-glycaemia can occur with isolated ACTH deficiency and in isolated growth hormone deficiency, but in the latter condition it is never symptomatic.

Rather surprisingly, in view of its role in carbohydrate metabolism, lack of adrenaline in patients who have undergone bilateral adrenalectomy and who are maintained on cortical hormone replacement, neither causes hypoglycaemia nor interferes with the ability to recover from artificially induced hypoglycaemia.

Inherited metabolic disease

Fasting hypoglycaemia is an important feature of glycogen storage disease Type I, discussed in more detail on pages 247 and 248.

Insulinoma

Insulinoma is an uncommon tumour of the insulin-secreting β-cells of the pancreatic islets. Because the presentation often is not typical of hypoglycaemia (patients may, for example, present to psychiatrists with behavioural disturbances), there is frequently a delay in diagnosis. Once the diagnosis is considered and hypoglycaemia confirmed, the presence of an insulin-secreting tumour can be inferred from the demonstration of an inappropriately high serum insulin level at a time when the blood glucose is low. For this purpose hypoglycaemia can be provoked by fasting.

Blood samples should be collected three times after an overnight fast and when this is done, ninety per cent of patients with an insulinoma will demonstrate unequivocal, though often asymptomatic, hypoglycaemia. Clinically apparent hypoglycaemia is seen in almost all patients who are fasted for seventy-two hours. Blood is collected for glucose and insulin measurement at 4–6h intervals and when symptoms are present. Patients should be encouraged to exercise during this time.

Inappropriate insulin secretion is confirmed if the serum insulin level is greater than 10mu/l or more at a time when the blood glucose is less than 2.2mmol/l. Normal subjects may develop (women more frequently than men) hypoglycaemia during such a fast, but this is asymptomatic and serum insulin concentration is low or undetectable. Alternatively, inappropriate insulin secretion may be confirmed by performing an insulin tolerance test (see page 110) with the usual strict monitoring and measurement of C-peptide levels. C-peptide is released from pancreatic β-cells in equimolar quantities with insulin and its concentration provides an estimate of β-cell function (see page 167). If hypoglycaemia is achieved, a C-peptide concentration of greater than 1.5 μg/l indicates autonomous insulin secretion. This result reflects continuing insulin release at a time when it should be suppressed by the low blood glucose.

CASE HISTORY 13.2

A woman telephoned for an ambulance when she was unable to rouse her husband one morning; she noticed that his left leg and arm were jerking. In the hospital emergency room, he was seen to be pale and sweaty, with a rapid poor-volume pulse. His blood glucose concentration was 0.8mmol/l. He regained consciousness when given a bolus of glucose intravenously, but then became confused and required a continuous glucose infusion for several hours to prevent hypoglycaemia.

His wife revealed that she had been becoming increasingly worried about her husband. Formerly a man of equable temperament, over the past six months he had frequently arrived home in a bad mood, taken little notice of his wife and young child and sat in a sullen silence until his evening meal. After eating, he would behave quite normally, apparently with no recollection of his previous behaviour. On the two mornings immediately prior to admission, she had found him sitting up in bed, apparently conscious but staring vacantly at the wall and not speaking; she had managed to get him to drink his usual cup of sweet tea and he had rapidly recovered.

A presumptive diagnosis of insulinoma was made and was confirmed by the finding of a serum insulin concentration of 80mu/l at a time when he was hypoglycaemic. He had hepatomegaly and the alkaline phosphatase level was raised. A coeliac axis angiogram demonstrated a large filling defect in the liver; at laparotomy, the liver was found to have extensive tumour deposits, shown on histological examination to be characteristic of an insulinoma. A single, small tumour was present on the pancreas. No operative treatment was possible; he initially responded well to cytotoxic drugs but relapsed and died six months later.

Comment

This case illustrates the sometimes bizarre symptomatology of patients with insulinomas, who may be chronically hypoglycaemic. A provocative test was clearly not required to establish the diagnosis in this case. When a definitive test is required, the insulin tolerance test with measurement of C-peptide is preferable to the many other tests that have been described, using, for example, glucagon or alcohol as a secretagogue for insulin. The majority of insulinomas (approximately ninety percent) are, unlike the tumour in this case, benign. In some ten percent of cases there are multiple pancreatic tumours and there may be associated adenomas in other endocrine organs (multiple endocrine adenomatosis Type I, see pages 279–280).

The treatment of choice is surgical resection, when possible, and with benign tumours the prognosis is good. Diazoxide may be used to prevent hypoglycaemia pre-operatively. Its main action is to reduce insulin secretion from both normal and neoplastic β-cells. Streptozotocin, a drug which is specifically cytotoxic to β-cells, is valuable in the management of malignant insulinomas when surgical treatment is either not possible or has failed.

Non-pancreatic neoplasms

Hypoglycaemia is also seen in association with non-pancreatic neoplasms, particularly large mesen-chymal tumours such as retro-peritoneal sarcomas, with hepatocellular and adrenal carcinomas, and with carcinoid tumours. In the majority of cases, the cause of the hypoglycaemia is uncertain. Except with some carcinoid tumours, serum insulin levels are not elevated. Increased glucose uptake by the tumour may be a factor but is unlikely ever to be the sole cause. Hepatic glucose output is often reduced although there is a normal glucogenic response to glucagon. In some cases, there appear to be increased serum levels of factors having an insulin-like action. These are discussed in more detail in *Chapter 21*.

Alcohol-induced fasting hypoglycaemia

CASE HISTORY 13.3

An elderly man was found to be unrousable one morning by fellow inmates of a derelict house in which they slept. He had been drunk the previous evening and although this was not uncommon he had never before been so stuporose in the morning. An ambulance was called and he was admitted to hospital and found to be profoundly hypoglycaemic. He responded rapidly to intravenous glucose and did not then appear inebriated. He refused further treatment and discharged himself later the same day.

Comment

Alcohol-induced fasting hypoglycaemia is caused mainly by the inhibitory effect of alcohol on gluconeogenesis. However, acquired ACTH deficiency may be present in some alcoholics and poor nutrition and liver disease may also contribute. As in this case, hypoglycaemia characteristically develops several hours after alcohol ingestion (compare alcohol and reactive hypoglycaemia, page 184), when hepatic glycogen stores become exhausted. Signs of inebriation may not be present at this time and blood alcohol levels are unremarkable. Although first described, and most commonly seen, in poorly nourished chronic alcoholics, hypogly-caemia can be readily precipitated by alcohol in healthy subjects whose hepatic glycogen reserves have been depleted by lack of food.

HYPOGLYCAEMIA IN CHILDHOOD

Neonatal hypoglycaemia

In the newborn, as in adults, the definition of hypoglycaemia is somewhat arbitrary. It is generally accepted that, during the first three days of life, a blood glucose concentration of less than 1.6mmol/l is abnormal for a normal weight baby born at term and less than 1.1mmol/l is abnormal for a premature or small-for-dates infant.

Hypoglycaemia may occur transiently in apparently normal babies, but is particularly common in those who have respiratory distress, severe infection, brain damage or who are small-for-dates. Premature and small-for-dates babies are particularly at risk of developing neonatal hypoglycaemia because they are born with low hepatic glycogen stores and are more likely to have feeding problems. Extensive physio-logical changes occur at birth and, in terms of glucose metabolism, there is a sudden interruption of the maternal glucose supply and glycogenolysis must span the period until feeding starts. Babies born to diabetic mothers may have islet cell hyperplasia which increases the risk of hypoglycaemia developing in the immediate post-natal period, though this does not persist thereafter.

Hypoglycaemia in infancy

Any of the conditions discussed above may cause hypoglycaemia in infancy. A variety of other conditions may cause hypoglycaemia at this time (Fig.13.3); these are discussed below. The inherited metabolic diseases associated with hypoglycaemia are particularly likely to present during the first few weeks of life.

The commonest form of hypoglycaemia in infancy is ketotic hypoglycaemia. This may be either primary (idiopathic) or secondary to conditions leading to relative carbohydrate deficiency, such as starvation or generalized illness in children who were small-for-dates at birth. There is a reduced availability of glucogenic precursors, especially alanine, but the reason for this is unknown. The low blood glucose suppresses insulin secretion and this is the cause of the ketosis (not vice versa as used to be thought). Hypoglycaemia can be prevented by frequent feeding and the attacks become less frequent as the child grows older.

Nesidioblastosis is a developmental abnormality of the pancreas in which there is overgrowth of ducts and islet cells. Leucine may precipitate the hypoglycaemia in this condition, but there is doubt as to whether 'leucine-induced hypoglycaemia' is a distinct clinical syndrome.

The inherited metabolic diseases associated with hypoglycaemia include galactosaemia, glycogen storage disease Type I and hereditary fructose intolerance, discussed below.

CASE HISTORY 13.4

A healthy two-month-old female infant, who had previously been exclusively breast-fed, vomited when given supplementary feeds of sweetened cow's milk. She reacted similarly when given fruit juice and sometimes became quiet and sleepy

---

Causes of hypoglycaemia in childhood
Transient neonatal hypoglycaemia
Ketotic hypoglycaemia idiopathic secondary
Hyperinsulinaemia islet cell hyperplasia nesidioblastosis insulinoma
Inherited metabolic disorders, including: glycogen storage diseases galactosaemia hereditary fructose intolerance
Other causes prematurity small-for-dates endocrine disorders starvation drugs

Fig.13.3   Causes of hypoglycaemia in childhood.

after such a feed. Her mother experimented with various feeds and learnt to avoid those that made her child ill. The child grew up with an aversion to sweet foodstuffs and fruit. Her brother born three years later had a similar history.

Later, when both became medical students, they wondered if their aversion was due to hereditary fructose intolerance. Fructose tolerance tests were performed and were found to induce hypoglycaemia, vomiting and the other metabolic changes characteristic of this condition.

Comment

If the link between a child's illness and dietary fructose (and sucrose) is not made, there may be serious long-term consequences: failure to thrive, cirrhosis and renal tubular dysfunction, for example. If fructose is avoided and irreversible liver or kidney damage has not occurred, patients with hereditary fructose intolerance remain symptom-free. The cause of the intolerance is a lack of the B isoenzyme of fructose-1-phosphate aldolase, which catalyzes the conversion of fructose-1-phosphate and fructose 1,6-bisphosphate to trioses.

In the absence of the B isoenzyme, the other isoenzymes (A and C), which account for fifteen percent of the catalytic activity, convert much less fructose-1-phosphate than normal but are still sufficiently active to convert fructose 1,6-bisphosphate to trioses in the glycolytic pathway. When fructose is ingested, however, it is still converted by fructokinase to fructose-1-phosphate in the normal way, but because the A and C isoenzymes have insufficient activity to metabolize it, the fructose-1-phosphate accumulates. The clinical manifestations stem from the accumulation of fructose-1-phosphate, which inhibits glucose synthesis, and the depletion of ATP and phosphate as fructose is phosphorylated but not further metabolized.

SUMMARY

The causes of hypoglycaemia fall into two groups according to whether the condition either occurs in the fasting state (fasting hypoglycaemia) or is provoked by a specific stimulus, which can include food intake (reactive hypoglycaemia). Causes of fasting hypoglycaemia include insulin-secreting tumours (insulinomas) and certain other tumours producing insulin-like substances, pituitary and adrenal failure, severe liver disease and glycogen storage diseases, notably Type I (glucose-6-phosphatase deficiency). Apart from insulinomas, hypoglycaemia is rarely the only clinical feature of any of these conditions. The diagnosis of an insulinoma depends upon the finding of inappropriately high insulin (and C-peptide) levels in the blood at a time when the patient is hypoglycaemic. This is often demonstrable after an overnight fast, precipitated if necessary by exercise. Provocative tests to stimulate insulin secretion are rarely required in patients with an insulinoma.

Reactive hypoglycaemia may be caused by drugs. Most patients with insulin-dependent diabetes experience occasional episodes of hypoglycaemia, for example, as a result of a delayed meal following an insulin injection, an error in insulin dosage or unaccustomed exercise. Oral hypoglycaemic drugs (sulphonylureas) can also cause hypoglycaemia. This is more common in patients treated with the longer acting drugs, particularly the elderly, whose capacity to metabolize or excrete the drugs may be impaired. Following gastric surgery, rapid transit of food into the small intestine may cause inappropriate insulin secretion and lead to hypoglycaemia. Such post-prandial hypoglycaemia also occurs occasionally in normal subjects. Patients with hereditary fructose intolerance and galactosaemia develop hypo-glycaemia after ingesting fructose and galactose, respectively.

Hypoglycaemia may occur following alcohol ingestion and several distinct syndromes of alcohol-related hypoglycaemia have been described. Alcohol potentiates insulin- and drug-induced hypoglycaemia and may enhance any tendency to post-prandial reactive hypoglycaemia. The hypoglycaemia that can develop 12–24h after alcohol ingestion, particularly in chronic alcoholics, is due in part to impairment of gluconeogenesis but the presence of liver disease, poor nutrition and depletion of hepatic glycogen reserves may also be important.

Hypoglycaemia is particularly common in neonates who are small-for-dates and is a risk in those born to diabetic mothers. In addition to the conditions described, ketotic hypoglycaemia may occur in infancy. This is a condition of unknown aetiology in which there appears to be a decreased supply of glucogenic substrates.

Acutely, hypoglycaemia causes clinical features related to increased activity of the sympathetic nervous system (sweating, tachycardia) and decreased substrate supply to the central nervous system (paraesthesiae, fits, coma). These usually respond rapidly to the administration of glucose. Patients who are chronically hypoglycaemic, for example, due to an insulinoma, often present with behavioural disturbance or frank psychosis and the acute manifestations of hypoglycaemia may be absent.

FURTHER READING

Marks V & Rose FC (1981) *Hypoglycaemia*. 2nd edition. Oxford: Blackwell Scientific Publications.

14. Calcium, Phosphate, Magnesium and Bone

INTRODUCTION

Calcium is the most abundant mineral in the human body. The average adult body contains approximately 25,000mmol (1kg), of which ninety-nine percent is bound in the skeleton. The total calcium content of the extracellular fluid (ECF) is only 22.5mmol, of which about 9mmol is in the plasma. Bone is not metabolically inert; some of its calcium is rapidly exchangeable with the ECF, the turnover between bone and ECF being approximately 500mmol/day (Fig.14.1). In the kidneys, ionized calcium is filtered by the glomeruli (240mmol/day). Most of this is reabsorbed in the tubules and normal renal calcium excretion is 2.5–7.5mmol/day. Gastrointestinal secretions contain calcium, some of which is reabsorbed together with dietary calcium. Since calcium in the ECF pool is effectively exchanged through the kidneys, gut and bone about thirty-three times every twenty-four hours, a small change in any of these fluxes will have a profound effect on the concentration of calcium in the ECF and, hence, in the plasma.

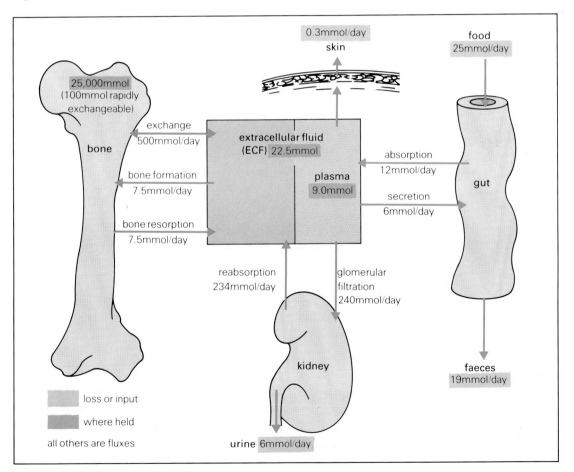

Fig.14.1 Calcium exchange in the body.

Functions of calcium	
Function	**Example**
structural	bone teeth
neuromuscular	control of excitability release of neurotransmitters initiation of muscle contraction
enzymic	coenzyme for coagulation factors
hormonal	intracellular second messenger

Fig.14.2 Functions of calcium.

Calcium has many important functions in the body (Fig.14.2). Its effect on neuromuscular activity is of particular importance in the symptomatology of hypocalcaemia and hypercalcaemia, as is described later in this chapter.

BONE

Bone consists of osteoid, a collagenous organic matrix, on which is deposited complex inorganic hydrated calcium salts known as hydroxyapatites. These have the general formula:

$$Ca^{2+}{}_{10-x} \ (H_3O^+)_{2x} \ (PO_4{}^{3-})_6 \ (OH^-)_2$$

Even when growth has ceased, bone remains biologically active. Continuous turnover occurs and bone reabsorption (mediated by osteoclasts) is followed by new bone formation (mediated by osteoblasts); the control of this process is poorly understood. Bone formation requires osteoid synthesis and adequate calcium and phosphate for the laying down of hydroxyapatite. Alkaline phosphatase, secreted by osteoblasts, is essential to the process, probably acting by releasing phosphate from pyrophosphate. Bone provides an important reservoir of calcium, phosphate and, to a lesser extent, magnesium and sodium.

SERUM CALCIUM

In the serum, calcium is present in three forms (Fig.14.3): bound to protein (mainly albumin); complexed with citrate and phosphate; and free ions. Only the latter form is physiologically active and it is the concentration of ionized calcium which is maintained by homoeostatic mechanisms.

In alkalosis, the decrease in hydrogen ion concentration results in increased binding of calcium to protein as hydrogen ions dissociate from albumin and increases the proportion of complexed calcium. The effect is a decrease in the concentration of ionized calcium which may be sufficient to produce clinical symptoms and signs of hypocalcaemia. In the short-term, there is no change in the total calcium concentration; it is the proportion of physiologically active calcium that is decreased. In chronic alkalosis, however, homoeostatic mechanisms may return ionized calcium levels to normal, so that the total calcium concentration increases. In an acute acidosis, the reverse effect is observed, the ionized calcium concentration being increased, although in chronic acidosis it is usually normal.

The most commonly used methods for determining serum calcium concentration measure the total calcium, although technology for the measurement of ionized calcium using ion-selective electrodes is

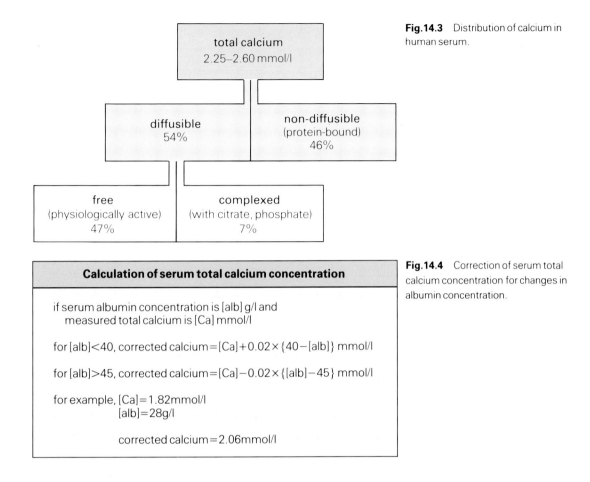

Fig. 14.3 Distribution of calcium in human serum.

Fig. 14.4 Correction of serum total calcium concentration for changes in albumin concentration.

becoming more widely available. No clear superiority has yet been demonstrated for measurements of ionized, as opposed to total, calcium concentration in the management of the majority of cases of disordered calcium metabolism. Considerable care is necessary in the collection and preservation of blood samples if ionized calcium is to be measured.

Changes in plasma albumin concentration will affect total calcium levels independently of the ionized calcium concentration, leading to possible misinterpretation of results in both hypoproteinaemic and hyperproteinaemic states. Various formulae have been devised to indicate the total calcium concentration to be expected if the albumin concentration were normal. One widely used formula is given in Fig. 14.4, but such estimates of corrected calcium concentration should be interpreted with caution, especially when serum hydrogen ion concentration is abnormal. This is an instance where a direct measurement of ionized calcium may be helpful. A common cause of apparent hyperproteinaemia, and hence hypercalcaemia, is venous stasis during blood sampling; this must be avoided when determinations of serum calcium are to be made; for example, a tourniquet should not be used. Although globulins bind calcium to a much lesser extent than albumin, the increase in γ-globulin in patients with myeloma can also increase the total serum calcium concentration. In myeloma, however, hypercalcaemia is commonly present with an increased ionized calcium concentration.

Actions of parathyroid hormone		
Organ	**Action**	**Effect**
bone	increases resorption	release of calcium and phosphate into ECF
kidney	stimulates 1α-hydroxylation of 25-hydroxycholecalciferol	increased uptake of calcium and phosphate from gut
	decreases phosphate reabsorption	increased phosphate excretion
	increases calcium reabsorption	increased calcium excretion
	decreases bicarbonate reabsorption	acidosis

Fig.14.5 Actions of parathyroid hormone. The effect on tubular reabsorption of calcium is small in comparison with the effect of hypercalcaemia on the amount of calcium filtered. The overall effect of PTH is to increase serum calcium, decrease serum phosphate and increase renal calcium excretion.

CALCIUM-REGULATING HORMONES

Calcium concentration in the ECF is normally maintained within narrow limits by a control system involving two hormones, parathyroid hormone (PTH) and calcitriol (1,25-dihydroxycholecalciferol). These hormones also control the inorganic phosphate concentration of the ECF.

Parathyroid hormone

This hormone is a polypeptide, comprising eighty-four amino acids; as with many hormones, it is synthesized as a larger precursor, pre-pro-PTH (115 amino acids). Prior to secretion, two amino acid sequences are lost; the removal of a twenty-five amino acid chain produces pro-PTH, a further six amino acids being lost to form PTH itself. The pre-peptide and pro-peptide are thought to be involved in the intracellular transport of the hormone. The biological activity of PTH resides in the N-terminal 1–34 amino acid sequence of the hormone. PTH is secreted by the parathyroid glands in response to a fall in serum (ionized) calcium concentration. Its actions tend to increase the serum calcium concentration and reduce serum phosphate (Fig.14.5).

Although PTH increases the tubular reabsorption of calcium in the kidneys, its action on bone releases calcium into the ECF and thus increases the amount of calcium filtered by the glomeruli. Although the fraction of the filtered load that is reabsorbed is increased by PTH, the overall effect is to produce hypercalciuria.

Despite the importance of PTH in the control of phosphate excretion, phosphate levels do not affect secretion of the hormone. Mild hypomagnesaemia stimulates PTH secretion, but more severe hypomagnesaemia reduces it as the secretion of PTH is magnesium-dependent.

PTH can be measured by immunoassay, but considerable problems attend both the measurement itself and the interpretation of results. PTH is rapidly metabolized in the blood stream into a number of polypeptide fragments. The C-terminal fragments, which are biologically inactive, are more immunogenic than the active N-terminal fragments and have a longer half-life. Thus, the fraction measured may not be biologically active. Assays which have been designed to measure only intact PTH molecules may reflect the activity of the parathyroid glands, but not necessarily the biological activity in the plasma.

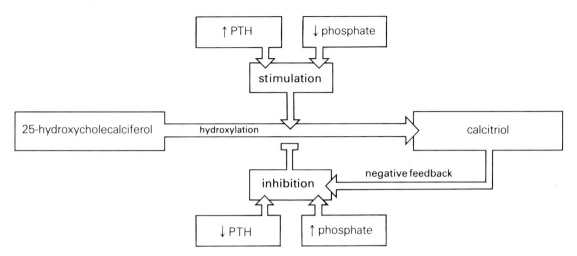

Fig.14.6   Synthesis of calcitriol, which in the long-term is also stimulated by growth hormone, prolactin and oestrogens.

Actions of calcitriol		
Gut	**Kidney**	**Bone**
increases calcium absorption and phosphate absorption via calcium-binding protein in intestinal cells	increases calcium reabsorption	increases resorption (PTH dependent)

Fig.14.7   Actions of calcitriol. The main physiological action is to increase calcium absorption from the gut, by increasing the synthesis of a calcium-binding protein in enterocytes.

Calcitriol

This hormone is derived from vitamin D by successive hydroxylation in the liver (25-hydroxylation) and kidney (1α-hydroxylation). Hydroxylation in the liver is not subject to feedback control, but that in the kidney is usually closely regulated (Fig.14.6). When the 1α-hydroxylation of 25-hydroxycholecalciferol is inhibited, there is an increase in 24-hydroxylation. The product of this reaction, 24,25-dihydroxy-cholecalciferol, has no known physiological function. Both this metabolite and calcitriol undergo further metabolism in the kidney but the products are physiologically inactive.

The actions of calcitriol are listed in Fig.14.7; they tend to increase both serum calcium and phosphate concentration. The presence of adequate calcium and phosphate is a prerequisite for the normal mineralization of bone, but calcitriol does not appear to have any direct effect on osteoblastic activity. The way in which calcitriol and PTH act together to control serum calcium and phosphate concentrations is considered later in this chapter.

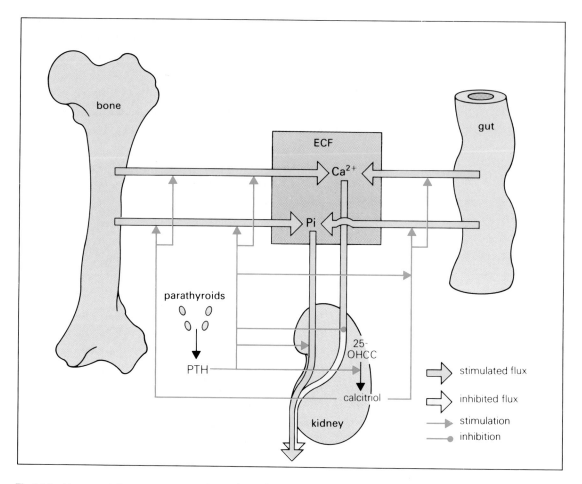

Fig.14.8 Homoeostatic responses to hypocalcaemia. Hypocalcaemia stimulates the release of PTH which in turn stimulates calcitriol synthesis. These hormones act together to restore the calcium concentration to normal, independently of phosphate.

Calcitonin

This polypeptide hormone, produced by the C cells of the thyroid, can be shown experimentally to inhibit osteoclast activity, and thus bone resorption, but it is not known if this is of any physiological significance. Subjects who have had a total thyroidectomy develop no clinical syndrome that can be ascribed to calcitonin deficiency. Also, calcium homoeostasis is normal in patients with medullary carcinoma of the thyroid, a tumour which secretes large quantities of calcitonin. Serum calcitonin levels are elevated during pregnancy and lactation, and it has been suggested that in these states it blocks the action of calcitriol on bone and permits increased calcium uptake from the gut to take place without loss of mineral from bone.

CALCIUM AND PHOSPHATE HOMOEOSTASIS

The response of the body to a fall in serum calcium concentration, provided that this is not due to

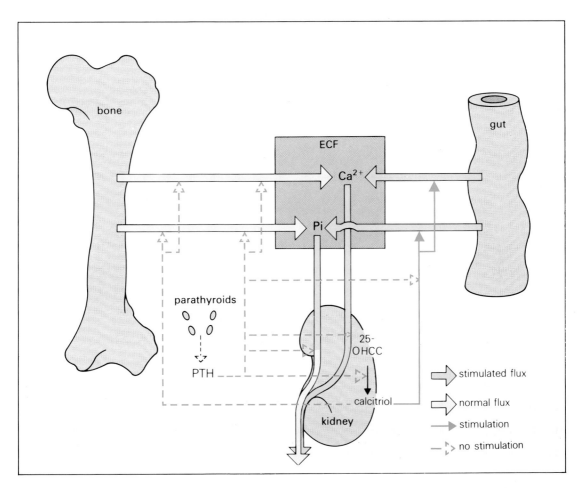

Fig.14.9 Homoeostatic responses in hypophosphat-aemia. In the absence of PTH (secretion is not affected by phosphate), an increase in calcitriol production due to stimulation of 1α-hydroxylase tends to increase the serum phosphate independently of calcium concentration.

disordered homoeostasis in the first instance, is illustrated in Fig.14.8. Hypocalcaemia stimulates the secretion of PTH and, through this, increases the production of calcitriol. There will be an increase in the uptake of both calcium and phosphate from the gut and in their release from bone. PTH is phosphaturic, so the excess phosphate is excreted but the fractional reabsorption of calcium by the kidney is increased; some of the mobilized calcium is retained and the serum calcium concentration tends to rise towards normal.

In hypophosphataemia (Fig.14.9), calcitriol secretion is increased but PTH is not. Calcium and phosphate absorption from the gut are stimulated. Calcitriol has a much smaller effect on renal calcium reabsorption than PTH so that, in the absence of PTH, the excess calcium absorbed from the gut is excreted in the urine. The net result is the restoration of the phosphate concentration towards normal, independently of calcium. In the absence of PTH, calcitriol alone has little effect on bone.

DISORDERS OF CALCIUM, PHOSPHATE AND MAGNESIUM METABOLISM

Hypercalcaemia

The causes of hypercalcaemia are listed in Fig.14.10. Hypercalcaemia may be discovered during the investigation of an illness in which it is known to be a potential complication, during the investigation of clinical features suggestive of hypercalcaemia (Fig.14.10) and frequently as an incidental finding because of the use of multichannel autoanalyzers for biochemical analysis.

Malignant disease

This is a very common cause of hypercalcaemia and may be seen in patients both with and without obvious metastases in bone. Non-metastatic hypercalcaemia is discussed on page 276; it is almost certainly due to humoral factors. Although PTH has been implicated, it has not been found in the majority of cases examined critically, and some other substance, or substances, with similar actions is probably responsible. In patients with metastases in bone, there is often no relationship between the extent of metastasis and the severity of the hypercalcaemia.

Hypercalcaemia	
Causes	**Clinical features**
Common	weakness, tiredness, lassitude, weight loss and muscle weakness
malignant disease, with and without metastasis to bone	mental changes (impaired concentration, drowsiness, personality changes, coma)
primary hyperparathyroidism	anorexia, nausea, vomiting and constipation
Less common	abdominal pain (rarely peptic ulceration and pancreatitis)
thyrotoxicosis	polyuria, dehydration and renal failure
vitamin D intoxication	renal calculi and nephrocalcinosis (mainly associated with primary hyperparathyroidism)
thiazide diuretics	short QT interval on ECG
sarcoidosis	cardiac arrhythmias and hypertension
idiopathic hypocalciuric hypercalcaemia	corneal calcification and vascular calcification
renal transplantation (tertiary hyperparathyroidism)	there may also be features of the underlying disorder, such as bone pain in malignant disease and hyperparathyroidism
Uncommon	
milk-alkali syndrome	
lithium	
tuberculosis (very rare)	
immobilization (especially in Paget's disease)	
acromegaly	
adrenal failure	
idiopathic hypercalcaemia of infancy	
diuretic phase of acute renal failure	

Fig.14.10   Causes and clinical features of hypercalcaemia.

Thus humoral factors may be involved in the pathogenesis of hypercalcaemia in malignant disease whether or not metastases are present.

With certain tumours, prostaglandins produced by the metastatic deposit may act as local hormones to stimulate osteoclast activity. The hypercalcaemia of multiple myeloma (see *Case history 15.1*) is due to the activation of osteoclasts by humoral factors (lymphokines) secreted by the tumour cells.

Primary hyperparathyroidism

The prevalence of this condition is of the order of one case per thousand persons. It can occur at any age and affects both men and women but is most common in post-menopausal women. It is usually due to a parathyroid adenoma, less often to diffuse hyperplasia of the glands, and only rarely to a parathyroid carcinoma. Adenomas may be multiple and the condition is sometimes familial; it may also be part of one of the multiple endocrine adenomatosis syndromes.

CASE HISTORY 14.1

A fifty-one-year-old woman was investigated after two episodes of ureteric colic, shown on radiological examination to be due to calcium-containing calculi. She also complained of constipation, although she previously had normal bowel movements, but was otherwise well. No abnormality was found on physical examination.

Investigations

serum:	calcium	2.95 mmol/l
	phosphate	0.70 mmol/l
	total CO_2	19 mmol/l

plasma immunoreactive PTH
(N-terminal assay):   110 pmol/l
(reference range < 120 pmol/l)

bone radiographs:	normal
urea, albumin and alkaline phosphatase:	all normal

Comment

Hyperparathyroidism may present in many ways (see Fig.14.10), including renal or ureteric colic due to calculi which are themselves a result of hypercalciuria. Only about ten percent of patients have clinical evidence of bone disease at presentation, although biochemical and radiological evidence is present in more than twenty percent. Many patients with hyperparathyroidism have no or few symptoms and are detected as a result of biochemical screening.

The serum calcium concentration is nearly always raised. Exceptions to this occur if there is concomitant renal disease, vitamin D deficiency or hypothyroidism; occasionally the calcium is only raised intermittently. The phosphaturic action of PTH causes hypophosphataemia but this is not invariable; the serum phosphate concentration may be normal or raised, particularly when there is renal damage. The serum alkaline phosphatase is raised in only twenty to thirty percent of cases. Hypercalciuria is a reflection of the hypercalcaemia and is of no diagnostic importance.

The serum PTH concentration may be frankly elevated but the results of PTH assays must be interpreted with caution. In the presence of hypercalcaemia, PTH production from normal parathyroid glands should be suppressed. A PTH concentration in the normal range, as in this case, is therefore inappropriate and suggests autonomous PTH secretion.

Before the advent of assays for PTH, measurements of the theoretical renal phosphate threshold, assessment of the response of the raised serum calcium to corticosteroids (serum calcium concentration is reduced by steroids in most patients with hypercalcaemia due to causes other than hyperparathyroidism) and multivariate analysis were all used as tools in the diagnosis of hyperparathyroidism. Direct assay of PTH, despite its limitations, has tended to render these obsolete. However, it is important to consider and exclude other common causes of hypercalcaemia, such as malignancy, before making the diagnosis.

MANAGEMENT

The definitive treatment for hyperparathyroidism is surgery, but patients whose serum calcium remains below 3.0mmol/l, who are symptom free and have no renal impairment, often stay healthy for many years without an operation. A high fluid intake should be maintained, to discourage renal calculus formation, and regular reassessment is necessary. Parathyroid adenomas are usually small and rarely palpable. Isotopic imaging techniques may help to localize the tumour preoperatively. A tumour may be localized by measuring PTH in blood samples obtained by selective catheterization of neck veins. This is a highly specialized technique, but is particularly useful if hypercalcaemia recurs and a second operation is required, since the normal anatomical relationships will have been distorted by the previous surgery.

Secondary and tertiary hyperparathyroidism

Serum PTH concentrations are also raised in many patients with chronic renal disease and vitamin D deficiency. Both these conditions are associated with decreased synthesis of calcitriol, which causes hypocalcaemia, and the increase in PTH secretion is an appropriate physiological response. This is termed secondary hyperparathyroidism. The increase in PTH may not, however, normalize the serum calcium; in the absence of adequate calcitriol, there is resistance to the calcium-mobilizing effect of PTH on bone. Occasionally, patients with end-stage renal failure become hypercalcaemic, due to the development of autonomous PTH secretion, presumably as a result of the prolonged hypocalcaemic stimulus. It may be seen for the first time in a patient given a renal transplant, who can then metabolize vitamin D normally. This is termed tertiary hyperparathyroidism.

Parathyroid hormone is, in part, metabolized and excreted by the kidney. Increased serum levels of PTH in renal failure reflect impairment of these processes as well as increased secretion. Much of the excess consists of C-terminal fragments, which are inactive in calcium homoeostasis.

Other causes of hypercalcaemia

Malignancy and hyperparathyroidism account for the majority of cases of hypercalcaemia. However, it is sometimes a feature of thyrotoxicosis; thyroid hormones have no specific role in calcium homoeostasis and the hypercalcaemia is due to the increased osteoclastic activity that may be present in this condition. Coincidental thyrotoxicosis may provoke symptomatic hypercalcaemia in a patient with mild, subclinical hyperparathyroidism.

Vitamin D intoxication is a preventable cause of hypercalcaemia and all patients being treated with vitamin D or its derivatives should have their serum calcium concentration measured regularly. Even though serum concentrations of calcitriol may be normal in patients treated with vitamin D, the serum level of 25-hydroxycholecalciferol may be increased sufficiently for it to have a direct hypercalcaemic effect.

In the milk-alkali syndrome, hypercalcaemia is associated with the ingestion of milk and antacids for the control of dyspeptic symptoms. The ingestion of alkali is important in the pathogenesis of the hypercalcaemia; it is thought that it decreases renal excretion of calcium but the precise mechanism is unknown. This syndrome is uncommon and is becoming more so with the widespread use of more specific remedies, for example, histamine-receptor (H_2) antagonists, for the treatment of patients with dyspepsia. It should also be remembered that dyspepsia itself may be a feature of hyperparathyroidism since calcium stimulates gastrin release. Occasionally, patients with parathyroid adenomas may have associated gastrin-secreting tumours (multiple endocrine adenomatosis Type I).

Thiazide diuretics are a common cause of mild hypercalcaemia, as they interfere with renal calcium excretion. Chronic lithium therapy is also occasionally associated with it possibly due to stimulation of PTH secretion; it is also seen in approximately ten percent of patients with sarcoidosis and in other chronic granulomatous disorders, including tuberculosis, as a result of 1-hydroxylation of 25-hydroxycholecalciferol by the macrophages in the granulomatous tissue. Hypercalcaemia (and hyperphosphataemia) is occasionally seen in acromegaly, probably due to the stimulation of renal 1α-hydroxylase by growth hormone and, very rarely, in adrenal failure, particularly when it is acute.

During a period of immobilization, the decreased stimulus to bone formation and resorption (a part of normal bone turnover) results in hypercalciuria. Hypercalcaemia is usually only seen if bone turnover has previously been higher than normal, and it may therefore develop in a patient with Paget's disease of bone who is immobilized for some reason.

Familial hypocalciuric hypercalcaemia is a condition of unknown cause, inherited as an autosomal dominant; it has only recently been recognized and is probably underdiagnosed. Chronic hypercalcaemia develops from childhood and is usually asymptomatic. Hypophosphataemia is sometimes present; PTH levels are usually normal but may be slightly elevated. The diagnosis may be made only when hypercalcaemia persists after parathyroid-ectomy, but it may be inferred from a low rate of calcium excretion in the urine. The cause of idiopathic hypercalcaemia of infancy is unknown. It is associated with characteristic elfin facies and supravalvar aortic stenosis.

Investigation

The way in which hypercalcaemia is investigated is dependent upon the clinical setting. The most common causes of hypercalcaemia are hyper-parathyroidism and malignancy, and preliminary investigations should be directed towards these diagnoses. The serum phosphate is of limited discriminative value; although low in most uncomplicated cases of primary hyperpara-thyroidism, it may also be decreased in the hypercalcaemia of malignancy but may be raised in either condition if there is renal impairment. The serum alkaline phosphatase is increased if there is increased osteoblastic activity, but this may occur, or be absent, in both conditions.

Radiographic examination may reveal the characteristic subperiosteal bone reabsorption and bone cysts of hyperparathyroidism, but these are only present in a minority of cases; their absence does not exclude the diagnosis. A primary lung tumour or bony metastases will often, but not always, be revealed by radiography but other tumours may not be so easily diagnosed.

The limitations of serum PTH measurements have been discussed, but as long as these are borne in mind this is a valuable investigation if available. The measurement of urinary calcium excretion is of no diagnostic value, except in the diagnosis of familial hypocalciuric hypercalcaemia.

If hyperparathyroidism and malignancy are ex-cluded, reassessment of the history, for both drugs and features of other conditions associated with hypercal-caemia, may prompt appropriate investigations.

CASE HISTORY 14.2

A thirty-eight-year-old man developed thirst and polyuria while on holiday in Spain. He had no other symptoms. He consulted his family doctor when he arrived home. The urine was tested but there was no glycosuria. Blood was taken for biochemical investigations.

Investigations

serum:		
	calcium	3.24mmol/l
	phosphate	1.20mmol/l
	alkaline phosphatase	90iu/l
	urea	10.0mmol/l
	creatinine	150μmol/l

The patient, a non-smoker, was admitted to hospital for investigation. He had previously been well, apart from some joint pain and a painful rash on his legs several months before which had resolved spontaneously. The chest radiograph showed some increased hilar shadowing but was otherwise normal. No bony abnormality was seen on skeletal radiographs. He was slightly dehydrated and was given an intravenous saline infusion. Despite a good diuresis, the serum calcium was unchanged. He was then given hydrocortisone, 40mg three times daily, and a week later the serum calcium was 2.80mmol/l. At this time, the result of the PTH assay on blood taken on admission became available; no PTH could be detected.

Comment

The patient presents with acute, symptomatic hypercalcaemia. The diagnosis could be hyperparathyroidism, an occult malignancy, or some other condition. In view of the slight renal impairment, the normal serum phosphate is not helpful. PTH is undetectable, implying suppression of the parathyroids by hypercalcaemia, rather than autonomous PTH secretion, and the dramatic response to hydrocortisone also militates against hyperparathyroidism. The hypercalcaemia of malignancy responds unpredictably to steroids.

199

The clue to the diagnosis is provided by the chest radiograph and the previous history, which are suggestive of sarcoidosis. The hypercalcaemia in this condition is characteristically sensitive to steroids and is often more severe in the summer, due to increased synthesis of vitamin D by the action of ultraviolet light on the skin. The diagnosis of sarcoidosis was later confirmed by a positive Kveim test and the finding of a raised serum angiotensin-converting enzyme activity.

Management

When possible, the underlying cause should be treated but the hypercalcaemia itself may require treatment in the short-term. Dehydrated patients should be rehydrated with intravenous saline. Once this has been achieved, frusemide may be used; this stimulates a diuresis and inhibits the renal tubular reabsorption of calcium, thus promoting calcium excretion. Various other drugs may be used, for example, calcitonin, bisphosphonates, corticosteroids and mithramycin. Sodium phosphate, given intravenously, is potentially very dangerous, particularly in a patient with renal impairment, since it may cause extensive metastatic calcification. Life-threatening, resistant hypercalcaemia may require treatment by dialysis or, exceptionally, emergency parathyroidectomy.

Hypocalcaemia

The causes of hypocalcaemia are listed in Fig.14.11. The importance of interpreting a low serum calcium concentration in the light of the albumin concentration has already been stressed. The clinical features relate to increased neural and muscular excitability (Fig.14.11).

Vitamin D deficiency

The causes of this condition, which causes osteomalacia in adults and rickets in children, are discussed in *Chapter 23*. Deficiency may be dietary, due to inadequate endogenous synthesis and dietary supply of vitamin D, or to malabsorption. Whatever the cause, the effect is to decrease the amount of 25-hydroxycholecalciferol available for calcitriol synthesis, leading to decreased absorption of calcium and phosphate from the gut (see *Case history 7.2*). Although the 1α-hydroxylation of 25-hydroxycholecalciferol is stimulated in hypocalcaemia, with severe deficiency of the vitamin, lack of the substrate will prevent sufficient calcitriol being formed.

Vitamin D deficiency is a cause of secondary hyperparathyroidism. This further lowers the serum phosphate concentration and patients with vitamin D deficiency usually show hypocalcaemia, hypophosphataemia and a raised serum alkaline phosphatase activity. The serum concentration of 25-hydroxycholecalciferol is reduced.

Hypocalcaemia and bone disease are occasionally seen in epileptic patients treated with phenobarbitone or phenytoin. Both drugs are inducers of hepatic microsomal hydroxylating enzymes and it is thought that this enzyme induction interferes in some way with vitamin D metabolism, although anticonvulsants may also have other effects on calcium homoeostasis. This condition is much less common than was once thought, early studies having been conducted in institutionalized patients who may have had a concomitant degree of dietary vitamin D deficiency. In some forms of chronic liver disease, particularly primary biliary cirrhosis, hypocalcaemia and a metabolic bone disease with some features of osteomalacia develop. The cause of this syndrome is complex, defective 25-hydroxylation of vitamin D playing only a minor role.

Vitamin D-dependent rickets is a recessively inherited disorder in which renal 1α-hydroxylation is defective. It responds to massive doses of vitamin D or small doses of 1-hydroxylated derivatives. Renal tubular phosphate leaks, which may be isolated (familial X-linked hypophosphataemic rickets) or part of the Fanconi syndrome, also cause rickets and osteomalacia but are not associated with hypocalcaemia. They do not respond to vitamin D alone and phosphate supplements are also necessary.

Renal disease

Hypocalcaemia is common in patients with end-stage renal disease (see *Case history 5.3*) but is rarely symptomatic. It is often associated with a complex metabolic bone disease known as renal osteodystrophy. The hypocalcaemia is due primarily to decreased synthesis of calcitriol. This is itself due partly to the decrease in renal substance and partly to

Hypocalcaemia	
Causes	**Clinical features**

vitamin D deficiency:
  dietary
  malabsorption
  inadequate exposure to ultraviolet light
disordered vitamin D metabolism:
  renal failure
  anticonvulsant treatment
  1α-hydroxylase deficiency
hypoparathyroidism
pseudohypoparathyroidism
magnesium deficiency
acute pancreatitis
treatment of metabolic bone disease
hyperphosphataemia (rare)
alkalosis
neonatal hypocalcaemia
massive transfusion with citrated blood
	behavioural disturbance and stupor
	numbness and paraesthesiae
	muscle cramps and spasms
	laryngeal stridor
	convulsions
	cataracts (chronic hypocalcaemia)
	basal ganglia calcification (chronic
	  hypocalcaemia)
	Chvostek's sign positive
	Trousseau's sign positive
	prolonged QT interval on ECG

Fig.14.11 Causes and clinical features of hypocalcaemia. Chvostek's sign (contraction of facial muscles on tapping facial nerve) and Trousseau's sign (carpal spasm when sphygmomanometer cuff applied to upper arm is inflated to midway between systolic and diastolic blood pressures for three minutes) may be positive before other signs are present (latent tetany). Additional features in patients with vitamin D deficiency include myopathy and bone pain.

inhibition of the enzyme by the high serum phosphate levels. A direct effect of uraemic toxins may also be important. It can be treated (and prevented) with 1-hydroxylated derivatives of vitamin D. As has been mentioned, the hypocalcaemia stimulates parathyroid hormone synthesis (secondary hyperparathyroidism) and this may persist after successful renal transplantation giving rise to hypercalcaemia (tertiary hyperparathyroidism). Aluminium retention from dialysis fluid may also be an important factor in the pathogenesis of renal osteodystrophy.

Hypoparathyroidism

This may be congenital or acquired. Acquired causes are listed in Fig.14.12. The congenital form may be associated with thymic aplasia and immune deficiency, the Di George syndrome.

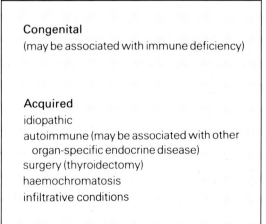

Causes of hypoparathyroidism
Congenital (may be associated with immune deficiency)
Acquired idiopathic autoimmune (may be associated with other   organ-specific endocrine disease) surgery (thyroidectomy) haemochromatosis infiltrative conditions

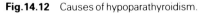

Fig.14.12 Causes of hypoparathyroidism.

201

CASE HISTORY 14.3

A fifty-six-year-old lady was admitted to hospital for cataract extraction, in good health apart from her failing vision. She had undergone thyroidectomy for a multinodular goitre twenty years before. Routine preoperative investigations were carried out.

Investigations

serum: calcium   1.60mmol/l
       phosphate   2.53mmol/l
       albumin   44g/l
       alkaline phosphatase   76iu/l

Comment

The combination of hypocalcaemia, hyperphosphataemia and a normal alkaline phosphatase is typical of hypoparathyroidism, probably due in this case to inadvertent removal of the glands. It is not uncommon for patients with chronic hypocalcaemia to be symptom free. In this patient, both Chvostek's and Trousseau's signs (see Fig.14.11) were positive. Cataracts are a recognized complication of chronic hypocalcaemia

Patients with hypoparathyroidism are treated with vitamin D or its hydroxylated analogues with or without calcium supplements, but care must be taken to avoid hypercalcaemia.

Pseudohypoparathyroidism superficially resembles hypoparathyroidism, but serum levels of PTH are elevated. There is more than one type of pseudo-hypoparathyroidism, but the common factor is end-organ resistance to the action of PTH, a result of a receptor or post-receptor defect. The effects of PTH on the kidneys are mediated through cyclic 3,5-AMP; administration of PTH to patients with true hypoparathyroidism causes an increase in urinary cyclic AMP excretion. In pseudohypoparathyroidism there is no such response. There is a strong hereditary element. In addition to any features due to hypocalcaemia, certain skeletal abnormalities are characteristic of this condition, including a rounded face, short stature, shortening of the fourth and fifth metacarpals and metatarsals, and a tendency for exostoses to form. Patients may be mentally retarded. In the condition pseudopseudohypoparathyroidism, similar skeletal abnormalities are present but the serum calcium concentration is normal. Both conditions are rare.

Magnesium deficiency

Since magnesium is required for both PTH secretion and its action on target tissues, magnesium deficiency can cause hypocalcaemia or render patients insensitive to the treatment of hypocalcaemia with vitamin D or calcium, or both.

CASE HISTORY 14.4

An elderly woman who presented with weight loss and malabsorption due to amyloidosis of the small intestine was found to have osteomalacia and was hypocalcaemic. She was given parenteral nutritional support but despite what was considered to be adequate calcium and vitamin D supplementation, she remained hypocalcaemic.

Investigations

serum magnesium   0.35mmol/l

Comments

Patients with malabsorption may develop magnesium deficiency and while this patient's parenteral feeds contained magnesium there was presumably an insufficient amount to correct her deficit. When given additional magnesium supplements her serum calcium rapidly returned to normal.

Pancreatitis

The causes of hypocalcaemia in acute pancreatitis are discussed in *Chapter 7*.

Causes of hyperphosphataemia
renal failure
hypoparathyroidism
pseudohypoparathyroidism
acromegaly
excessive phosphate intake/ administration
vitamin D intoxication
catabolic states

Fig.14.13 Causes of hyperphosphataemia. This can develop *in vitro* if there is a delay in separating serum from cells prior to analysis.

Causes of hypophosphataemia
vitamin D deficiency
primary hyperparathyroidism
enteral/parenteral nutrition with inadequate phosphate (particularly in malnourished patients) and following intravenous glucose therapy
diabetic ketoacidosis (recovery phase)
alcohol withdrawal (rare)
hypophosphataemic rickets
phosphate binding agents, such as magnesium and aluminium salts (rare)

Fig.14.14 Causes of hypophosphataemia.

Hyperphosphataemia

By far the most common cause of hyperphosphataemia is renal insufficiency; other causes are presented in Fig.14.13. Hyperphosphataemia is a hazard if infants are fed undiluted cow's milk but excessive intake is an uncommon cause in adults, only occurring if excessive phosphate is given intravenously, for example, during parenteral feeding. Increased tissue catabolism, for example, in the treatment of malignant disease (particularly haematological malignancy), can cause hyperphosphataemia. Tissue catabolism is also one of its causes in diabetic ketoacidosis, but in these patients renal impairment is often also present.

Hyperphosphataemia is important clinically because it results in inhibition of the 1α-hydroxylation of 25-hydroxycholecalciferol in the kidney; phosphate may also combine with calcium, resulting in metastatic calcium deposits in the tissues and hypocalcaemia.

Management

Management should be directed at the underlying cause but, in practice, the most effective treatment is to give aluminium or magnesium salts by mouth to bind phosphate in the gut and prevent its absorption.

Hypophosphataemia

This is a common biochemical finding. When mild it is probably of little consequence, but severe hypophosphataemia (<0.3mmol/l) may have important consequences on the function of all cells, particularly the red and white blood cells, by limiting the formation of essential phosphate-containing compounds such as adenosine triphosphate (ATP) and 2,3-diphosphoglycerate (2,3-DPG).

Causes of hypophosphataemia are given in Fig.14.14. Although hyperphosphataemia is usual in diabetic ketoacidosis, hypophosphataemia is seen during the recovery phase when there is increased uptake of phosphate into depleted tissues. This is also the mechanism of hypophosphataemia seen in patients with malnutrition who are given a high calorie intake either enterally or parenterally. Hypophosphataemia is common during alcohol withdrawal and is multifactorial in origin. The causes include decreased intake, magnesium deficiency and, when it occurs, ketoacidosis.

Management

Hypophosphataemia should be anticipated, and prevented, in conditions where it may occur. It is treated by the administration of phosphate, either

203

Serum concentration				
Condition	**Calcium**	**Phosphate**	**Alkaline phosphatase**	**Other**
osteoporosis	N	N	N	–
osteomalacia	↓ or N	↓	↑ (↑ ↑ †)	–
Paget's disease	N(↑ *)	N	↑ ↑ ↑	–
renal osteodystrophy	↓ or N	↑	↑	↑ creatinine
primary hyperthyroidism	↑	N or ↓	N or ↑	–
secondary tumour deposits	N or ↑	↓ N or ↑	↑	–
* during immobilization † during early phase of recovery				

Fig.14.15 Biochemical changes in the serum in metabolic bone disease.

enterally or parenterally as appropriate but intravenous phosphate should not be given to a patient who is hypercalcaemic (except, very rarely, to treat the hypercalcaemia itself) or oliguric.

Metabolic bone disease

Various bone diseases are characterized by disordered metabolism of bone. They include osteomalacia and rickets, due usually to vitamin D deficiency, Paget's disease (see page 243), renal osteodystrophy (see page 65) and osteoporosis. In the latter there is a, usually generalized, loss of bone substance; serum calcium and phosphate concentrations and alkaline phosphatase activity are normal, in distinction to certain other conditions (Fig.14.15). However, osteoporosis and osteomalacia may coexist, particularly in the elderly.

Primary and secondary osteoporosis

Primary osteoporosis, that is, bone loss which is not related to other pathology, is a feature of normal ageing and tends to be more severe in women, particularly after the menopause. Secondary osteoporosis can occur in young people in association with a wide spectrum of disorders (Fig.14.16). The cause of primary osteoporosis is unknown. The rate of

new bone formation declines with age but in post-menopausal women, there is also an increase in bone resorption. This increased resorption may be related to decreased intestinal calcium absorption and increased calcium requirement. Oestrogen, and possibly calcitonin, deficiency appears to contribute to this. Established osteoporosis cannot be reversed, but oestrogen replacement in post-menopausal women and calcium supplementation in both men and women reduce the rate of bone loss.

Magnesium

Magnesium is the fourth most abundant cation in the body. The adult human body contains approximately 1000mmol, with about half in bone and the remainder distributed equally between muscle and other soft tissues. Only 15–29mmol is found in the ECF, the serum concentration being 0.8–1.2mmol/l. The normal daily intake of magnesium (10–12mmol) is greater than is necessary to maintain magnesium balance (approximately 8mmol/day) and the excess is excreted through the kidneys.

Urinary magnesium excretion is increased by ECF volume expansion, hypercalcaemia and hyper-magnesaemia, and decreased in the opposite of these states. Various hormones, including PTH and aldosterone, affect the renal handling of magnesium; the effects of aldosterone are probably secondary to

Causes of osteoporosis
Ageing especially post-menopausal
Endocrine premature ovarian failure thyrotoxicosis Cushing's syndrome diabetes mellitus hypogonadism
Drugs prolonged heparin treatment glucocorticoids alcoholism
Others immobilization malabsorption of calcium weightlessness

Fig.14.16   Causes of osteoporosis.

Magnesium deficiency
Causes

malabsorption, malnutrition and fistulae
alcoholism (chronic alcoholism and alcohol
    withdrawal)
cirrhosis
diuretic therapy (especially loop diuretics)
renal tubular disorders (in advanced renal
    disease hypermagnesaemia is usual)
chronic mineralocorticoid excess

Clinical features
tetany (with normal or decreased calcium) agitation ataxia, tremor, choreiform movements and   convulsions muscle weakness

Fig.14.17   Causes and clinical features of magnesium deficiency.

changes in ECF volume but PTH, which increases the tubular reabsorption of filtered magnesium, appears to act directly.

Magnesium has an important role as a cofactor in many enzyme reactions, including some involved in protein synthesis, ATPases and creatine kinase. It has an essential role in muscle contraction since a magnesium complex with ATP is the substrate for myosin ATPase.

Hypermagnesaemia

Significant hypermagnesaemia is uncommon. Cardiac conduction is affected at concentrations of 2.5–5.0mmol/l; very high concentrations (>7.5mmol/l) cause respiratory paralysis and cardiac arrest. Such extreme hypermagnesaemia may occasionally be seen in renal failure.

Intravenous calcium may give short-term protection against the adverse effects of hypermagnesaemia but in renal failure, dialysis may be necessary.

Hypomagnesaemia

The more common causes of magnesium deficiency are presented in Fig.14.17. Significant magnesium deficiency with hypomagnesaemia is also uncommon, although less so than hypermagnesaemia. Its clinical features are also summarized in Fig.14.17. Hypocalcaemia, due to decreased PTH secretion is an important consequence of hypomagnesaemia. Hypokalaemia is also sometimes present, but all three abnormalities usually respond to magnesium supplementation. Serum magnesium concentration tends to be measured far less frequently than calcium, even in conditions where deficiency might be expected, and it should always be measured when hypocalcaemia, or hypocalcaemic signs and symptoms, fail to respond to calcium supplementation alone.

Mild magnesium deficiency is treated by oral supplementation; in severe deficiency, and with malabsorption, magnesium may be given by slow intravenous infusion.

SUMMARY

Calcium has many functions in the body in addition to its obvious structural role in bones and teeth. It is, for example, essential for muscle contraction and its concentration affects the excitability of nerves; it is a second messenger, involved in the action of several hormones; and it is required for blood coagulation. About half the calcium in the serum is bound to protein; it is the unbound fraction that is physiologically active and whose concentration is closely regulated.

Two hormones have a central role in calcium homoeostasis. The main action of calcitriol, the hormone derived from vitamin D by successive hydroxylations in liver and kidney, is to stimulate calcium (and phosphate) uptake from the gut. Parathyroid hormone, secreted in response to a fall in serum ionized calcium concentration, stimulates calcitriol formation; it also stimulates calcium reabsorption from bone and by the renal tubules, and has a powerful phosphaturic action. These two hormones also regulate extracellular phosphate concentration. The role of calcitonin in calcium homoeostasis has yet to be established.

The common causes of hypercalcaemia are primary hyperparathyroidism, due to parathyroid adenomas or hyperplasia, and malignant disease, with or without metastasis to bone, including myeloma. Less common causes include sarcoidosis and overdosage with vitamin D or its derivatives. Mild hypercalcaemia is often asymptomatic; when more severe, clinical features may include bone and abdominal pain, renal calculi, polyuria, thirst and behavioural disturbances.

Hypocalcaemia causes hyperexcitability of nerve and muscle, leading to muscle spasm (tetany) and, in severe cases, to convulsions. Causes include vitamin D deficiency and hypoparathyroidism. Vitamin D deficiency may be either dietary in origin, often exacerbated by poor exposure to sunlight (and hence reduced endogenous synthesis), or due to malabsorption.

Vitamin D deficiency causes osteomalacia in adults and rickets in children, both diseases being characterized by defective bone mineralization. In osteoporosis, a condition particularly common in post-menopausal women, there is a generalized loss of both bone matrix and mineral. Paget's disease of bone is common in the elderly in both sexes; there is greatly increased osteoclastic bone reabsorption which stimulates new bone formation, but this is structurally abnormal and patients present with bone pain and deformity. End-stage renal disease also leads to a metabolic bone disease. This is multifactorial in origin; features of osteomalacia and hyperparathyroidism are usually present.

Magnesium is an essential cofactor for many enzymes. Its concentration in the extracellular fluid is controlled primarily through regulation of its urinary excretion. Hypomagnesaemia can cause clinical features similar to those of hypocalcaemia and indeed can cause hypocalcaemia since the secretion of parathyroid hormone is magnesium-dependent. Deficiency of magnesium can occur with prolonged diarrhoea and malabsorption. Hypermagnesaemia is common in renal failure but it appears to be tolerated well by the body and increased concentrations rarely give rise to obvious clinical disturbances.

FURTHER READING

Nordin BEC (ed.) (1984) *Metabolic Bone and Stone Disease*. 2nd edition. Edinburgh: Churchill Livingstone.

Heath D & Marx SJ (eds) (1982) *Calcium Disorders*. London: Butterworth.

15. Plasma Proteins

INTRODUCTION

Proteins are present in all body fluids, but it is the proteins of the blood plasma that are examined most frequently for diagnostic purposes. Over one hundred individual proteins have a physiological function in the plasma (Fig.15.1). Quantitatively, the single most important protein is albumin. The other proteins are known collectively as globulins. Changes in the concentrations of individual proteins occur in many conditions and their measurement can provide useful diagnostic information.

MEASUREMENT OF PLASMA PROTEINS

Total plasma protein

In very general terms, variations in plasma protein concentrations can be due to three changes: in the rate of protein synthesis; rate of removal, and in the volume of distribution.

The concentration of proteins in plasma is affected by posture; an increase in concentration of ten to twenty percent occurs within thirty minutes of becoming upright after a period of recumbency. Further, if a tourniquet is applied before vene-puncture, a significant rise in protein concentration can occur within a few minutes. In both cases, the fluctuation in protein concentration is caused by an increased diffusion of fluid from the vascular into the interstitial compartment. These effects must be borne in mind when blood is being drawn for the determination of protein concentration.

Only changes in the more abundant plasma proteins will have a significant effect on the total protein concentration; in practice, this means albumin and the immunoglobulins.

When patients have not been given blood or proteins intravenously, a short-term increase in the total plasma protein is always due to a decrease in the volume of distribution (in effect, to dehydration). A rapid decrease in concentration is most usually due to an increase in plasma volume. Thus, changes in

Functions of plasma proteins	
Function	**Example**
transport	thyroxine-binding globulin (thyroid hormones) apolipoproteins (cholesterol, triglyceride) transferrin (iron)
humoral immunity	immunoglobulins
maintenance of oncotic pressure	all proteins, particularly albumin
enzymes	renin clotting factors
protease inhibitors	α_1-antitrypsin (acts on proteases)
buffering	all proteins

Fig. 15.1 Functions of plasma proteins.

Causes of changes in total plasma protein concentration			
Increase		**Decrease**	
hypergammaglobulinaemia paraproteinaemia	↑ protein synthesis	malnutrition and malabsorption liver disease humoral immunodeficiency	↓ protein synthesis
artefactual	haemoconcentration due to stasis of blood during venepuncture	over-hydration increased capillary permeability	↑ volume of distribution
dehydration	↓ volume of distribution	protein-losing states catabolic states	↑ excretion/ catabolism

Fig. 15.2   Causes of changes in total plasma protein concentration.

plasma protein concentration can provide a valuable aid to the assessment of a patient's state of hydration.

The total protein concentration of plasma can also fall rapidly if capillary permeability increases, since protein will diffuse out into the interstitial space. This is seen, for example, in some patients with septicaemia and generalized inflammatory conditions. Causes of increased and decreased total plasma protein concentration are summarized in Fig.15.2.

Protein electrophoresis

This technique is widely used for the semi-quantitative assessment of serum proteins and is essential for the detection of paraproteins. Electrophoresis is usually performed on serum rather than plasma since the fibrinogen present in plasma produces a band in the β_2 region which might be mistaken for a paraprotein.

Electrophoresis, on cellulose acetate or agarose gel, separates the proteins into distinct bands: albumin, α_1- and α_2-globulins, β-globulins, and γ-globulins. The principal proteins comprising these groups are listed in Fig.15.3. A band due to prealbumin may be visible depending on the technique used. When present in excess, C-reactive protein and α-fetoprotein may occasionally be seen as discrete bands. Paraproteins, if present in the serum, characteristically migrate as discrete bands. The electrophoretic mobility of proteins, in particular albumin, may change when they bind drugs and other ligands, such as bilirubin, thereby producing additional bands.

Some of the most frequently seen electrophoretic patterns are shown in Fig.15.4. However, whether or not the demonstration of these patterns is of any value

in the diagnosis and management of the underlying disorders is debatable, with the important exception of paraproteins. For instance, pattern b in Fig.15.4 shows decreases in albumin and α_1- and γ-globulins, and increases in α_2-globulin and β-globulin (due to α_2-macroglobulin and apolipoprotein B). While this pattern is characteristic of the nephrotic syndrome, it is only seen in very severe cases and may also be seen in other protein-losing states.

Pattern c in Fig.15.4 shows a decrease in the γ-globulin band which may be seen in defects of humoral immunity involving IgG, the major component of the γ-globulins. However, because IgA and IgM are quantitatively minor components of the total γ-globulin, their concentrations can be low without the γ-globulin band appearing diminished.

In liver cirrhosis, a characteristic electrophoretic pattern may be seen (pattern d in Fig.15.4), with lowered albumin, a diffuse increase in γ-globulins, and β–γ-fusion (a merging of the β and γ bands due to the increase in IgA that occurs in some forms of the condition). This pattern is often only present in advanced cases and is of little diagnostic value.

A deficiency of α_1-antitrypsin may be detectable by electrophoresis (pattern e in Fig.15.4). This protein is the major component of the α_1-globulin band and, with deficiency, the band may be characteristically faint. Almost complete absence of α_1-antitrypsin may be seen, mainly in homozygotes for the Z gene, but the α_1-globulin band may appear normal in heterozygotes. Thus, a technique such as isoelectric focusing, which specifically identifies the variant proteins, must be used for the determination of phenotypes when screening the relatives of patients, in genetic counselling and in antenatal diagnosis.

Principal serum proteins		
Class	**Protein**	**Approximate mean serum concentration (g/l)**
	prealbumin	0.25
	albumin	40
α_1-globulin	α_1-antitrypsin α_1-acid glycoprotein	2.9 1.0
α_2-globulin	haptoglobins α_2-macroglobulin caeruloplasmin	2.0 2.6 0.35
β-globulin	transferrin low density lipoprotein complement components (C3)	3.0 1.0 1.0
γ-globulins	IgG IgA IgM IgD IgE	14.0 3.5 1.5 0.03 trace

Fig. 15.3 Principal serum proteins. Many other important proteins are present in only very low concentrations, for example, thyroxine-binding globulin, transcortin and vitamin-D binding globulin.

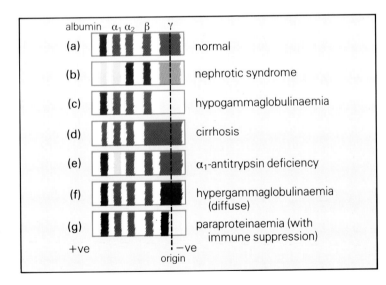

Fig. 15.4 Some typical serum electrophoretic abnormalities.

Pattern f in Fig.15.4 shows the electrophoretic appearance in diffuse hypergammaglobulinaemia, discussed later in this chapter as is the use of serum protein electrophoresis to detect paraproteins (pattern g in Fig.15.4), for which it is an essential technique.

The plasma proteins of clinical interest can all be measured using specific assays and, in general, such measurements will provide more precise diagnostic information. Preliminary electrophoresis is not helpful and may even be misleading. For example, IgA deficiency, the most frequently occurring congenital immunodeficiency disorder, is often not apparent on electrophoresis.

SPECIFIC PLASMA PROTEINS

Albumin

Albumin, the most abundant plasma protein, makes the greatest contribution to the oncotic pressure of plasma. Oncotic pressure is the osmotic pressure due to the presence of proteins and is an important factor in the distribution of extracellular fluid (ECF) between the intravascular and extravascular compartments.

In hypoalbuminaemic states, the decreased plasma oncotic pressure disturbs the equilibrium between plasma and interstitial fluid so that there is a decrease in the movement of interstitial fluid back into the blood at the venular end of the capillaries (Fig.15.5). The accumulation of interstitial fluid is seen clinically as oedema. The relative decrease in plasma volume results in a fall in renal blood flow. This stimulates the secretion of renin, and hence of aldosterone, through the formation of angiotensin (secondary aldosteronism, see page 134). This results in sodium retention and thus an increase in ECF volume which potentiates the oedema.

There are many possible causes of hypoalbuminaemia (Fig.15.6), a combination of which may be important in individual cases. For example, in a patient with malabsorption due to Crohn's disease, a low albumin may reflect both decreased synthesis (decreased supply of amino acids due to malabsorption) and increased loss (directly into the gut from ulcerated mucosa).

Hyperalbuminaemia may be either an artefact, for instance, as a result of venous stasis during blood collection, or due to an over-infusion of albumin or due to dehydration. There are no significant pathological causes of increased albumin synthesis.

Plasma albumin measurements are often used to assess a patient's response to nutritional support. Albumin is useless for this purpose in the short-term, though it may help with the assessment of fluid balance, since it has a half-life in the plasma of approximately twenty days. However, it is of use in the assessment of patients receiving long-term (several weeks or more) nutritional support.

The plasma albumin concentration is also a useful test of liver function (see page 80); it is often decreased in patients with chronic liver disease but, because of the relatively long half-life, the concentration is usually normal in acute hepatitis.

Albumin is a high capacity, low affinity transport protein for many substances, such as thyroid

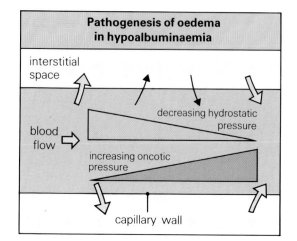

Fig. 15.5 Pathogenesis of oedema in hypoalbuminaemia. The normal balance of hydrostatic and oncotic pressures is such that there is net movement of fluid out of the capillaries at their arteriolar ends and net movement in at their venular ends (indicated here by arrows). Oedema can thus be due to: an increase in capillary hydrostatic pressure; a decrease in plasma oncotic pressure; or an increase in capillary permeability.

hormones, calcium and fatty acids. The influence of a low plasma albumin on measurement of thyroid hormones and calcium is considered on pages 142 and 191, respectively.

Many drugs are bound to albumin in the blood stream and a decrease in albumin concentration can have important pharmacokinetic consequences. Additionally, the binding of drugs may displace endogenous substances from albumin; for example, salicylates displace bilirubin and could precipitate kernicterus in an infant with unconjugated hyperbilirubinaemia (see page 78).

A number of molecular variants of albumin exist. In bisalbuminaemia, the variant protein has a slightly different electrophoretic mobility from normal albumin and a pair of albumin bands are seen on electrophoresis; there are no clinical consequences. Analbuminaemia is a rare, inherited condition where the plasma albumin concentration is 250mg/l or less. People with this condition tend to suffer episodic mild oedema but are otherwise well.

α_1-Antitrypsin

This α_1-globulin is a naturally occurring inhibitor of proteases. Its significance is related to the clinical consequences of inherited disorders of α_1-antitrypsin synthesis. Associated conditions are emphysema, occurring at a younger age (third and fourth decades) than is usual for this condition, and neonatal hepatitis which usually progresses to cirrhosis.

Homozygotes for the normal protein are termed Pi (protease inhibitor) MM, and over thirty alleles of the gene have been described. α_1-Antitrypsin deficiency is most frequently due to homozygosity for the Z allele (Pi ZZ), this genotype having a frequency of about one in three thousand in the United Kingdom. In affected individuals, the serum α_1-antitrypsin level is reduced to between ten and fifteen percent of normal. The defect is due to a single amino acid substitution which interferes with the attachment of carbohydrate residues to the protein; it is probably this structural modification that impedes the secretion of the protein from the liver and causes liver damage.

The development of emphysema is believed to be due to a lack of natural inhibition of the enzyme, elastase, which results in destructive changes in the lung. It should be emphasized, however, that not all PiZZ homozygotes develop liver or lung disease, so other factors must obviously be involved in the pathogenesis of these conditions.

PiMZ heterozygotes have serum α_1-antitrypsin levels which are about sixty percent of normal; there is probably only a very slightly increased tendency for these individuals to develop lung disease, when compared with the normal PiMM homozygotes. Neither homozygotes for the other relatively common alleles, F and S (that is, PiFF and PiSS) nor heterozygotes (PiMF, PiMS), appear to be at increased risk of developing liver or lung disease, although PiSZ heterozygotes seem to show some susceptibility.

Accurate phenotyping is required for the screening of an affected individual's family members and for antenatal diagnosis. This involves the use of special techniques, such as isoelectric focusing, to allow identification of individual proteins.

α_1-Antitrypsin is an acute phase protein. Its concentration increases in acute inflammatory states and this may be sufficient to bring a genetically determined low level of the protein, for example, in a PiMZ heterozygote, into the normal range. However, even with an acute phase response, the α_1-antitrypsin level in PiZZ homozygotes never rises above fifty percent of the lower limit of the normal range.

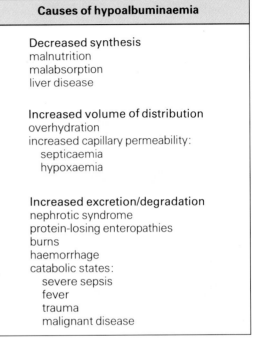

Causes of hypoalbuminaemia

Decreased synthesis
malnutrition
malabsorption
liver disease

Increased volume of distribution
overhydration
increased capillary permeability:
    septicaemia
    hypoxaemia

Increased excretion/degradation
nephrotic syndrome
protein-losing enteropathies
burns
haemorrhage
catabolic states:
    severe sepsis
    fever
    trauma
    malignant disease

Fig. 15.6   Causes of hypoalbuminaemia.

Haptoglobin

Haptoglobin is an α_2-globulin. Its function is to bind free haemoglobin released into the plasma during intravascular haemolysis. The haemoglobin–haptoglobin complexes formed are removed by the reticuloendothelial system and the concentration of haptoglobin falls correspondingly. Thus, a low serum haptoglobin level is indicative of intravascular haemolysis. However, low levels due to decreased synthesis are also seen in chronic liver disease, metastatic disease and severe sepsis.

Haptoglobin is an acute phase protein and its concentration also increases in hypoalbuminaemic states such as the nephrotic syndrome. It demonstrates considerable genetic polymorphism; the molecule consists of pairs of two types of subunit, α and β, and whilst the β-chain is constant, there are three alleles for the α-chain. However, as far as is known, these different proteins are functionally similar and their existence is of no clinical significance.

α_2-Macroglobin

α_2-Macroglobin is a high molecular weight protein (820,000 daltons) that constitutes approximately one-third of the α_2-globulins. Its increased synthesis in hypoalbuminaemic states contributes to the increased α_2-globulin band seen on electrophoresis (see pattern b in Fig.15.4).

Caeruloplasmin

A deficiency of this copper-carrying α_2-globulin is characteristic of Wilson's disease. Its concentration is increased in pregnancy and by oestrogen-containing oral contraceptives. It is also an acute phase protein.

Transferrin

This β-globulin is the major iron-transporting protein in the plasma; normally about thirty percent saturated with iron, it is characteristically one hundred percent saturated with iron in haemochromatosis. It is useful in the assessment of a patient's response to nutritional support (see *Chapter 23*).

Transferrin and ferritin are discussed in more detail in *Chapter 19*. Ferritin is also an iron-carrying protein and measurement of its serum concentration is the best simple test now available for determining body iron stores.

Other acute phase proteins

Characteristic changes in some serum proteins occur in clinical conditions where there is an acute inflammatory response, for example, trauma, burns, myocardial infarction and exacerbations of inflammatory joint disease. Characteristically, there is an increase in α_1- and α_2-globulins as a result of increases in the synthesis of α_1-antitrypsin, α_1-acid glycoprotein and haptoglobin. At the same time, there is often a rapid fall in serum albumin concentration, due primarily to redistribution as a result of increased vascular permeability. Falls in the concentrations of prealbumin and transferrin are also characteristic.

Similar changes may be seen in malignancy and in chronic inflammatory conditions, but more usually the α_1-globulin remains normal, the α_2-globulin is somewhat elevated and there is a diffuse increase in γ-globulins.

Although these changes, if gross, are readily apparent on electrophoresis of serum, this method is an insensitive means of monitoring the acute phase response. During this response, the concentration of C-reactive protein (a glycoprotein with α_2 mobility, quantitatively a minor component of the serum proteins) may increase by as much as thirty-fold. Sufficiently sensitive methods of measuring C-reactive protein are now available for it to be the test of choice in monitoring the acute phase response, of particular value in monitoring patients with inflammatory joint disease such as rheumatoid arthritis.

Other plasma proteins

Measurement of other plasma proteins may provide useful information in particular circumstances. Measurement of clotting factors (fibrinogen, factor VIII and others) is usually carried out in haematology laboratories and is essential in the investigation of some bleeding disorders. Measurement of the proteins of the complement system is of considerable value in the investigation of some diseases with an immunological basis. The apolipoproteins and tumour markers such as α-fetoprotein are considered in detail in *Chapters 16* and *21*, respectively. The importance of hormone-binding proteins, such as cortisol-binding globulin and sex hormone-binding globulin, is considered in *Chapters 9* and *11*, respectively. Serum proteins used in the assessment of nutritional status are discussed in *Chapter 23*.

IMMUNOGLOBULINS

The immunoglobulins are a group of plasma proteins that function as antibodies, recognizing and binding foreign antigens. This facilitates the destruction of these antigens by elements of the cellular immune system.

Since every immunoglobulin molecule is specific for one antigenic determinant, or epitope, there are vast numbers of different immunoglobulins. All share a similar basic structure (Fig.15.7) consisting of two identical 'heavy' polypeptide chains and two identical 'light' chains, linked by disulphide bridges. There are five types of heavy chain (γ, α, μ, δ, ϵ) and two types of light chain ($\varkappa$, λ), the immunoglobulin class being determined by the type of heavy chain that the molecule contains (Fig.15.8).

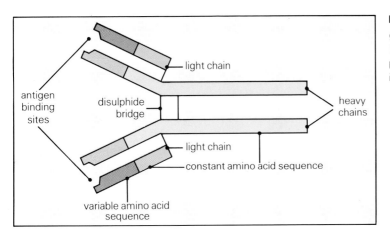

Fig. 15.7 Structure of immuno-globulins. It is basically similar in all immunoglobulins. IgM consists of a pentamer of the basic structure and IgA is secreted as a dimer.

		Characteristics of the immunoglobulins		
Class	**Heavy chain**	**Mean serum concentration (g/l)**	**Molecular weight (daltons)**	**Function**
IgG	γ	14.0	146,000	the major antibody of secondary immune responses
IgA	α	3.5	160,000	secreted as a dimer (molecular weight 385,000 daltons) the major antibody in seromucous secretions, e.g. saliva, bronchial mucus
IgM	μ	1.5	970,000	a pentamer, confined to the vascular spaces the major antibody of the primary immune response
IgD	δ	0.03	184,000	present on the surface of B lymphocytes ? involved in antigen recognition
IgE	ε	trace	188,000	present on surface of mast cells and basophils probable role in immunity to helminths and associated with immediate hyper-sensitivity reactions

Fig. 15.8 Characteristics of the immunoglobulins. Immunoglobulins of each class contain either ϰ- or λ-light chains. In IgA and IgG, slight variations in the structure of the constant regions of the heavy chains give rise to different subclasses.

The N-terminal amino acid sequence of both the heavy and light chains shows considerable variation between individual immunoglobulin molecules; these form the part of the immunoglobulin molecule responsible for recognition of the antigen (the antigen binding site). The amino acid sequence of the rest of the chain varies little within one immunoglobulin class; this constant part of the molecule is concerned with complement activation and interaction with the cellular elements of the immune system. The characteristics and functions of the immunoglobulins are summarized in Fig.15.8.

On electrophoresis the immunoglobulins behave mainly as γ-globulins but IgA and IgM may migrate with the β- or α₂-globulins. Because the normal serum concentration of IgG is much higher than that of the other immunoglobulins, the γ-globulin band seen on electrophoresis of normal serum is largely due to IgG.

Abnormal immunoglobulin concentrations in the plasma may show an increase or a decrease, which may be either physiological or pathological in origin.

Hypogammaglobulinaemia

Physiological causes

At birth IgA and IgM levels are low and rise steadily thereafter (Fig.15.9), although IgA may not reach the normal adult level until the end of the first decade. IgG is transported across the placenta during the last trimester of pregnancy and levels are high at birth (except in premature infants). The IgG concentration then declines, as maternal IgG is cleared from the body, before rising again as it is slowly replaced by the infant's own IgG.

Physiological hypogammaglobulinaemia is one of the reasons for the susceptibility of infants (especially the premature) to infection.

Pathological causes

Various inherited disorders of immunoglobulin synthesis are known, ranging in severity from X-linked agammaglobulinaemia (Bruton's disease), in which there is a complete absence of immunoglobulins and affected children develop recurrent bacterial infections, to milder dysgammaglobulinaemias, in which there is a defect or partial defect of only one or two immunoglobulins. The commonest of these, IgA deficiency, has an incidence of about one in four hundred.

Hypogammaglobulinaemia may also be acquired. It commonly occurs in haematological malignancies, such as chronic lymphatic leukaemia, multiple myeloma and Hodgkin's disease. It can be a complication of the use of cytotoxic drugs and is a feature of severe protein losing states, for example, the nephrotic syndrome.

Measurement of the specific class of immunoglobulin is essential for the diagnosis of hypogammaglobulinaemia. As previously discussed, electrophoresis is not sufficient for this purpose since the

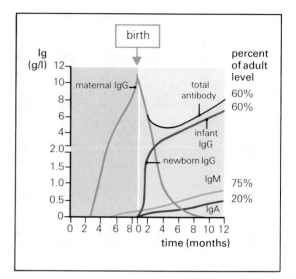

Fig. 15.9 Changes in serum immunoglobulin concentrations with age.

normal concentrations of the immunoglobulins, with the exception of IgG, are low and the effect of any decrease on the γ-globulin peak is too small to be detectable. IgG deficiency can be inferred if the γ-globulin band is faint, but possible coexistent deficiencies of other immunoglobulins will be missed.

Hypergammaglobulinaemia

Physiological causes

Increased levels of immunoglobulins are seen in both acute and chronic infections. Measurement of a particular immunoglobulin class, such as IgM, is of no value diagnostically. However, the detection and measurement of an immunoglobulin directed against a specific antigen provides an important aid to the diagnosis of many infectious diseases. Such investigations are usually performed in departments of medical microbiology.

Pathological causes

Increases in serum immunoglobulin concentrations are common in autoimmune diseases, for example, rheumatoid disease and systemic lupus erythematosus

Diagnostic criteria of benign paraproteinaemia
no clinical features of myeloma or associated disorder
no suppression of normal immunoglobulins
no lytic lesions in bone on radiography
normal bone marrow
paraprotein concentration <10g/l
no Bence–Jones proteinuria
no increase in paraprotein concentration with age
no positive evidence of malignancy on follow-up (at least three years)

Fig. 15.10   Diagnostic criteria for benign paraproteinaemia.

(SLE), and in chronic liver diseases, some of which have an autoimmune basis.

The quantification of immunoglobulin classes is rarely of diagnostic use in such conditions, although the measurement of specific auto-antibodies, such as the rheumatoid factor, an IgM directed against the body's own IgG, is of immense diagnostic value in many autoimmune diseases. Many different immunoglobulins are produced in these conditions and they give rise to a diffuse (polyclonal) increase in the γ-globulin band on electrophoresis (see Fig.15.4, pattern f).

Paraproteins

A paraprotein is an immunoglobulin produced by a single clone of B-cells. Since all the molecules are identical, the paraprotein is seen on electrophoresis as a discrete band, usually in the γ-region (see Fig.15.4, pattern g). The band may migrate elsewhere, particularly if the protein is an IgA or IgM, or if complex formation with another serum protein has occurred. More than one paraprotein band may occasionally be seen; this may be due to dimerization, as frequently occurs with IgA paraproteins, or to the presence of complexes and fragments of paraproteins in addition to the intact molecule.

If plasma is electrophoresed, the presence of a fibrinogen band may mimic or mask a paraprotein. Even in serum, a paraprotein may be missed on electrophoresis if, as occasionally happens, it coincides exactly with a normal band, for example, α_2-globulin.

Paraproteins are characteristic of malignant diseases of B-cells, the most common of which is multiple myeloma. Others include Waldenström's macroglobulinaemia and Franklin's (heavy chain) disease. Paraproteins are occasionally seen in other malignancies of B-cell origin, such as chronic lymphatic leukaemia.

While serum protein electrophoresis is essential for the detection of paraproteins, the urine must also be examined. In approximately twenty percent of cases of myeloma, the tumour produces light chains only. Since these are of low molecular weight they are cleared rapidly from the blood stream and may be undetectable in serum. They are, however, detectable in urine; immunoglobulin light chains found in the urine are known as Bence–Jones protein and this is present in some fifty percent of all cases of myeloma.

Paraproteins can also be benign, that is, not associated with malignant disease. The incidence of benign paraproteinaemias increases with age and has been reported to be as high as three percent in people over the age of seventy.

Although benign paraproteins occur frequently, especially in the elderly, this diagnosis should not be made without vigorous investigation to exclude malignancy (Fig.15.10). The most definitive diagnostic criterion is a failure of the paraprotein concentration to increase with time and this necessitates a regular follow-up of the patient. There are no absolute criteria; the diagnosis of a benign paraprotein is essentially one of exclusion.

CASE HISTORY 15.1

A seventy-year-old man presented with back pain and loss of weight. Although a non-smoker, he had had several recent infections and was increasingly short of breath on exercise. On examination, he was anaemic but there were no other obvious abnormalities.

Investigations

serum: sodium   130mmol/l
      urea   15.3mmol/l
      creatinine 212μmol/l
      calcium   2.75mmol/l
      total protein   85g/l
      albumin   30g/l
      urate   0.51mmol/l

ESR (in first hour)   >100mm
haemoglobin   8.5g/dl

A blood film showed normochromic, normocytic anaemia; rouleaux were present on the blood film and there was increased background staining.

Serum protein electrophoresis revealed a paraprotein in the γ-globulin region (Fig.15.4, pattern g); this was typed by immunofixation and shown to be IgG-ϰ. There was a decrease in the normal γ-globulin band. Bence–Jones protein was present in the urine and identified as the ϰ type.

Radiological examination showed the typical punched-out lytic lesions of myeloma in the lumbar vertebrae, ribs and pelvis.

Comment

This is a typical presentation of multiple myeloma. The paraprotein is an IgG in fifty percent of cases (Fig.15.11). Replacement of normal bone marrow by malignant plasma cells frequently results in anaemia and in decreased synthesis of normal immunoglobulins.

The diagnosis rests on the demonstration of any two of the following: the presence of a paraprotein; typical radiological appearances; the presence of increased numbers of abnormal plasma cells in the bone marrow. However, it is normal practice to confirm the diagnosis by examination of a bone marrow smear, even if the diagnosis is already obvious. Occasionally, if the marrow involvement is not widespread, an aspirate may not contain any abnormal cells.

The presence of the paraprotein causes red cells to adhere to each other (rouleaux formation) and may be

Paraproteins in myeloma	
Protein	**Incidence (%)**
IgG	55
IgA	22
IgD	1.5
Bence–Jones	50
Bence–Jones only	20

Fig. 15.11 Paraproteins in myeloma. IgE and IgM myelomas occur, but are very rare. In about 1% of all cases, no paraprotein can be detected.

sufficient to cause an increase in the background staining of the blood film. Hyponatraemia often occurs in sick people and in hyperproteinaemic states it may be 'spurious' in character (see page 20).

Renal failure is the cause of death in approximately one-third of patients with myeloma. This is often multifactorial in origin; contributory factors include obstruction of nephrons by protein, hypercalcaemia, hyperuricaemia, pyelonephritis and amyloid. Hypercalcaemia is common in myeloma; its cause is discussed elsewhere (see pages 196–200 and 276). Hyperuricaemia may be due partly to increased uric acid synthesis from tumour nucleic acids and partly to renal impairment.

Despite the extensive lytic lesions of bone, there is no increase in osteoblastic activity and the serum alkaline phosphatase is usually normal. The laboratory findings in myeloma are summarized in Fig.15.12.

The amount of paraprotein produced in myeloma is a reflection of tumour mass and may be used as a tumour marker. It should be appreciated that metabolic abnormalities may not be present when the condition is first diagnosed; they may develop subsequently so patients should be periodically monitored for these complications.

Myeloma is treated using cytotoxic drugs but the prognosis is poor. Local radiotherapy may be useful for isolated lesions (plasmacytomas) and for localized bone pain.

Waldenström's macroglobulinaemia is also a B-cell tumour. The paraprotein is an IgM and a hyperviscosity syndrome, causing sludging of red cells

---

Laboratory findings in multiple myeloma
Biochemical
paraprotein in serum
paraprotein in urine (Bence–Jones protein)
↑ serum:
urea, creatinine,
calcium and urate
normal alkaline phosphatase
Haematological
↑ erythrocyte sedimentation rate (ESR)
anaemia (usually normochromic, normocytic)
rouleaux formation

Fig. 15.12 Laboratory findings in multiple myeloma.

in capillaries and predisposing to thrombus formation, is a prominent feature. It is much less common than myeloma.

Rarer still is Franklin's (heavy chain) disease, in which the paraprotein produced is immunoglobulin heavy chain only. This is usually an α-chain, but may also be a γ- or μ-chain. Patients with α-chain disease present with malabsorption due to infiltration of the gut by malignant cells.

Some paraproteins precipitate out of solution when cooled to 4°C and redissolve on warming. These proteins are known as cryoglobulins and are associated with Raynaud's phenomenon, although the majority of patients with this condition do not have cryoglobulinaemia. The latter may also occur in other conditions where there are abnormalities of immunoglobulin production, for instance, systemic lupus erythematosus.

PROTEINS IN OTHER BODY FLUIDS

Cerebrospinal fluid

Cerebrospinal fluid (CSF) is usually obtained for diagnostic purposes by lumbar puncture. The protein concentration is normally 0.1–0.4g/l and the protein is predominantly albumin; higher concentrations are found in neonates (up to 0.9g/l) and the elderly. It is important that the CSF is not contaminated with blood since the presence of plasma proteins will completely invalidate the results of CSF protein measurement.

Examination of the CSF is most often performed in cases of suspected meningitis. The diagnosis of this is primarily the concern of the medical microbiologist, but it is usual also to request biochemical analysis for glucose and protein. The significance of the CSF glucose concentration is considered in *Chapter 12*. In meningitis, there is secretion of IgG into the CSF but this has little effect on the total amount of protein. However, meningeal inflammation may lead to an increase in capillary permeability and, therefore, a marked increase in CSF protein content. It is important to note that, in suspected meningitis, a normal CSF protein does *not* exclude the diagnosis.

CSF protein concentration is increased in patients with tumours of the central nervous system and may exceed 5g/l in patients with tumours which obstruct the normal circulation of the CSF (spinal block or Froin's syndrome).

Examination of the CSF can be of great value in the diagnosis of multiple sclerosis. Although the total protein concentration is usually only slightly raised, the ratio of IgG to albumin is increased from less than ten percent to as much as fifty percent. Greater sensitivity is provided if the IgG/albumin ratio of the CSF is compared with that of plasma. The ratio is abnormal in approximately eighty percent of cases of multiple sclerosis but may also be abnormal in neurosyphilis, with tumours of the central nervous system and after cerebrovascular accidents.

An even more sensitive test is provided by electrophoresis of CSF on polyacrylamide gel. In multiple sclerosis, only a small number of clones of B-cells produce IgG which is seen as discrete 'oligoclonal' bands when CSF is electrophoresed. The methodology is technically demanding and considerable experience is necessary for interpretation of the results. Oligoclonal bands may be detected in over ninety-five percent of cases of multiple sclerosis, although they may also be seen in other, less common, demyelinating diseases, such as subacute sclerosing panencephalitis, and in neurosyphilis.

Transudates and exudates

The protein concentration of pleural fluid or abdominal ascites is occasionally measured to determine whether the sample is a transudate (fluid with a low protein content derived by filtration across capillary endothelium) or an exudate (fluid with a high protein content actively secreted in response to infection). A value of 30g/l is usually taken as the dividing line between the two types of fluid, but this is not a reliable criterion as the protein content of both is very variable.

The important diagnostic differentiation is whether the ascites or pleural fluid is infected or if it is related to the presence of a tumour. This can only be determined by microbiological and cytological examination, so protein measurement is of little value.

Urine

The investigation and significance of proteinuria is discussed in *Chapter 5*.

SUMMARY

The most abundant protein in plasma is albumin, which is synthesized in the liver. Through its contribution to the colloid osmotic pressure, albumin has an important role in determining the distribution of the extracellular fluid between the vascular and extravascular spaces. It is also an important transport protein for several hormones, drugs, free fatty acids, unconjugated bilirubin and various ions. Its concentration is, however, affected by so many pathological processes (decreases occur in chronic liver disease, protein-losing states, malabsorption, following trauma and when capillary permeability is increased) that measurements must be interpreted with caution.

Most of the other plasma proteins are classified as globulins. The immunoglobulins are synthesized by plasma cells and constitute the humoral arm of the immune system. Five main classes are known of which the most abundant are IgG, IgM and IgA. IgM is the main antibody of the primary immune response and is largely confined to the vascular compartment; IgG is involved in the secondary response and is distributed throughout the extracellular fluid; IgA is secreted onto mucosal surfaces. An increase in total immunoglobulins is characteristic of chronic inflammatory conditions, for example, chronic infections and chronic autoimmune disease. The measurement of specific immunoglobulins is of value in the investigation of immunodeficiency syndromes and certain autoimmune diseases.

Myelomas are malignant tumours of plasma cells which produce large amounts of identical, monoclonal, immunoglobulin molecules or fragments thereof, known as paraproteins. Serum urine and protein electrophoresis is essential for the detection of paraproteins but other abnormalities of plasma proteins are better investigated by specific measurement of the protein, or proteins, in question. Metabolic features of myeloma include renal impairment, hypercalcaemia and hyperuricaemia. Patients are frequently anaemic and may have an immune paresis. Causes of death include infection and renal disease.

Other plasma proteins include the clotting factors, complement components and various transport proteins, for example, thyroxine-binding globulin, transcortin, sex hormone-binding globulin, transferrin and caeruloplasmin. Increases in the concentration of certain proteins occur in association with acute inflammatory reactions. These 'acute phase proteins' include α_1-antitrypsin, C-reactive protein and the haptoglobins. Measurement of C-reactive protein is valuable in following the course of conditions characterized by episodes of acute inflammation, such as rheumatoid arthritis and Crohn's disease. α_1-antitrypsin is a protease inhibitor; inherited deficiency of the protein can cause neonatal hepatitis, which may progress to cirrhosis, and to emphysema in adults, particularly those who smoke. The condition can now be detected antenatally, by examination of fetal blood.

The investigation of proteins in cerebrospinal fluid is a valuable technique in the diagnosis of multiple sclerosis; the presence of oligoclonal immunoglobulin bands is characteristic, although not specific, to this condition. Total cerebrospinal fluid protein is frequently measured in patients with suspected meningitis but is of limited diagnostic value.

FURTHER READING

Keyser JW (1979) *Human Plasma Proteins*. Chichester: John Wiley & Sons Ltd.

Whicher JT (1983) Abnormalities of plasma proteins. In *Biochemistry in Clinical Practice*. Edited by Williams DL & Marks V. pp. 221–251. London: Heinemann.

16. Lipids and Lipoproteins

INTRODUCTION

The major lipids present in the plasma are fatty acids, triglycerides, cholesterol and phospholipids. Other lipid-soluble substances, present in much smaller amounts but of considerable physiological importance, include steroid hormones and fat-soluble vitamins but these are discussed in *Chapters 9 and 23* respectively.

TRIGLYCERIDES, CHOLESTEROL AND PHOSPHOLIPIDS

Triglycerides consist of glycerol esterified with three long-chain fatty acids, such as stearic (18 carbon atoms) and palmitic (16 carbon atoms) acids. These, together with the free fatty acids (FFA), are important energy sources in the body. Although the majority of fatty acids are saturated, certain unsaturated fatty acids are important as precursors of prostaglandins and in the esterification of cholesterol. Both saturated and unsaturated fatty acids are important components of cell membranes.

Cholesterol is also important in membrane structure and is the precursor of steroid hormones and bile acids.

Phospholipids are compounds similar to the triglycerides but with one fatty acid residue replaced by phosphate and a nitrogenous base.

Because they are not water-soluble, lipids are transported in the plasma in association with proteins. Albumin is the principal carrier of FFA while the other lipids circulate in complexes known as lipoproteins. These consist of a non-polar core of triglyceride and cholesterol esters surrounded by phospholipids, cholesterol and proteins known as apolipoproteins (Fig.16.1). The latter are important both structurally and in the metabolism of lipoproteins (Fig.16.2).

CLASSIFICATION OF LIPOPROTEINS

Lipoproteins are classified on the basis of their densities as demonstrated by their ultracentrifugal separation. In order of increasing density, these are

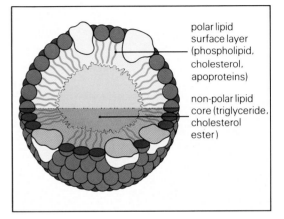

Fig.16.1 Diagram showing the composition of a lipoprotein particle. A segment has been removed to reveal the non-polar core of cholesterol ester and triglyceride surrounded by phospholipids and apoprotein.

Apolipoprotein	Function
AI	cofactor for LCAT structural (in HDL)
AII	inhibitor of LCAT and hepatic lipase? structural (in HDL)
AIV	?
B-100	structural (in LDL and VLDL) receptor binding
B-48	structural (in chylomicrons)
CI	cofactor for LCAT
CII	activator of LPL
CIII	inhibitor of LPL?
E	receptor binding

Fig.16.2 Functions of the major apolipoproteins.

Classification of apolipoproteins			
Svedberg coefficient (SF)		**Density**	**Diameter (A)**
>400	CM	<0.96	750–10,000
20–400	VLDL	0.960–1.006	300–800
0–20	LDL	1.006–1.063	200–220
	HDL	1.063–1.210	75–100

Fig.16.3 Classification of lipoproteins.

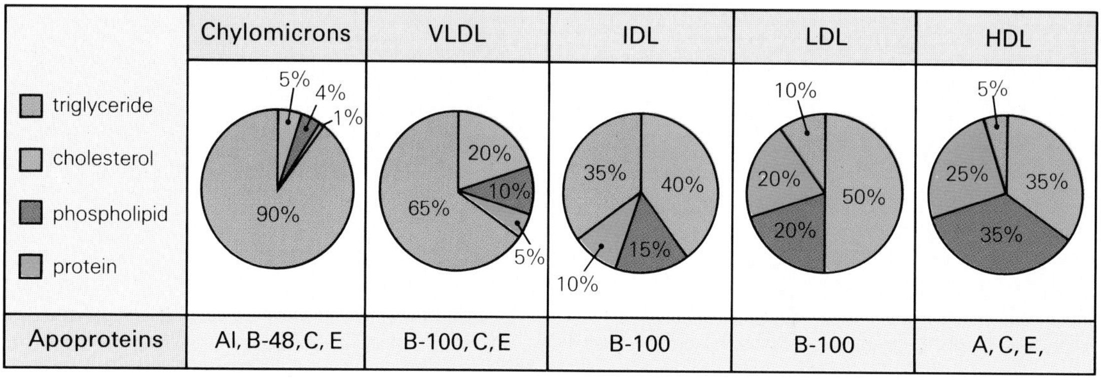

Fig.16.4 Composition of lipoproteins; although the composition in each class is similar, the particles are heterogeneous so that the percentages given are approximate. Only the principal apoproteins are shown.

Lipoprotein	Function
chylomicrons	transport of exogenous triglyceride
VLDL	transport of endogenous triglyceride
LDL	transport of cholesterol from liver to other tissues
HDL	transport of cholesterol from peripheral tissues and other lipoproteins to liver

Fig.16.5 Principal functions of the lipoproteins.

chylomicrons (CM) and lipoproteins of very low density (VLDL), intermediate density (IDL), low density (LDL), and high density (HDL). HDL can be separated, on the basis of density, into two metabolically distinct subtypes, HDL2 (density 1.063–1.125) and HDL3 (density 1.126–1.210). IDL are normally present in the blood stream in only small amounts but can accumulate in pathological disturbances of lipoprotein metabolism. This classification is illustrated in Fig.16.3 and the approximate lipid and apolipoprotein contents in Fig.16.4. However, it is important to appreciate that the composition of the circulating lipoproteins is not static. They are in a dynamic state with continuous exchange of components between the various types. The principal functions of these lipoproteins are summarized in Fig.16.5 and discussed in greater detail in the next section.

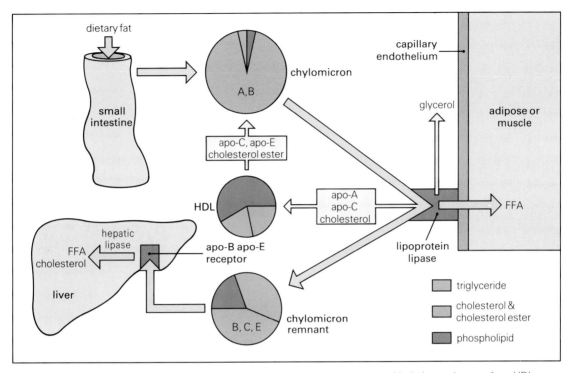

Fig.16.6 Chylomicrons transport dietary triglyceride to tissue where it is removed by the action of lipoprotein lipase. The resulting remnant particles are removed by the liver. They bind to apo-E/apo-B receptors on hepatic cells, are internalized and catabolized. Apolipoproteins A and B are synthesized in intestinal cells; apo-C and apo-E are acquired, together with cholesterol esters, from HDL. Apolipoprotein CII activates lipoprotein lipase. As triglyceride is removed from chylomicrons, apo-A, apo-C, cholesterol and phospholipids are released from their surfaces and transferred to HDL where the cholesterol is esterified.

LIPOPROTEIN METABOLISM

Chylomicrons

Chylomicrons (Fig.16.6) are formed from dietary fat (principally triglyceride, but also cholesterol) in enterocytes; they enter the lymphatics and reach the systemic circulation via the thoracic duct. Chylomicrons are the major transport form of exogenous (dietary) fat. Triglyceride constitutes about ninety percent of the lipid. Triglyceride is removed from chylomicrons by the action of the enzyme lipoprotein lipase (LPL), located on the luminal surface of the capillary endothelium of adipose tissue, skeletal and cardiac muscle and lactating breast, so that free fatty acids are delivered to these tissues either to be used as energy substrates or, after re-esterification to triglyceride, for energy storage.

LPL is activated by apolipoprotein C (apo-CII). Apolipoproteins A and B (apo-A, apo-B) are synthesized in enterocytes and are present in newly formed chylomicrons. Apolipoproteins C and E (apo-C, apo-E), together with cholesterol esters, are transferred to chylomicrons from HDL. As triglyceride is stripped off the chylomicrons, they become smaller; cholesterol, phospholipids, apo-A and some apo-C are released from the surfaces of the particles and taken up by HDL. The resultant particles, known as chylomicron remnants, are cleared from the circulation mainly by the liver; this process involves recognition of apo-E and apo-B by specific hepatic receptors. Dietary cholesterol transported in chylomicrons reaches the liver in the remnant particles. Under normal circumstances, chylomicrons cannot be detected in plasma after a twelve-hour fast.

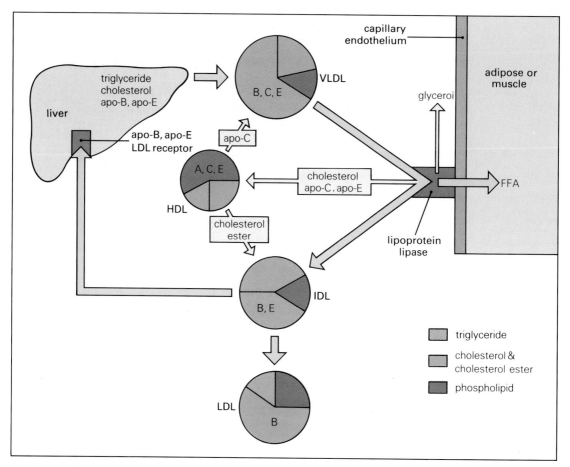

Fig.16.7 VLDL are synthesized in the liver and transport endogenous triglyceride from the liver to other tissues where it is removed by the action of lipoprotein lipase. At the same time, cholesterol, phospholipids and apo-C and apo-E are released and transferred to HDL. By this process VLDL are converted to IDL. Cholesterol is esterified in HDL and cholesterol ester is transferred to IDL, increasing their density so that they become LDL. Thus the triglyceride-rich VLDL are the precursors of LDL, which comprise mainly apo-B and cholesterol ester. In addition, some IDL is removed from the circulation by the liver.

Very low density lipoproteins

Very low density lipoproteins (Fig.16.7) are formed from triglyceride synthesized in the liver either *de novo* or by re-esterification of free fatty acids. VLDL also contain some cholesterol, apo-B, apo-C and apo-E; the apo-E and some of the apo-C is transferred from circulating HDL.

VLDL are the principal transport form of endogenous triglyceride and share a similar fate to chylomicrons, triglyceride being stripped off by the action of LPL. As the VLDL particles become smaller, phospholipids, free cholesterol and apolipoproteins are released from their surfaces and taken up by HDL, thus converting the VLDL to denser particles, IDL. Cholesterol that has been transferred to HDL is esterified and the cholesterol ester is transferred back to IDL, some of which become converted to LDL, composed largely of cholesterol esters and apo-B, together with some phospholipid. Some IDL are not converted to LDL but taken up by the liver via LDL receptors. Under normal circumstances, there are very few IDL in the circulation, because of their rapid removal or conversion to LDL.

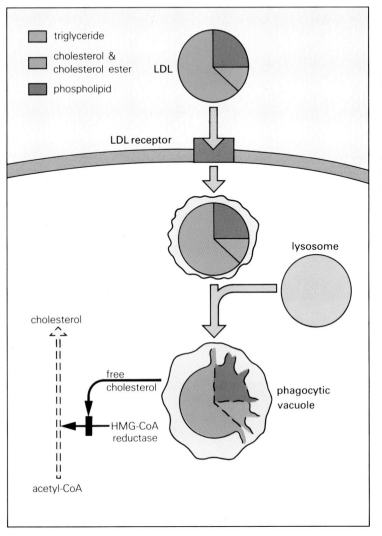

triglyceride

cholesterol & cholesterol ester

phospholipid

LDL

LDL receptor

lysosome

cholesterol

free cholesterol

phagocytic vacuole

HMG-CoA reductase

acetyl-CoA

Fig.16.8 LDL are derived mainly from VLDL (via IDL) although some is produced in the liver, this amount being greater in patients with familial hypercholesterolaemia. LDL are taken up by the liver and other tissues in a receptor-dependent process and hydrolyzed by lysosomal enzymes. Free cholesterol regulates cholesterol synthesis by inhibiting the rate-limiting enzyme, HMG-CoA reductase. LDL are also removed by a non-receptor-mediated process, particularly when present in high concentration.

Low density lipoproteins

Low density lipoproteins (Fig.16.8) are the vehicle for the transport of cholesterol from the liver to peripheral tissues. LDL can pass through the junctions between capillary endothelial cells and attach to specific LDL-receptors on cell membranes that recognize apo-B. This is followed by internalization and lysosomal degradation with release of free cholesterol. Cholesterol can also be synthesized in these tissues but the rate-limiting enzyme, HMG-CoA reductase (hydroxymethylglutarate-CoA reductase),

is in fact inhibited by cholesterol so, in the average adult, cholesterol synthesis in peripheral cells probably does not occur.

The LDL receptors are saturable and subject to down-regulation, but non-receptor-mediated LDL degradation by macrophages can also occur and has been shown to account for two-thirds of cholesterol uptake at a 'normal' serum cholesterol and even more at higher levels. This uncontrolled uptake of cholesterol is thought to be a factor of importance in the pathogenesis of atherosclerosis. When macrophages become overloaded with cholesterol esters, they are converted to 'foam cells', the classic components of atheromatous plaques. In human neonates, serum LDL levels are much lower than in adults and cellular cholesterol uptake is probably all receptor-mediated and controlled.

High density lipoprotein

High density lipoprotein (Fig.16.9) is synthesized primarily in the liver and, to a lesser extent, in small intestinal cells as a precursor ('nascent HDL') comprising phospholipid, cholesterol, apo-E and apo-A. Nascent HDL is disc-shaped; in the circulation, it acquires apo-C and more apo-A from other lipoproteins and free cholesterol both from other lipoproteins and from effete cells, and in doing so assumes a spherical conformation. The free cholesterol is esterified by the enzyme lecithin-cholesterol acyltransferase (LCAT), which is present in nascent HDL and activated by its cofactor, apo-AI. This increases the density of the HDL particles which are thus converted from HDL3 to HDL2.

While some of the cholesterol ester remains in HDL and is transported therein to the liver, much is transferred to chylomicrons, VLDL and LDL, and reaches the liver indirectly. Cholesterol is excreted by the liver in the bile, both as cholesterol and after metabolism to

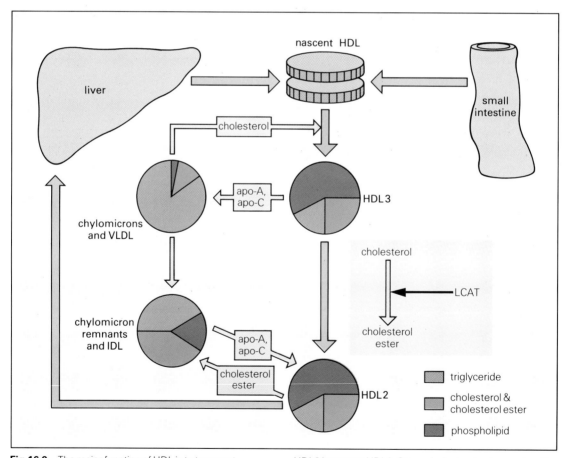

Fig.16.9 The major function of HDL is to transport cholesterol to the liver, whence it is excreted. Nascent HDL, synthesized in the liver and intestine, are converted to HDL3 as they acquire cholesterol from effete cells and other lipoproteins. This cholesterol is esterified by the enzyme lecithin-cholesterol acyltransferase (LCAT) and the HDL3 becomes HDL2. Some cholesterol ester reaches the liver directly in HDL2 but most is transported to the liver indirectly in chylomicron remnants, IDL and LDL. HDL is also an important source of apolipoproteins, particularly apo-A and apo-C, for chylomicrons and VLDL.

bile acids. Thus HDL has an important function in removing cholesterol from the circulation and in facilitating its excretion. It is also a source of apo-C for chylomicrons and VLDL, supplying it to newly synthesized lipoproteins and acquiring it from them as their triglyceride is removed.

The essential features of lipoprotein metabolism are as follows:

- Dietary triglyceride is transported in chylomicrons to tissues where it may be used as an energy source or stored.
- Endogenous triglyceride, synthesized in the liver, is transported in VLDL and is also available to tissues as an energy source or for storage.
- Cholesterol synthesized in the liver is transported to tissues in LDL, derived mainly from VLDL; dietary cholesterol reaches the liver in chylomicron remnants.
- HDL acquire cholesterol from effete cells and other lipoproteins and this is esterified by LCAT. Cholesterol ester for excretion is transported to the liver in HDL and also after transfer to chylomicron remnants, IDL and LDL.

REFERENCE RANGES AND LABORATORY INVESTIGATIONS

There is a problem in defining a reference range for serum cholesterol concentration. At birth, the concentration is very low (total cholesterol less than 2.6mmol/l, LDL-cholesterol less than 1.0mmol/l). There is a rapid increase in concentration in the first year of life, but in childhood the total does not usually exceed 4.1mmol/l. In affluent societies particularly, concentrations rise further in early adulthood. The reference range for serum cholesterol varies for different populations, but in the United Kingdom, the upper limit of the reference range for total cholesterol is usually taken as 6.5mmol/l. However, epidemiological studies indicate that the risk of coronary artery disease increases with cholesterol levels of more than 5.5mmol/l so that 'normal' levels are in fact associated with a significant risk.

There is an undoubted association between high serum cholesterol levels (and in particular, high LDL cholesterol) and an increased risk of coronary artery disease, while there is an inverse correlation with HDL cholesterol. Many physiological factors influence LDL and HDL cholesterol levels, of which some are indicated in Fig.16.10.

Total triglyceride and cholesterol concentrations may easily be measured in the laboratory; HDL cholesterol may be determined by first using a simple precipitation technique to separate HDL from LDL. LDL cholesterol can be calculated using the formula:

$$\text{LDL.CHOL} = \text{TOTAL CHOL} - \left(\text{HDL.CHOL} + \frac{\text{TRIG}}{2.2} \right)$$

where all quantities are expressed in mmol/l. This formula should not be used if the triglyceride concentration exceeds 4.5mmol/l.

Influences on plasma lipoproteins		
Variable	HDL cholesterol	LDL cholesterol
sex	F>M	M=F
age	slight ↑ in F	↑
high P:S ratio	– or ↓	↓
exercise	↑	↓
obesity	↓	↑
alcohol	↑	–
exogenous oestrogens	↑	–

Fig.16.10 Some physiological influences on plasma lipoproteins. P:S is the ratio of polyunsaturated to saturated fats in the diet.

Ultracentrifugation and electrophoresis

Ultracentrifugation is not a convenient technique for routine use but separation of lipoproteins by electrophoresis is a simple procedure. An example of normal fasting lipid electrophoresis is included in Fig.16.11. The β-band on electrophoresis corresponds to LDL, pre-β to VLDL and the α-band to HDL. Chylomicrons, if present, remain at the origin. Simple visual inspection of serum, after standing overnight at 4°C, is useful (Fig.16.12); chylomicrons, being less dense than serum, float to the top giving a creamy supernatant layer; VLDL remain in suspension and impart an opalescence or turbidity to the serum, while both LDL and HDL are too small to scatter light, so that even when they are present in excess the serum remains clear.

Blood for lipid studies

Blood for lipid studies should be drawn after an overnight fast (14h), when chylomicrons, being derived from dietary fat, should normally have been cleared so that a pathological disturbance may be inferred if they are present. The patient should have kept to his own normal diet for two weeks before the blood is taken. Alcohol should not have been taken on the evening before blood sampling. It is a common cause of hypertriglyceridaemia even in patients who have otherwise fasted. If lipid studies are to be done on a patient who has had a myocardial infarct, blood should either be taken within 24h or after an interval of three months, since the metabolism of lipoproteins is disturbed during the convalescent period and analytical results may be misleading. Lipid studies are indicated in the following circumstances:

- premature coronary artery disease (less than age forty years);
- a family history of coronary artery disease;
- the presence of stigmata of hyperlipoproteinaemia, (such as xanthomata or corneal arcus at age below forty years);
- when a patient's fasting serum is seen to be lipaemic.

DISORDERS OF LIPID METABOLISM

There are a number of rare, inherited metabolic diseases associated with the accumulation of lipids in tissues and others in which serum lipoprotein

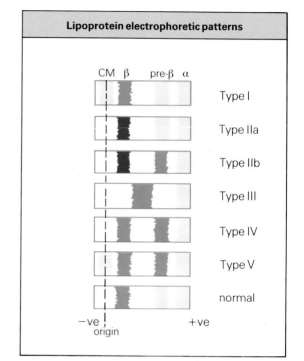

Fig.16.11 Diagrammatic representation of electrophoretic patterns characteristic of various lipoprotein disorders, based on the WHO classification.

concentrations are reduced. By far the commonest disorders, however, are the hyperlipoproteinaemias and it is to these conditions that the rest of this chapter is devoted.

Classification

The WHO classification of hyperlipoproteinaemias, based on the work of Fredrickson, is essentially a phenotypic classification, being dependent on the type of lipoprotein involved (Fig.16.13). As these WHO types do not correspond to specific disease entities, and since hyperlipoproteinaemia may be secondary to other conditions (see below), different patients with the same condition may manifest different WHO types: for example, both Types IIa and IIb may be seen in hypothyroidism. Further, even with primary inherited hyperlipoproteinaemias, relatives may show

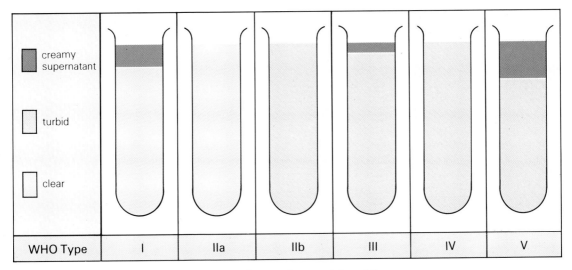

Fig.16.12   Appearance of fasting serum samples, after standing overnight at 4°C, in various hyperlipoproteinaemias.

WHO classification of hyperlipoproteinaemias					
Type	**CM**	**VLDL**	**LDL**	**Cholesterol**	**Triglyceride**
I	↑	N	N	N	↑ ↑
IIa	–	N	↑ ↑	↑ ↑	N
IIb	–	↑	↑	↑	↑
III	–	'broad β band'		↑	↑
IV	–	↑	N	N (↑)	↑
V	↑	↑	N	N (↑)	↑ ↑

Fig.16.13   WHO classification of hyperlipoproteinaemias. An indication of the concentrations of cholesterol and triglyceride characteristic of these types (N = normal, ↑ = raised) is shown but is not part of the basis of the classification. LDL levels, though usually normal, may be reduced in Types I, IV and V; total cholesterol may be slightly increased in types IV and V due to the cholesterol in VLDL. Chylomicrons (CM) are not usually present in fasting serum.

different WHO types, while the same pattern of lipoprotein excess may occur in more than one distinct inherited condition, for instance Type IIa in both familial monogenic and polygenic hypercholesterolaemia. The inadequacy of the WHO classification has another drawback in that the WHO type of an individual patient may change in response to dietary or drug treatment. A further drawback of this classification is that it takes no account of HDL cholesterol.

Secondary hyperlipoproteinaemias

These are common and, since resolution of the lipid abnormality should follow successful treatment of the underlying condition, management should be directed towards the cause. It follows that although the presence of a primary disorder may be inferred from a relevant family history, it is always important to exclude secondary causes in the investigation of patients with hyperlipoproteinaemias (common causes are shown in Fig.16.14). It should be appreciated that the presence of any of these may exacerbate a coexistent, inherited tendency to hyperlipoproteinaemia.

CASE HISTORY 16.1

A fifty-five-year-old man presented with a history of lethargy, loss of concentration and constipation. He had suffered from angina for two years, but this had become less of a problem recently since he had become much less active. On examination, he appeared myxoedematous.

Investigations

serum:
thyroxine	30nmol/l
TSH	more than 100mu/l
cholesterol	12.2mmol/l
triglyceride	1.5mmol/l

He was treated cautiously with tri-iodothyronine, his angina was controlled effectively with nitrates and, once euthyroid, with β-blockers. His serum cholesterol fell to only 8.2mmol/l on treatment and electrophoresis confirmed that the excess was in the β-lipoprotein band.

Comment

Hypothyroidism is commonly associated with hypercholesterolaemia due to decreased removal of LDL from the circulation. The persistence of a raised cholesterol despite adequate treatment of the hypothyroidism is suggestive, in this case, of the presence of an additional, genetically determined, predisposition to hypercholesterolaemia.

CASE HISTORY 16.2

A forty-five-year-old obese barman who complained of recurrent epigastric pain underwent gastroscopy. Because he admitted to heavy alcohol ingestion, blood was taken for liver function tests prior to the precedure. Gastroscopy revealed a duodenal ulcer.

In the laboratory, the technician noticed that the serum looked opalescent and analyzed it for lipids.

Investigations

serum:
cholesterol	7.5mmol/l
triglyceride	8.4mmol/l

lipid electrophoresis:

excess pre-β-lipoprotein
slight excess of α-lipoprotein
normal β-lipoprotein

Comment

Alcohol causes hypertriglyceridaemia by increasing triglyceride synthesis and the insulin resistance seen in obesity has a similar effect. Massive hypertriglyceridaemia (more than 20mmol/l) may occur in patients with a high alcohol intake when there is an additional, inherited tendency to hypertriglyceridaemia. Alcohol increases HDL cholesterol levels (accounting for the slight excess of α-lipoprotein in this case), although levels are commonly reduced when the high triglyceride level has another cause.

Various abnormalities of serum lipids may be seen in liver disease; with hepatocellular damage, triglyceride levels may be low due to decreased synthesis and chylomicron remnants and IDL accumulate because of decreased hepatic uptake. In cholestasis, an abnormal lipoprotein (lipoprotein X) may appear in the serum.

Condition	WHO Type	Lipid predominantly increased: cholesterol	triglyceride
obesity	IV		*
excessive alcohol intake	IV, V		*
diabetes mellitus	IIb, IV, V	*	*
hypothyroidism	IIa, IIb, III	*	
nephrotic syndrome	IIa, IIb, IV, V	*	*

Fig.16.14   Common causes of secondary hyperlipoproteinaemias.

Primary hyperlipoproteinaemias

Familial monogenic hypercholesterolaemia (FH)

This condition is characterized by very high serum cholesterol levels which are present from early childhood and do not depend upon the presence of environmental factors (see polygenic hypercholesterolaemia, below). Its inheritance is characteristic of a single gene defect and heterozygotes occur at a frequency of about one in 500 births. There is a defect of LDL receptors; in one form, these are reduced in number (to fifty percent), and in another there is defective internalization of receptor-bound LDL. Consequently, there is a decrease in LDL catabolism and serum levels of LDL are increased.

In the very rare homozygotes, no receptors are present. These individuals develop coronary artery disease in childhood and, if untreated, rarely survive into adult life; heterozygotes tend to develop coronary artery disease some twenty years earlier than the general population; more than half of those untreated die before the age of sixty.

CASE HISTORY 16.3

A thirty-two-year-old man requested screening for possible ischaemic heart disease after the death of his brother from a myocardial infarct at the age of thirty-six. He was not obese and was a non-smoker. On examination his blood pressure was normal and the only abnormalities were tendon xanthomata, arising from the Achilles tendons, and bilateral arcus senilis. An ECG taken at rest was normal but ischaemic changes developed on exercise. Analysis of fasting blood for lipids showed the following.

Investigations

serum: cholesterol 14.2mmol/l
      triglyceride 1.3mmol/l

lipid electrophoresis massive excess of β-lipoprotein

Comment

This is a characteristic picture of familial monogenic hypercholesterolaemia. Tendon xanthomata are common and may develop in the third decade; they are accumulations of cholesterol, but deep-seated so that the overlying skin has a normal colour. Arcus senilis and xanthelasmata are frequently present, but, unlike tendon xanthomata, may also occur in people who have no disturbance of lipid metabolism. Patients are not usually obese. Even though this patient is normotensive and a non-smoker the hypercholesterolaemia alone considerably increases his risk of dying of ischaemic heart disease and indeed he has an abnormal exercise ECG. Familial hypercholesterolaemia is ten times more common in victims of myocardial infarctions than in the rest of the population.

Patients with familial monogenic hypercholesterolaemia require rigorous treatment, usually with lipid-lowering drugs as well as diet (see below) and if other risk factors are present these, of course, must also be tackled. Children and most adults with familial monogenic hypercholesterolaemia show a Type IIa phenotype, but in some adults the phenotype is IIb.

Polygenic hypercholesterolaemia

In this condition, which is more common than familial monogenic hypercholesterolaemia, serum cholesterol levels are not as high and do not usually exceed 10mmol/l. Its expression is very dependent on environmental factors, such as diet. Though the tendency to hypercholesterolaemia is inherited, the pattern of inheritance suggests that several genes are involved, and in affected relatives there may be a continuous distribution of serum cholesterol concentrations, whereas with familial monogenic hypercholesterolaemia there is a clear distinction between normals, heterozygotes and homozygotes.

The significance of this condition lies again in its relationship to the risk of coronary artery disease and the principles of management are similar to those for familial monogenic hypercholesterolaemia; in polygenic hypercholesterolaemia, however, dietary treatment alone is often successful so that the use of lipid-lowering drugs may not be required.

Familial dysbetalipoproteinaemia (broad β-disease)

This condition is characterized clinically by the presence of fat deposits in the palmar creases and tuberous xanthomata; the latter tend to occur over bony prominences and, unlike tendon xanthomata, are reddish in colour. However, neither of these cutaneous stigmata are invariably present. In some patients eruptive xanthomata are present. Biochemically, the condition is characterized by the presence of a 'broad' band on electrophoresis of serum, due to the presence of an excess of IDL and chylomicron remnants; chylomicrons are sometimes also present. The total cholesterol and triglyceride levels are elevated and the molar ratio VLDL-cholesterol/total triglyceride characteristically

exceeds 0.68 (although this cannot be determined routinely since an ultracentrifuge is required to separate out VLDL).

Patients have an increased risk not only of coronary artery disease but also of peripheral and cerebral vascular disease. The molecular basis of this hyperlipoproteinaemia is the presence of a variant form of apo-E, which decreases the rate of IDL removal from the circulation by the liver. However, the fact that this variant apo-E is present in one in 100 of the normal population, while dysbetalipoproteinaemia is an uncommon disorder, implies a role for other factors in its expression, and in this context it is noteworthy that although the variant apoprotein is present from birth, the condition does not appear clinically until adult life. Such factors include obesity, alcohol, hypothyroidism and diabetes.

Although the diagnosis can be inferred from the clinical and biochemical findings, it should ideally be confirmed by apo-E phenotyping.

CASE HISTORY 16.4

A middle-aged man was referred by his family doctor to a dermatologist because of extensive yellowish papules, with an erythematous base, on his buttocks and elbows. The dermatologist recognized these as eruptive xanthomata and noticed that there were yellow, fatty streaks in the palmar creases. Blood was drawn after an overnight fast for lipid analysis and the serum was seen to be slightly turbid.

Investigations

| serum: | cholesterol | 8.5mmol/l |
| | triglyceride | 6.4mmol/l |

Serum protein electrophoresis showed a broad β band and a trace of chylomicrons.

Comment

This patient was treated with a low fat diet and bezafibrate and after three months serum lipid concentrations had become normal as had the electrophoretic appearance. There was also considerable regression of the xanthomata. When the bezafibrate was stopped, the abnormality returned but resolved again on restarting the drug. Familial dysbetalipoproteinaemia characteristically responds very well to treatment.

Familial chylomicronaemia

Fasting chylomicronaemia is a feature of two rare hyperlipoproteinaemias both having an autosomal recessive inheritance; in one there is a deficiency of the enzyme lipoprotein lipase and in the other of apo-CII which is required for activation of this enzyme. The result in each case is a failure of chylomicron clearance from the blood stream. Presentation is usually in childhood, with eruptive xanthomata, recurrent abdominal pain due to pancreatitis and sometimes hepatosplenomegaly.

Chylomicronaemia may also be seen in other patients with a genetic predisposition to hypertriglyceridaemia when this is exacerbated by obesity, diabetes mellitus, hyperuricaemia or alcohol ingestion; some drugs, for example thiazides, may also have this effect.

Management involves giving a low fat diet, with substitution of some fat by triglycerides based on medium-chain fatty acids; these are absorbed directly from the gut into the blood stream and therefore do not produce chylomicrons. The major complication of the chylomicronaemic syndromes is the recurrent pancreatitis and since this is uncommon with triglyceride levels below 20mmol/l, it is not usually necessary to achieve normalization of serum triglyceride concentration.

Familial hypertriglyceridaemia

In this condition the serum triglyceride level is elevated and the phenotypic expression is WHO Type IV or V according, it seems, to environmental factors and the presence of other causes of hypertriglyceridaemia. The basic abnormality appears to be increased production of triglyceride leading to increased VLDL secretion by the liver. Although usually only VLDL are present in excess, patients may also have chylomicronaemia, presumably because the excess VLDL block chylomicron catabolism.

Hypertriglyceridaemia is often asymptomatic and not usually diagnosed before the fourth decade, when it is commonly detected during the investigation of an associated condition such as diabetes mellitus. Mild hypertriglyceridaemia in itself is not an important risk factor for coronary artery disease, but the low HDL that is often present, together with the slight overall increase in serum cholesterol due to the cholesterol in VLDL, may be significant in this respect.

Familial combined hyperlipidaemia

This form of hyperlipoproteinaemia is due to hepatic overproduction of apo-B leading to increased VLDL secretion and production of LDL from VLDL. Either the serum cholesterol or triglyceride, or both, may be elevated; typically, in affected relatives, one-third have an increase in LDL, one-third in VLDL and one-third have an excess of of both lipoproteins, so hat the phenotypic expression may be IIa, IIb or IV. Cutaneous manifestations of hyperlipidaemia may be seen and in all cases there is an increased risk of coronary artery disease. Recent work suggests that this risk is associated with the increased apo-B in LDL, even though LDL cholesterol may be normal.

MANAGEMENT OF LIPID DISORDERS

The decision as to whether to treat a patient with a hyperlipoproteinaemia may not be straightforward. There is no doubt that treatment is vital for patients with familial monogenic hypercholesterolaemia, but in those with mild hyperlipoproteinaemia the possible risks of the use of drugs for treatment may outweigh those associated with the disease *per se*. In general, hypercholesterolaemia is more sinister than hypertriglyceridaemia alone.

It is important to treat any condition known to exacerbate hyperlipidaemia and in the context of coronary artery disease, attention to other risk factors such as smoking, hypertension and lack of exercise is vital. Overall, a more aggressive approach to management should be dictated by the presence of a poor family history or a personal history of arterial disease, especially in a younger patient; by the presence of hypercholesterolaemia either alone or in association with hypertriglyceridaemia; and by the presence of reduced levels of HDL. All patients should be encouraged to achieve a normal body weight.

Hypercholesterolaemia

In hypercholesterolaemia, a diet low in cholesterol and saturated fat should be prescribed; unsaturated fats may be substituted for saturated. In monogenic hypercholesterolaemia dietary treatment alone is seldom sufficient to normalize cholesterol levels. A total cholesterol concentration of >7.8mmol/l or a ratio (HDL. CHOL/TOTAL CHOL–HDL) of less than 0.2 is often taken as an indication for drug treatment. Bile acid sequestrants, such as cholestyramine and colestipol, though sometimes poorly tolerated, are very effective; other drugs which may be valuable, particularly if there is associated hypertriglyceridaemia, include fibric acid derivatives such as clofibrate, gemfibrozil and bezafibrate. The very rare homozygotes may require surgical treatment (partial ileal bypass) or plasmapheresis to control the disorder.

Familial dysbetaproteinaemia

In familial dysbetalipoproteinaemia, normalization of body weight and substitution of unsaturated fat for saturated fat is often very successful, but lipid-lowering drugs may be required. Although nicotinic acid is effective, it is often poorly tolerated; fibric acid derivatives are to be preferred. Bile acid sequestrants aggravate the hyperlipidaemia and are contra-indicated in this condition.

Hypertriglyceridaemia

Hypertriglyceridaemia may respond well to control of body weight and any coexistent exacerbating factors, such as diabetes, obesity or hyperuricaemia. Except in the case of the chylomicronaemic syndromes, lipid-lowering drugs are seldom required for hypertriglyceridaemia alone but nicotinic acid, clofibrate and bezafibrate are all effective.

The bile salt sequestering agents act by binding bile salts in the gut thus reducing the enterohepatic circulation and, by depleting the pool of bile salts in the liver, encouraging conversion of cholesterol to bile acid. Bloating and constipation are frequent side-effects and there may be decreased absorption of drugs taken at the same time. Nicotinic acid works by decreasing VLDL secretion; flushing is a frequent side-effect and tolerance is poor. Fibric acid

derivatives probably act at more than one point in lipoprotein metabolism; there is an increased risk of gallstones with clofibrate but not, apparently, with bezafibrate or gemfibrozil.

LIPOPROTEIN DEFICIENCY

There are two rare inherited lipoprotein deficiencies.

Abetalipoproteinaemia

In abetalipoproteinaemia, there is a defect in the synthesis of apo-B; CM, VLDL and LDL are absent from the plasma. Clinically, there is malabsorption of fat, acanthocytosis, retinitis pigmentosa and an ataxic neuropathy.

Tangier disease

In Tangier disease, HDL levels are reduced; clinically, the condition is characterized by hyperplastic, orange tonsils and the accumulation of cholesterol esters in other reticuloendothelial tissues. The molecular basis of this disorder is unknown.

SUMMARY

The main lipids in the blood are triglyceride, an important energy substrate, and cholesterol, a component of the membranes of cells and their organelles. Cholesterol and triglycerides are insoluble in water and are transported in the blood in lipoproteins, complexes of lipids with specific proteins known as apolipoproteins. There are four major classes of lipoprotein: (i) chylomicrons, which carry exogenous, that is, dietary, fat (mainly triglyceride) from the gut to peripheral tissues; (ii) very low density lipoproteins (VLDL), which carry endogenous triglyceride from the liver to these tissues; (iii) low density lipoproteins, which transport cholesterol from the liver to peripheral tissues; and (iv) high density lipoproteins (HDL), which scavenge cholesterol and transport it to the liver for metabolism and excretion. These particles are in a dynamic state and there is considerable exchange of lipid and proteins between them.

Hypercholesterolaemia, when due to an increase in LDL, is an important risk factor for coronary heart disease; an excess of cholesterol in HDL appears to confer some protection from this condition. Hypertriglyceridaemia is not on its own a major risk factor for coronary heart disease but, when very severe, can cause pancreatitis. Both hypercholesterolaemia and hypertriglyceridaemia are associated with various types of cutaneous fat deposition, or xanthomata.

Hyperlipoproteinaemias may be classified into six distinct phenotypes (the WHO classification) according to which lipoprotein particles are present in excess. They may be either primary, that is, genetically determined, or occur secondarily to a variety of other conditions, including diabetes mellitus, hypothyroidism, obesity, alcoholism, renal disease and certain drugs. The diagnosis of a primary hyperlipoproteinaemia is supported when such conditions can be excluded, especially if there is a family history; often, however, an underlying genetic tendency to hyperlipoproteinaemia is exacerbated by the presence of one of these conditions.

The most important primary hyperlipoproteinaemia is familial (monogenic) hypercholesterolaemia. The molecular basis of this condition is a functional defect in, or a decrease in the number of, LDL receptors which leads to decreased clearance of these lipoproteins from the blood and an increase in cholesterol synthesis. Heterozygotes for the condition occur with a frequency of approximately 0.2% and have a greatly increased risk of coronary heart disease. Homozygotes are fortunately very rare; affected patients develop coronary heart disease in their teens. Other inherited hyperlipoproteinaemias include familial hypertriglyceridaemia, combined hyperlipidaemia and dysbetalipoproteinaemia, in which a particle with a density intermediate between the low and very low density lipoproteins accumulates.

A case can be made for screening all adults for hypercholesterolaemia but this is essential in those with premature coronary heart disease (<40 years of age) or a family history of this condition or of a hyperlipoproteinaemia; in patients with xanthomata; and in patients whose serum is observed to be lipaemic.

The management of secondary hyperlipoproteinaemias is to treat the underlying condition; primary disorders are treated with diet, often with a lipid-lowering drug such as an ion exchange resin, to sequester bile salts in the gut (for hypercholesterolaemia) or a fibrate derivative (for hypercholesterolaemia and hypertriglyceridaemia). In patients with hypercholesterolaemia, it is essential to identify, and if possible eliminate, any other risk factors for coronary heart disease, such as smoking and hypertension.

FURTHER READING

Lewis B (1986) The appropriate use of diagnostic services: (viii) the investigation of hyperlipidaemia: why, how and for whom? *Health Trends*, **18**, 1–4.

Schaefer EJ & Levy RI (1985) Pathogenesis and management of lipoprotein disorders. *New England Journal of Medicine*, **312**, 1300–1310.

Stanbury JB, Wyngaarden JB, Fredrickson DS, Goldstein JL & Brown MS (eds) (1983) *The Metabolic Basis of Inherited Disease*. 5th edition. New York: McGraw-Hill Book Company.

Rifai N (1986) Lipoproteins and apolipoproteins. *Archives of Pathology and Laboratory Medicine*, **110**, 694–701.

17. Clinical Enzymology

INTRODUCTION

Enzymes are present throughout the body and their measurement can provide valuable diagnostic information. Tissue enzyme measurements are important in the diagnosis of inherited metabolic disease (*Chapter 18*), but this chapter is devoted mainly to the use of enzyme assays in conditions where a change in the concentration of an enzyme is a reflection, rather than the cause, of a disease process. Such measurements are usually made on serum, although enzyme assays in other body fluids, such as urine and pancreatic juice, can also provide useful information. Some enzymes present in the serum (such as renin and the clotting factors) have a physiological function there. However, diagnostic clinical enzymology is particularly concerned with intracellular enzymes that can be detected in the serum.

Small amounts of intracellular enzymes are present in the blood as a result of normal cell turnover. When pathological damage to cells occurs, increased amounts of enzymes will be released and their concentration in the blood will rise. However, such increases are not always due to tissue damage. Other possible causes include an increase in cell turnover, or in intracellular enzyme levels as a result of enzyme induction, leakage into the blood of an enzyme present in exocrine secretions when there is obstruction of the duct, and decreased clearance of an enzyme from the serum.

Little is known about the mechanisms by which enzymes are removed from the circulation. Small molecules, such as amylase, are filtered at the glomerulus but most enzymes are probably removed by reticuloendothelial cells. Serum amylase activity rises in acute renal failure but, in general, changes in clearance rates are not known to be important as causes of changes in serum enzyme levels.

Enzyme activity

Enzyme assays usually depend on the measurement of the catalytic *activity* of the enzyme, rather than the *concentration* of the enzyme protein itself. Since each enzyme molecule can catalyze the reaction of many molecules of substrate, measurement of activity provides great sensitivity. It is, however, important that the conditions of the assay are optimized and standardized to give reliable and reproducible results. Reference ranges for serum enzymes are dependent on assay conditions, for example, temperature, and may also be subject to physiological influences. It is thus important to be aware of both the reference range for the laboratory in question and the physiological circumstances when interpreting the data provided. Ranges quoted in this book (see page 315) are from the author's own laboratory and may not necessarily agree with those of the reader's.

Disadvantages of enzyme assays

A major disadvantage in the use of enzymes for the diagnosis of tissue damage is their lack of specificity to a particular tissue or cell type. Many enzymes are common to more than one body tissue, so that an increase in the serum level of a particular enzyme could reflect damage to any one of these tissues. This problem may be obviated to some extent in two ways: first, different tissues may contain (and thus release when they are damaged) two or more enzymes in different proportions; thus alanine and aspartate transaminase are both present in cardiac muscle and hepatocytes, but there is relatively more alanine transaminase in the liver; secondly, some enzymes exist in different forms or isoenzymes; these isoenzymes are often characteristic of a particular tissue, in that they have similar catalytic activities but differ in some other measurable property, such as heat stability, or sensitivity to inhibitors.

After a single insult to a tissue, the activity of intracellular enzymes in the serum rises as they are released from the damaged cells, and then falls as the enzymes are cleared. It is thus important to consider the time at which the blood sample is taken in relation to the insult. If taken too soon, there may have been insufficient time for the enzyme to reach the blood stream and if too late, they may have been completely cleared (see *Case history 17.2*). As with all diagnostic techniques, data acquired from measurements of enzymes in serum must always be assessed in the light of whatever clinical and other information is available, and their limitations borne in mind.

ENZYMES OF DIAGNOSTIC VALUE

Alkaline phosphatase

This enzyme is present in high concentrations in the liver, bone (osteoblasts), placenta and intestine. The hepatic content of alkaline phosphatase is increased in cholestatic disease as a result of enzyme induction, thus an increased serum alkaline phosphatase is usually a reflection of cholestasis. In bone, however, the enzyme is released from active osteoblasts so that an increase in the level of the enzyme is a reflection of new bone formation.

The causes of an increase in serum alkaline phosphatase are summarized in Fig.17.1. Physiological increases are seen in pregnancy, due to the placental isoenzyme, and in childhood (when bones are growing), due to the bone isoenzyme. The serum level is high at birth but falls rapidly thereafter. However, it remains two to three times the normal adult level and rises again during the adolescent growth spurt before falling to the adult level as bone growth ceases (Fig.17.2).

Serum alkaline phosphatase levels are commonly slightly higher than normal in apparently healthy elderly people. This may reflect the high incidence of mild, subclinical Paget's disease in the elderly. Levels of alkaline phosphatase as high as ten times the upper limit of normal (10 × ULN) may be seen in severe Paget's disease of bone, rickets and osteomalacia and occasionally in cholestatic liver disease. Levels of up to 5 × ULN are, however, more common in these conditions (see Fig.17.1).

In Paget's disease, increased osteoblastic activity is essential to the bone remodelling which occurs as a consequence of the underlying increased osteoclastic bone resorption. The serum alkaline phosphatase is thus a valuable guide to the activity of the disease and to the efficacy of treatment (see *Case history 17.3*). It should be noted that alkaline phosphatase levels are not increased in uncomplicated osteoporosis, unless the condition has been complicated by collapse or fracture of bone.

Serum alkaline phosphatase is commonly elevated in malignant disease; it may be of bony or hepatic origin and associated with the presence of both

---

Causes of an increased serum alkaline phosphatase

Physiological

pregnancy (last trimester)

childhood

Pathological

> 5 × ULN
  Paget's disease of bone
  osteomalacia and rickets
  cholestasis (intra- and extra-hepatic)
  cirrhosis

usually < 5 × ULN
  bone tumours (primary and secondary)
  renal bone disease
  primary hyperparathyroidism with bone
    involvement
  healing fractures
  osteomyelitis
  hepatic space-occupying lesions
    (tumour, abscess)
  infiltrative hepatic disease
  hepatitis
  inflammatory bowel disease

Fig.17.1 Causes of an increased serum alkaline phosphatase level.

---

primary and secondary tumours at these sites. A number of apparently tumour-specific alkaline phosphatases, secreted by tumour cells themselves, have also been described. The best known of these is the Regan isoenzyme, which has similar heat stability to placental alkaline phosphatase and is present in some patients with bronchial carcinoma.

Serum alkaline phosphatase is frequently measured as part of a biochemical profile and it is not uncommon to find a raised level in the absence of clinical evidence of bone, liver, or gut disease, and in the absence of other biochemical abnormalities. In establishing the cause of such an increase it is clearly helpful to determine the tissue of origin. This is done by measuring tissue-specific isoenzymes of alkaline phosphatase. These may be separated and quantitated using various techniques, including electrophoresis and differential heat inactivation. This is a more reliable approach, although more time-consuming

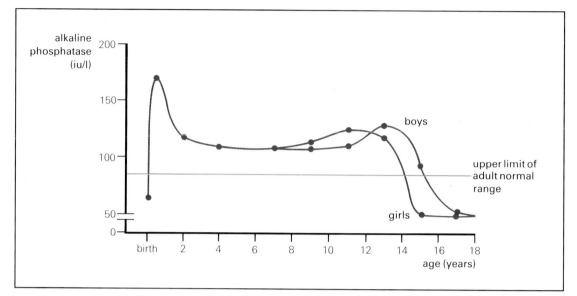

Fig.17.2 Serum alkaline phosphatase level as a function of age in childhood. Mean values are shown; the peaks between ten and sixteen years correspond to the pubertal growth spurt and levels of up to three times the upper limit of the adult normal range may be seen at this time.

than measuring serum γ-glutamyl transpeptidase which is found in the liver but not bone. The serum level of this enzyme is commonly, but not always, increased when there is an excess of hepatic alkaline phosphatase in the serum. Further, because the γ-glutamyl transpeptidase is such a sensitive indicator of hepatic dysfunction, it may be increased while hepatic alkaline phosphatase is normal (see *Case histories 6.2 and 6.4*). An alternative approach is the measurement of the enzyme 5′-nucleotidase, which is specific to the liver.

Acid phosphatase

This enzyme is present in high concentrations in the prostate gland and is elevated in the serum of some patients with prostatic cancer. It is of little value in the diagnosis of this condition, being raised in only about twenty percent of patients, in whom the tumour is confined to the gland. It is, however, raised in up to eighty percent of cases when there are metastases and its use as a tumour marker in such patients is discussed in *Chapter 21*. Acid phosphatase is also increased in some cases of prostatitis and occasionally in benign prostatic hypertrophy.

The prostate is examined clinically through the rectum, a procedure that may cause some release of acid phosphatase into the circulation producing a transient increase in its concentration. Although this happens much less frequently than was once thought, it is wise to take a blood sample for acid phosphatase determination before, rather than after, clinical examination. The enzyme is labile and blood for analysis must be transported to the laboratory rapidly and the serum deep-frozen until the assay is performed. Since the enzyme is present in red blood cells, haemolysis invalidates acid phosphatase measurements.

In addition to the prostate and red blood cells, acid phosphatase is also present in platelets, bone, liver and spleen. The prostatic isoenzyme is characteristically inhibited by tartrate but not formaldehyde. In most laboratories, acid phosphatase is measured as the prostate-specific 'tartrate-labile' or 'formol-stable' isoenzyme. Specific immunoassays for prostatic acid phosphatase exist but are not in general use. High concentrations of non-prostatic acid phosphatase may be seen in bone diseases (particularly in Paget's disease), in thrombocythaemia and in Gaucher's disease, an inherited disorder of lipid storage.

Transaminases

Two transaminases are of use in diagnostic enzymology. These are aspartate transaminase (AST, also known as glutamate-oxaloacetate transaminase, GOT) and alanine transaminase (ALT, or glutamate-pyruvate transaminase, GPT). Both enzymes are widely distributed in body tissues, the concentration of aspartate transaminase being lower in all tissues except the liver where the enzymes are present in approximately equal amounts.

The causes of increased serum aspartate transaminase are shown in Fig.17.3. Very high concentrations of AST, sometimes in excess of $100 \times$ ULN, are seen with severe tissue damage, such as acute hepatitis, crush injuries and tissue hypoxaemia. More usually in hepatitis the peak concentration is only $10-20 \times$ ULN; this peak may occur in the prodromal stage before the patient is jaundiced or at the time of onset. In myocardial infarction, serum AST begins to rise some twelve hours after the infarct, reaching a peak of up to $10 \times$ ULN at 24–36h and then declining over two to three days providing that there is no further cardiac damage (see page 241).

In most conditions where AST is elevated there is a concurrent, though proportionally smaller, rise in ALT. In hepatitis, however, serum levels of ALT may exceed those of AST. If, as in many laboratories, only one transaminase assay is available, this should be for AST. Its major use is in the management of liver disease, where a raised level suggests hepatocellular damage and the results of serial estimation can indicate persistence of, or recovery from, such damage (see page 79). Serum AST is often measured as part of a biochemical profile in multichannel autoanalyzers. It is very uncommon to find levels greater than $10 \times$ ULN unexpectedly; this is most likely to be seen in the prodromal phase of viral hepatitis. Levels of up to $2 \times$ ULN are sometimes found in patients who have no clinical evidence of tissue damage. There are no tissue-specific isoenzymes of aspartate transaminase and if there are no other biochemical changes, nor any clinical features which could explain the raised level, the wisest procedure is to repeat the analysis after an interval of one or two weeks.

γ-Glutamyl transpeptidase

This enzyme is present in high concentrations in the liver, kidney and pancreas. Its serum activity is commonly raised in liver disease, especially when

Causes of increased serum aspartate transaminase
> 10 × ULN acute hepatitis and liver necrosis major crush injuries severe tissue hypoxaemia (levels may sometimes exceed 100 × ULN in these conditions)
5–10 × ULN myocardial infarction following surgery or trauma skeletal muscle disease cholestasis chronic hepatitis
usually < 5 × ULN physiological (neonates) other liver diseases pancreatitis haemolysis (*in vivo* and *in vitro*)

Fig.17.3 Causes of an increased serum aspartate transaminase level. Serum alanine transaminase is raised to a similar extent in liver diseases but to a lesser degree, if at all, in the other conditions.

there is cholestasis, but this finding is not sufficiently consistent to be diagnostic.

γ-Glutamyl transpeptidase is, however, a sensitive index of hepatic dysfunction and the serum level may rise before that of alkaline phosphatase in early cholestasis.

Serum γ-glutamyl transpeptidase is raised in the absence of liver disease in many patients taking the anticonvulsant drugs, phenytoin and phenobarbitone; rifampicin, used in the treatment of tuberculosis, can have a similar effect. This is an example of enzyme induction where the increased serum activity of γ-glutamyl transpeptidase is not due to cell damage but to an increase in enzyme production within cells so that an increased amount is released during normal cell turnover.

The level of the enzyme in serum is also increased in forty to seventy percent of people who are heavy alcohol drinkers. Levels may remain elevated for up to three to four weeks after stopping alcohol even in the absence of liver damage. Serum γ-glutamyl transpeptidase tends to revert to normal in people with a history of more than five years of excessive alcohol

---

Some causes of an increased serum γ-glutamyl transpeptidase

> 10 × ULN
  cholestasis
  alcoholic liver disease

5–10 × ULN
  hepatitis (acute and chronic)
  cirrhosis (without cholestasis)
  other liver diseases
  pancreatitis

usually < 5 × ULN
  excessive alcohol ingestion
  enzyme-inducing drugs
  congestive cardiac failure

Fig.17.4 Causes of an increased serum γ-glutamyl transpeptidase level. Increases of less than 5 × ULN are seen in many conditions and probably reflect secondary effects on the liver. γ-Glutamyl transpeptidase is not usually increased with hepatic space-occupying lesions provided that liver function is normal.

intake. However, small increases are also seen in several other conditions (Fig.17.4), and a significant number of people who abuse alcohol have a normal serum level of the enzyme.

Lactate dehydrogenase

This enzyme exists in body tissues as a tetramer. There are two distinct monomers, H and M, which can combine in various proportions so that five isoenzymes of lactate dehydrogenase (LD) are known. The distribution of these isoenzymes in tissues is shown in Fig.17.5. The isoenzymes may be distinguished on the basis of several properties, including their sensitivity to heat and various inhibitors, and their electrophoretic mobility.

The major clinical uses of lactate dehydrogenase measurements are in suspected myocardial infarction and in the diagnosis of haemolytic crises in sickle cell disease. In both cardiac muscle and red blood cells LD_1 is the predominant isoenzyme. This shows much

greater catalytic activity with α-hydroxybutyrate (rather than lactate) as a substrate than the other isoenzymes. Consequently LD_1 is commonly measured by means of a reaction using this substrate and has the alternative name, α-hydroxybutyrate dehydrogenase (HBD).

α-Hydroxybutyrate dehydrogenase has a long half-life in the plasma; after myocardial infarction its activity rises slowly to reach a peak at two to three days, declining thereafter over a period of one week or more. Since it is present in red blood cells, levels increase after pulmonary embolism which may resemble myocardial infarction clinically. The presence of haemolysis invalidates the use of HBD_1 in diagnosis.

Measurements of total lactate dehydrogenase are less frequently performed due to the lack of the enzyme's specificity. Increases are seen in a wide variety of conditions including acute damage to the liver, skeletal muscle and kidneys, and also in megaloblastic and haemolytic anaemias.

Creatine kinase

The enzymatically active creatine kinase molecule consists of a dimer of two distinct monomers, M and B; three isoenzymes BB, MM and MB are found. BB is confined mainly to the brain and thyroid. There is little of the BB isoenzyme in the serum normally, and even with severe brain damage (due, for example, to a stroke) the concentration barely rises. Most of the creatine kinase normally present in the serum is the MM isoenzyme which originates from skeletal muscle. An increase in concentration is seen with skeletal muscle damage and with severe or prolonged exercise (Fig.17.6).

Creatine kinase present in cardiac muscle contains a considerably higher proportion of the MB isoenzyme (approximately thirty percent) than does skeletal muscle (less than one percent). A raised serum creatine kinase is characteristic of myocardial infarction (see page 241) and, in the absence of a possible contribution from skeletal muscle, separate measurement of the MB isoenzyme is not necessary. However, the finding that more than five percent of the total creatine kinase is due to the MB isoenzyme suggests that myocardial damage has been sustained if the suspected myocardial infarction follows exercise, if intramuscular injections, such as an analgesic, have been given, or if the patient has sustained trauma to skeletal muscle. In practice such measurements are not frequently required.

Isoenzymes of lactate dehydrogenase and their distribution					
Isoenzyme	1	2	3	4	5
subunit structure	H_4	H_3M	H_2M_2	HM_3	M_4
normal serum	+	++	+	trace	trace
heart muscle	++	++	trace	−	−
liver	−	−	−	+	++
skeletal muscle	−	−	−	+	++

Fig.17.5   The isoenzymes of lactate dehydrogenase and their distributions. + = present; − = absent.

Causes of an increased serum creatine kinase

> 10 × ULN
  myocardial infarction
  rhabdomyolysis
  malignant hyperpyrexia
 	
> 5 × ULN
  following surgery
  skeletal muscle trauma
  severe exercise
  grand mal convulsions
  myositis
  muscular dystrophy
 	
usually < 5 × ULN
  physiological (neonates)
  hypothyroidism (↓ catabolism)

Fig.17.6   Causes of an increased serum creatine kinase level. Concentrations as high as 100 × ULN may be seen in rhabdomyolysis and malignant hyperpyrexia.

Causes of an increased serum amylase

> 10 × ULN
  acute pancreatitis
 	
> 5 × ULN
  perforated duodenal ulcer
  intestinal obstruction
  other acute abdominal disorders
  acute oliguric renal failure
  diabetic ketoacidosis
 	
usually < 5 × ULN
  salivary gland disorders, e.g. calculi
    and inflammation (including mumps)
  chronic renal failure
  macroamylasaemia
  morphine administration (spasm of
    sphincter of Oddi)

Fig.17.7   Causes of an increased serum amylase level.

The MB isoenzyme may be separated from the others by electrophoresis; other available techniques include an immunoinhibition method in which the M subunit is selectively inhibited and various specific immunoassays for the B subunit.

Amylase

This enzyme is found in the salivary glands and exocrine pancreas and tissue-specific isoenzymes can be distinguished by means of electrophoresis or the use of inhibitors.

Causes of a high serum amylase are shown in Fig.17.7. The most important use of this enzyme is in the differential diagnosis of the acute abdomen. Its serum concentration is usually raised in acute pancreatitis, and levels greater than 10 × ULN are virtually diagnostic, being only very occasionally seen in other conditions. However, levels are not as high as this in all cases of pancreatitis and may be 5 × ULN or more in other abdominal emergencies, particularly with perforation of a duodenal ulcer. Acute pancreatitis is usually managed conservatively while urgent laparotomy and exploration is indicated for most other abdominal emergencies.

Extra-abdominal causes of a raised serum amylase level rarely cause increases of more than 5 × ULN. Macroamylasaemia is an example of a high serum enzyme level being due to reduced clearance. In this condition, amylase becomes complexed with another protein (in some cases, an immunoglobulin) to form an entity of much greater apparent molecular weight; renal clearance is reduced as a result. This has no direct clinical connotations but can misleadingly suggest the presence of pancreatic damage.

Separate measurement of the pancreas-specific amylase can improve the diagnostic specificity of serum amylase determinations. Early experience with pancreas-specific lipase suggests that this may be more specific for pancreatic damage than total serum amylase.

Cholinesterase

This enzyme is secreted by the liver into the blood stream and low serum levels are seen in chronic hepatic dysfunction. It is, however, rarely measured for this reason.

Interest in this enzyme derives from the fact that it hydrolyzes a muscle-relaxant drug, widely used in anaesthesia, called succinylcholine (Scoline). Occasionally, patients are found in whom the effect of this drug, which paralyzes respiration, persists for several hours after it has been administered (Scoline apnoea). Many of these patients have an abnormal cholinesterase level.

Four enzyme variants have been recognized based on the activity of the enzyme in the presence of inhibitors: normal, dibucaine-resistant, fluoride-resistant and inactive. Normal homozygotes account for ninety-five percent of the population, and heterozygotes for dibucaine resistance, four percent. Such individuals do not usually react abnormally to succinylcholine, but homozygotes for dibucaine resistance, (0.05%) are at risk of developing Scoline apnoea as are patients with some of the rarer genotypes. The relatives of patients found to have cholinesterase deficiency should be screened to identify those who might be at risk when undergoing surgery.

SERUM ENZYMES IN DISEASE

Liver disease

The use of serum enzyme measurements in liver disease is discussed in detail in *Chapter 6*. In a patient with jaundice, an increase in aspartate or alanine transaminase levels, often of 10 x ULN or more, suggests that the jaundice is due to hepatocellular damage. An increase in alkaline phosphatase of 2–3 × ULN or more is characteristic of cholestasis. Transaminases may be slightly increased with cholestasis, but usually not to more than 2–3 × ULN, while the alkaline phosphatase is not usually more than 2 × ULN in predominantly hepatocellular disease.

A high transaminase activity generally precedes any other biochemical change in early hepatitis, including the increase in serum bilirubin. An increase in liver-specific alkaline phosphatase is often the only biochemical abnormality in patients with compensated cirrhosis and with space-occupying lesions of the liver, such as secondary tumour deposits. The value of serial enzyme measurements in monitoring patients with liver disease is emphasized in *Chapter 6*.

γ-Glutamyl transpeptidase has become widely adopted as a useful 'liver enzyme', but in the jaundiced patient it provides little information in addition to that obtained from other enzyme and liver function tests.

Heart disease

Serum enzyme measurements are commonly made, in some cases routinely, in patients with suspected myocardial infarction, although it has been estimated that biochemical confirmation of the diagnosis is required in only fifteen to thirty percent of patients. Many patients with myocardial infarction will have a classical history of crushing central chest pain, perhaps radiating to the arm or face, and typical electrocardiographic (ECG) changes. Myocardial infarction can present atypically, however, or may even be clinically silent, particularly in the elderly (see *Case history 24.1*). The ECG changes may not always be typical, particularly with partial thickness infarcts, when there has been a previous infarction, and in left bundle branch block.

Significant changes in the serum activities of creatine kinase, aspartate transaminase and α-hydroxybutyrate dehydrogenase occur following myocardial infarction. The typical time course of

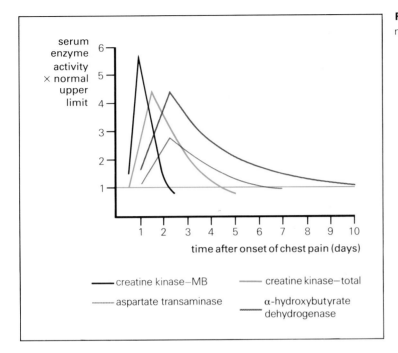

Fig.17.8 Serum enzyme levels after myocardial infarction.

these changes is shown in Fig.17.8. Also shown is the time course of the change in the MB isoenzyme of creatine kinase; this enzyme appears first and is rapidly cleared from the serum after myocardial infarction. It is obviously vital when interpreting serum enzyme changes that the time of the samples relative to the time of the suspected infarct is known and is appropriate. Thus the 'time-window' for creatine kinase is between twelve and thirty-six hours but no change in α-hydroxybutyrate dehydrogenase would be expected during this time.

Diagnosis

Given the lack of tissue specificity of aspartate transaminase, the most useful enzymes for diagnostic purposes in a patient with suspected myocardial infarction are creatine kinase, for early confirmation of the diagnosis, and α-hydroxybutyrate dehydrogenase, for those patients (often with atypical clinical features) who may present several days after the supposed infarct. However, if the patient has a typical history and unequivocal ECG changes, enzyme measurements add nothing either to the diagnosis or to the immediate management. This is because patients suspected of having infarcts are likely to be treated as if they have had an infarct, at least for the first critical twenty-four hours. However, the lack of a rise in enzyme levels should lead to a review of the diagnosis after this time. Even in a patient in whom there is no doubt as to the diagnosis, enzyme measurements may be considered to be worthwhile in that a persistently high level, or an increase after an initial fall, suggest an extension of the infarct which may not be obvious clinically or on the ECG.

The only circumstance in which the urgent measurement of serum creatine kinase is necessary is in patients with crescendo angina, in whom emergency coronary artery bypass grafting is considered. A raised creatine kinase would suggest that infarction has occurred and that surgery should not be performed.

Attempts have been made to correlate serial serum creatine kinase levels with infarct size, but this requires frequent measurements and is of little use in management.

CASE HISTORY 17.1

A recently retired lawyer was admitted to hospital with chest pain which had developed during the evening after a day spent digging in the garden.

There were no specific signs of myocardial infarction on the ECG. He was monitored in the acute coronary unit for twenty-four hours and then transferred to a general ward. His pain subsided rapidly and he was discharged after five days.

Investigations

	on admission	48h	72h
serum:			
creatine kinase (total)	300iu/l	80iu/l	40iu/l
creatine kinase (MB)	30iu/l	—	—
HBD	—	—	70iu/l

Comment

Although the total creatine kinase was raised, the MB isoenzyme was normal. It was concluded that no myocardial infarction had taken place and that the chest pain was musculoskeletal in origin, related to the unaccustomed exercise. Total creatine kinase may reach a peak of greater than $20 \times$ ULN after severe exercise, especially if unaccustomed. The normal HBD supported these conclusions.

CASE HISTORY 17.2

Two days after sustaining myocardial infarction, confirmed by ECG changes and the finding of a raised total creatine kinase, a fifty-four-year-old social worker complained of discomfort in the right epigastrum. On examination, the jugular venous pressure was elevated and the liver was slightly enlarged and tender.

Investigations

serum:	bilirubin	60μmol/l
	alkaline phosphatase	130iu/l

serum:	aspartate transaminase	125iu/l
	creatine kinase	80iu/l
		(280iu/l on admission)

Comment

The serum alkaline phosphatase is not usually increased after uncomplicated myocardial infarction but this patient has clinical evidence of right heart failure. Hepatic venous congestion is a consequence of this and can cause a mild, usually transient cholestatic jaundice, reflected here by the increase in alkaline phosphatase and bilirubin. If it is measured, the alanine transaminase is often elevated to an extent commensurate with aspartate transaminase, whereas in uncomplicated myocardial infarction the alanine transaminase is usually normal. In pulmonary embolism, which may mimic myocardial infarction clinically, aspartate transaminase (from an infarcted area of lung) and β-hydroxybutyrate dehydrogenase (from lysis of red blood cells in the embolus and infarct) may be elevated, but the creatine kinase is not. Note that the creatine kinase has become normal by forty-eight hours (see also Fig.17.8).

Bone disease

Alkaline phosphatase is secreted by osteoblasts and an increased serum concentration is seen in many diseases of bone (see *Chapter 14*); osteoporosis is an important exception to this. Unless there is coexisting osteomalacia or a fracture occurs, the alkaline phosphatase is normal. In multiple myeloma, despite extensive tumour deposition in bone, the alkaline phosphatase is not raised since there is no concomitant increase in osteoblastic activity.

Serial measurements of serum alkaline phosphatase are of particular value in the assessment of healing in osteomalacia, rickets, renal osteodystrophy and in Paget's disease.

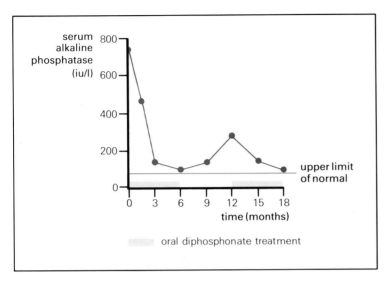

Fig.17.9 Serum alkaline phosphatase levels in a patient with Paget's disease of bone. Periods of treatment with oral diphosphonates are indicated. After a good response to the first period of treatment the serum alkaline phosphatase begins to rise, indicating recrudescence of the disease; a good response was again achieved when treatment was restarted.

CASE HISTORY 17.3

An elderly man who complained of severe pain in his pelvis and thighs was diagnosed on radiological evidence as having Paget's disease of bone. The serum alkaline phosphatase was 750iu/l. He was treated with oral diphosphonates and made a good clinical recovery, though when his medication was stopped his thighs became painful again.

Investigations

The serum alkaline phosphatase levels are shown in Fig.17.9.

Comment

Paget's disease is a condition in which osteoclastic activity is abnormally high. Bone destruction engenders osteoblastic activity which results in an increase in new bone formation. However, the new bone is abnormal and is laid down in a disorganized fashion so that bones become thickened, distorted and painful. The great increase in osteoblastic activity is reflected in the serum alkaline phosphatase levels and, as in this case, the enzyme may be used to monitor the condition and its response to treatment.

Paget's disease can also be monitored by serial measurements of the urinary excretion of hydroxyproline. This amino acid, a component of collagen, is not reutilized in the body and once released from bone by osteoclastic activity is excreted in the urine. However, the accurate measurement of urinary hydroxyproline is technically difficult; care must be taken to ensure that the dietary contribution is minimized (hydroxyproline is present in gravies and jellies, for example), while other sources of collagen, notably the skin, are also sources of hydroxyproline.

Muscle disease

Enzymes present within striated muscle cells appear in the serum in certain muscle diseases, including muscular dystrophies (particularly Duchenne type), polymyositis, toxic and other myopathies, and also after trauma and in ischaemia. Enzyme levels tend to be normal in neurogenic muscle disease, that is, syndromes of denervation such as motor neuron disease and lower motor neuron lesions.

The measurement of serum creatine kinase provides the most reliable evidence of muscle disease being more sensitive than, for example, aspartate transaminase or aldolase. However, no enzyme measurement is available which gives an indication of either the cause or nature of the disorder. Serial

243

measurements of serum creatine kinase are valuable in assessing the response to treatment (usually with corticosteroids) of patients with polymyositis.

Duchenne muscular dystrophy

Duchenne muscular dystrophy is a sex-linked recessive disorder of muscle which usually affects males. Serum creatine kinase may be very high in children with this condition though levels tend to fall in the later stages when most of the muscle has been destroyed. Female carriers of Duchenne muscular dystrophy are asymptomatic but some seventy-five percent have raised serum creatine kinase levels. It is helpful to be able to detect carriers to assist genetic counselling. However, in about one-third of cases of this condition there is no family history, nor is maternal serum creatine kinase raised, suggesting that there is a high incidence of spontaneous mutation giving rise to the disorder.

There are several other forms of muscular dystrophy and an increased serum creatine kinase is not always present. Increases that are present tend to be of a lesser magnitude than those seen in Duchenne muscular dystrophy.

Malignant disease

Changes in serum enzymes are common in malignant disease. They may be due to the release of normal or variant enzymes from the tumour itself (for example, acid phosphatase in carcinoma of the prostate, Regan isoenzyme in carcinoma of lung) or to the response of surrounding tissues to the presence of the tumour (for example, alkaline phosphatase in hepatic and osseous tumours).

A raised serum alkaline phosphatase may be caused by metastases in a patient known to have carcinoma, but other, non-biochemical techniques are necessary to confirm this, such as radiography and ultrasonography. The possibility of metastases from a clinically silent tumour should be considered when a raised alkaline phosphatase level is discovered incidentally. The use of acid phosphatase determinations in the management of prostatic cancer is discussed on page 236, but apart from these instances, enzyme measurements are of little value in either diagnosis or management of patients with malignant disease.

ENZYMES IN OTHER BODY FLUIDS AND TISSUES

Red blood cells

Many inherited defects of red cell enzymes have been described, most of which cause a haemolytic anaemia. Glucose-6-phosphate dehydrogenase (G6PD) deficiency is the commonest of these. It is transmitted on the X-chromosome so that most affected patients are males, and there is a particularly high incidence in American Negroes, and peoples of the Mediterranean littoral.

There is considerable variability in the clinical presentation, due to the presence of a large number of different abnormal enzymes. Some individuals bearing an abnormal enzyme are unaffected; others present with either the mild form or severe haemolytic anaemia induced by drugs, such as primaquine and sulphonamides, or a substance in fava beans (favism). Pyruvate kinase deficiency is another common defect of red cell enzymes. In many hospitals, the responsibility for the measurement of these and other red cell enzymes falls to the haematology rather than the clinical chemistry laboratory.

Measurements of red cell enzymes are used in the diagnosis of certain vitamin deficiencies (see *Chapter 23*) and in the diagnosis of some inherited metabolic diseases (see *Chapter 18*). Unfortunately, however, the definitive diagnosis of many inherited metabolic diseases requires the measurement of an enzyme in less readily available material, for example, liver or muscle. The measurement of enzymes in cultured amniotic cells and fetal blood for the antenatal diagnosis of inherited metabolic disease is also discussed in *Chapter 18*.

Urine

The measurement of enzymes in urine is technically difficult because of the presence of inhibitors, but some can usefully be measured, for example, amylase. The latter is excreted in urine and the presence of a high urinary amylase reflects an increase of its activity in the serum. Measurement of urinary amylase may occasionally be of value in the diagnosis of shortlived episodes of pancreatitis, when the serum amylase is increased only transiently. In the vast majority of cases of acute pancreatitis, however, the serum amylase remains elevated for several days.

Enzymes derived from renal tubular cells can also be detected in the urine and many have been investigated for a possible role in the diagnosis of tubular dysfunction (due, for example, to drugs) and renal graft rejection. The enzyme most widely used for this purpose is β-N-acetylglucosaminidase.

Intestinal secretions

The measurement of digestive enzymes in the investigation of malabsorption is considered in *Chapter 7*.

SUMMARY

The enzymes present in the plasma include those that have a physiological function there, for example, renin and the blood clotting factors, and those that have been released from cells as a result of damage or normal cell turnover. Diagnostic enzymology is principally concerned with the latter and the measurement of enzyme activity in the plasma can give useful diagnostic information concerning the site and extent of tissue damage. Examples of such enzymes include creatine kinase, which is released from cardiac muscle following myocardial infarction and after damage to skeletal muscle, and the transaminases, which are widely distributed and are released into the blood in a variety of conditions, including hepatitis, myocardial infarction and skeletal muscle injury.

Alkaline phosphatase is present in osteoblasts and an increase in its plasma activity occurs in conditions in which osteoblastic activity is increased, such as osteomalacia and Paget's disease of bone. Plasma alkaline phosphatase activity is also increased in patients with cholestatic jaundice.

Few enzymes that are measured for diagnostic purposes in plasma are tissue specific, but when the origin of increased plasma activity is not obvious either clinically or for other reasons, measurement of the isoenzymes (molecular variants of the enzymes which have similar catalytic activity but a different chemical structure so that they are distinguishable, for example, immunochemically or by electrophoresis) can often provide this information. Thus the measurement of alkaline phosphatase isoenzymes will distinguish between a hepatic, bony or other source for increased plasma activity of this enzyme, and the measurement of isoenzymes of creatine kinase will distinguish between a cardiac or skeletal muscle origin. Another method to improve specificity involves measuring more than one enzyme, since the concentration of different enzymes, and thus the amount released when cells are damaged, varies between different tissues.

Though tending to lack specificity, the measurement of plasma enzyme activity can provide a very sensitive means of detecting tissue damage and can be invaluable in following the course of an illness, such as hepatitis or Paget's disease, even though the diagnosis may have been established using another technique.

Other enzymes that are frequently measured for diagnostic purposes include amylase (elevated in acute pancreatitis), γ-glutamyl transpeptidase (a sensitive, but non-specific indicator of hepatobiliary disease and particularly useful in detecting alcohol abuse) and acid phosphatase (as a marker in prostatic carcinoma).

The measurement of enzymes in other body fluids can also provide useful diagnostic information and the measurement of enzymes in tissue samples may provide the definitive diagnosis in certain inherited metabolic diseases; for example, the enzyme galactose-1-phosphate uridyl transferase is deficient from red blood cells in galactosaemia. The measurement of enzymes in cultured amniotic cells, obtained by amniocentesis, and fetal tissue, obtained at fetoscopy, offers increasing opportunity for the antenatal diagnosis of inherited metabolic diseases and, if acceptable, termination of pregnancy when an infant is likely to be seriously affected.

FURTHER READING

Goldberg DM, Werner M & Zaidman JL (eds) (1987) *Enzymes and Isoenzymes in Pathogenesis and Diagnosis. Advances in Clinical Enzymology, volume 5.* Basel: Karger.

Moss DW (1982) *Isoenzymes.* London: Chapman and Hall.

Wilkinson JH (1976) *The Principles and Practice of Diagnostic Enzymology.* London: Edward Arnold.

18. Inherited Metabolic Diseases

INTRODUCTION

Many inherited diseases are known to be due to the genetically determined absence or modification of a specific protein. For example, in sickle cell anaemia, the protein is haemoglobin; in agammaglobulinaemia, antibody production is defective. However, in the majority of such diseases, the protein in question is an enzyme and as a result normal metabolism is disturbed. Other inherited metabolic diseases may be due to defective receptor synthesis (for example, monogenic hypercholesterolaemia, in which the receptor for low density lipoprotein is absent), or to defects involving carrier proteins (for example, cystinuria, in which renal tubular reabsorption of cystine is impaired). Whatever the cause, the clinical features of inherited metabolic diseases stem directly from the metabolic abnormalities to which they give rise. Although individually these conditions are rare, they are of considerable significance; the consequences of many of them are potentially disastrous but may in some cases be ameliorated if an early diagnosis is made and the appropriate treatment instituted.

The majority of these conditions have an autosomal recessive mode of inheritance. Certain of the porphyrias caused by disturbances of porphyrin metabolism (see *Chapter 19*) are an exception to this. Heterozygotes are usually phenotypically normal, but again there are exceptions such as familial monogenic hypercholesterolaemia.

EFFECTS OF ENZYME DEFECTS

Fig.18.1a shows a typical metabolic pathway involving the synthesis of product D from substrate A by successive, enzyme-catalyzed reactions through intermediates B and C. If the formation of B from A, catalyzed by enzyme a, is rate-limiting, as the first step unique to a metabolic pathway commonly is, then the concentrations of intermediates B and C will normally be low. The formation of product E from C, catalyzed by enzyme c', is normally a minor pathway, only a small amount of E being formed.

Three distinct sequelae of a lack of enzyme c, that could occur alone or in combination, can be envisaged.

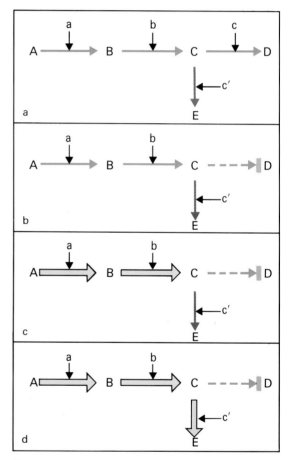

Fig.18.1 Effects of enzyme defects: (a) Product D is synthesized from A by a series of reactions catalyzed by enzymes a, b and c. Enzyme c' catalyzes the formation of a small amount of product E in a minor pathway. (b) In the absence of the enzyme c, no D is synthesized. (c) If the conversion of C to D is blocked, the concentration of the intermediate C, and possibly other precursors, may increase. (d) Increased formation of E may occur if the concentration of C increases and conversion of C to D is blocked.

246

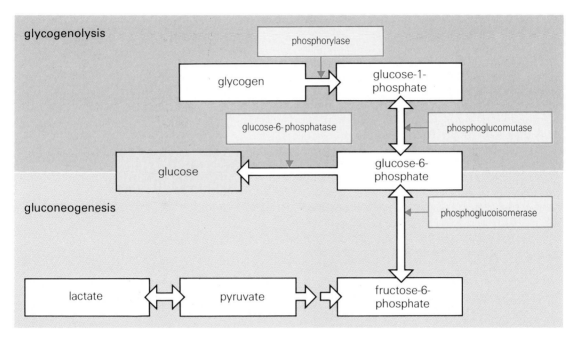

Fig.18.2 Glucose production by glycogenolysis or gluco-neogenesis. In the absence of glucose-6-phosphatase, glucose cannot be formed from glucose-6-phosphate.

Glucose-6-phosphate is an essential intermediate in the production of glucose by either glycogenolysis or gluconeogenesis, so in its absence hypoglycaemia results.

Decreased formation of the product

Decreased formation of the product of a reaction is the most obvious consequence (Fig.18.1b). If enzyme c is defective, D cannot by synthesized normally by this pathway. Clinical features will arise due to a lack of D if it is an essential product with no alternative pathway for its synthesis.

Accumulation of the substrate

Accumulation of the substrate (C) of the missing enzyme would also be expected (Fig.18.1c). If this is toxic to the body, clinical manifestations will result. Other, earlier substrates may also accumulate if the reactions prior to the one blocked are reversible. This will occur particularly if there is negative feedback by the product on an early reaction in the pathway so that with decreased formation of the product, the feedback is lost, thus releasing the inhibition and stimulating the formation of the intermediate substrates.

Increased formation of other metabolites

Increased formation of E, the product of a minor pathway, may occur if the concentration of C is increased as a result of the enzyme deficiency, the reaction being promoted by a mass action effect (Fig.18.1d). Again, if E is toxic, a clinical syndrome will result.

INHERITED METABOLIC DISORDERS

Glucose-6-phosphatase deficiency

Glucose-6-phosphatase deficiency (glycogen storage disease Type 1) exemplifies the production of a clinical syndrome due to lack of formation of the product of an enzyme-catalyzed reaction. Glucose synthesis from glycogen or by gluconeogenesis is blocked (Fig.18.2). Children with this disorder suffer severe fasting hypoglycaemia since their only source of glucose is dietary carbohydrate.

In this condition, blood glucose must be maintained by constant intragastric infusion of glucose or frequent ingestion of glucose and starch. Glucose-6-phosphatase deficiency also exemplifies the consequences of accumulation of a precursor other than the immediate substrate of the defective enzyme. Glycogen accumulates in the liver causing hepatomegaly. The block in gluconeogenesis results in an accumulation of lactate and lactic acidosis is a common finding. Hyperlipidaemia results from increased fat synthesis and hyperuricaemia is also frequently present. Accumulation of glycogen in platelets leads to disordered platelet function and a bleeding tendency. Due to the enzyme block, neither glucagon nor adrenaline increase the blood glucose in glucose-6-phosphatase deficiency but the definitive diagnosis is made by demonstrating lack of enzyme activity in a sample of liver obtained by biopsy. At least eight other glycogen storage diseases related to defects in glycogen metabolism are known.

Galactosaemia

Galactosaemia exemplifies the production of a clinical syndrome due to the accumulation of the substrate of the missing enzyme. The absence of the enzyme galactose-1-phosphate uridyl transferase, which is required for the conversion of galactose to glucose (Fig.18.3), results in the accumulation of galactose-1-phosphate, and the clinical features of the condition are thought to be due directly to the toxicity of this metabolite. Infants with galactosaemia present with failure to thrive, vomiting, hepatomegaly and jaundice. Cataracts may be present as a result of the conversion of excess galactose to galacticol in the lens. There may also be hypoglycaemia and impairment of renal tubular function. Galactose is a reducing sugar and a positive Clinitest in a child, diagnosed as galactosaemic on clinical grounds, merits withdrawal of galactose (and lactose) from the diet pending a definitive diagnosis, based on measurement of galactose-1-phosphate uridyl transferase in erythrocytes.

Galactokinase deficiency

In galactokinase deficiency, in contrast to galactosaemia, phosphorylation of galactose is impaired (see Fig.18.3). As a result, dietary galactose cannot be metabolized and its concentration in the blood rises. When the renal threshold is exceeded, galactose is

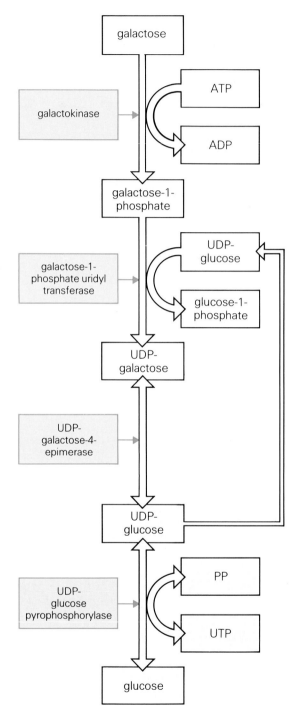

Fig.18.3 Metabolic pathway for the conversion of galactose to glucose. Sites of action of the potentially deficient galactokinase and galactose-1-phosphate uridyl transferase. UDP = uridine diphosphate.

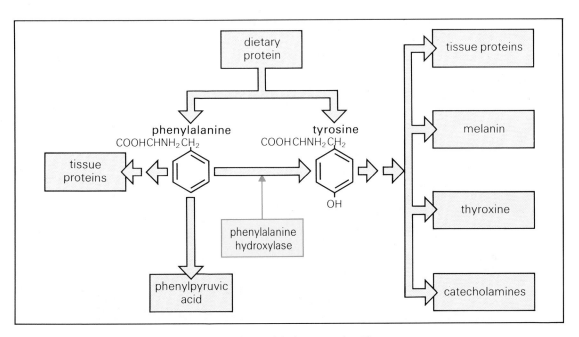

Fig.18.4 Metabolic pathway for the conversion of phenylalanine to tyrosine. The site of action of phenylalanine hydroxylase, the enzyme deficient in PKU, is shown.

excreted in the urine. However, it is not toxic and though cataracts may occur, the other clinical features characteristic of galactosaemia are not present in galactokinase deficiency.

Phenylketonuria

Phenylketonuria (PKU) is another condition in which the accumulation of the substrate of the missing enzyme gives rise to a clinical syndrome. The enzyme is phenylalanine hydroxylase, which hydroxylates phenylalanine to form tyrosine (Fig.18.4).

Phenylalanine accumulates in the blood and if the condition is untreated it results in severe mental retardation, thought to be due directly to the effect of excess phenylalanine on the developing brain. The name of the condition derives from the urinary excretion of phenylpyruvic acid, a phenylketone. This is normally a minor metabolite of phenylalanine but is produced in excess when the normal, major metabolic pathway is blocked. Many children with PKU have fair hair and blue eyes, due to defective melanin synthesis; tyrosine, the formation of which is blocked, is a precursor of this pigment. The diagnosis depends on the demonstration of an abnormally high concentration of phenylalanine in the blood; neonatal screening for the condition is discussed below.

The management involves restricting the dietary intake of phenylalanine by diets based on special proteins and pure amino acids. The serum level of phenylalanine should not be allowed to exceed 0.3mmol/l in the first year of life when there is rapid brain development, but may, without detriment, be allowed to rise to 0.5mmol/l by the age of four. By the tenth birthday, it is often possible to allow a normal diet but the patients must be followed up and the diet restarted if necessary. Strict dietary control is also required when a woman with PKU becomes pregnant, since maternal hyperphenylalaninaemia has been shown to affect the fetus *in utero* even if it does not itself have PKU.

Since phenylalanine is an essential amino acid, a certain amount must be provided in the diet and while tyrosine is not normally an essential amino acid, it becomes so when the intake of phenylalanine is limited; adequate quantities must therefore be provided. Thus treated, children in whom a diagnosis of PKU is made shortly after birth will grow and develop normally. Untreated, they rarely achieve an IQ of above 70, and many have to be cared for in institutions for the mentally subnormal for life.

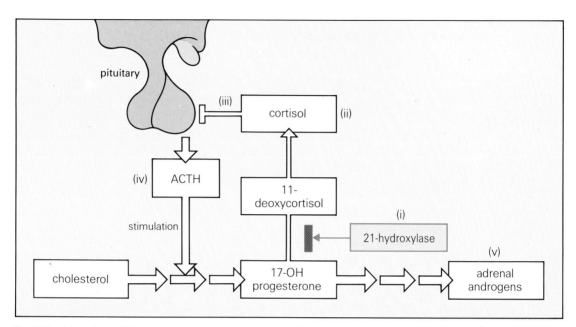

Fig.18.5   Adrenal steroid hormone synthesis, showing the increased synthesis of androgens when cortisol synthesis is blocked. Decreased 21-hydroxylase activity (i) leads to decreased cortisol synthesis (ii). Negative feedback to the pituitary (iii) is decreased leading to increased secretion of ACTH (iv). The conversion of cholesterol to 17-hydroxyprogesterone is stimulated, leading to increased synthesis of androgens (v).

Variants

The enzyme phenylalanine hydroxylase has tetrahydrobiopterin as a coenzyme. A number of variant forms of PKU has been described, some involving a defect in the metabolism of this coenzyme. Several other inherited metabolic diseases are associated with abnormalities of phenylalanine and tyrosine metabolism, including tyrosinosis and alcaptonuria.

Steroid 21-hydroxylase deficiency

Steroid 21-hydroxylase deficiency, the commonest cause of congenital adrenal hyperplasia, exemplifies the effects of increased activity of a normally minor metabolic pathway, in this case, the synthesis of adrenal androgens (Fig.18.5). Due to the defective synthesis of cortisol, there is decreased negative feedback to the pituitary and thus increased secretion of ACTH which further stimulates the synthesis of adrenal androgens.

DIAGNOSIS

The diagnosis of an inherited metabolic disease may be suggested by clinical features and the results of simple tests. However, the diagnosis will not be made if a possible metabolic origin of the symptoms is not considered. Although most inherited metabolic diseases are rare, as a group they are an important cause of the failure to thrive in neonates and infants. Effective treatment is now available for many of these conditions so that it would be tragic if treatment were not given because of a missed diagnosis.

The definitive diagnosis usually depends on the demonstration of a decreased activity of the enzyme responsible in an appropriate tissue. This may involve biopsy of an affected organ but in some cases the enzyme may be assayed in red or white blood cells. The identification of probable cases by neonatal or antenatal screening merits special consideration. Contrasting examples of the clinical presentation and management of two inherited metabolic diseases are provided by *Case histories 13.4* and *24.5*.

Criteria for effective screening tests	Criteria for prenatal diagnosis
reliable, cheap screening test available (no false negatives; some false positives acceptable)	reliable, safe diagnostic test available for use in early pregnancy
condition is treatable	disease sufficiently serious to justify termination of pregnancy if present
condition is fatal or leads to chronic disability if untreated	significant risk of disease occurring
condition is relatively common	disease not amenable to treatment
	parents are willing that pregnancy should be terminated if fetus shown to be affected

Fig.18.6 Criteria for effective screening tests and prenatal diagnosis of inherited metabolic disease.

SCREENING

The criteria for an effective screening programme are given in Fig.18.6. In many countries, including the United Kingdom and the United States, all babies are screened at birth for PKU (see also page 8) which has an incidence of approximately one in 10,000. The screening test involves measurement of the concentration of phenylalanine in a sample of capillary blood taken from a heel-prick six to ten days after birth. The most widely used test is the Guthrie test, a microbiological test using a strain of *Bacillus subtilis* in conditions such that growth is only seen if excess phenylalanine is present. Thin layer chromatography can also be used to screen for PKU. If the screening test is found to be positive, further definitive tests are then performed. Many babies are now also screened for congenital hypothyroidism (incidence of one in 4500). Economic considerations dictate that a screening test should be cost-effective. Even though it may be technically feasible, it is not economic to screen whole populations for very rare diseases.

In screening for PKU, the serum concentration of phenylalanine taken as positive is set such that the sensitivity of the test is virtually one hundred percent (all cases are detected). The specificity is greater than ninety-nine percent (there are few false positives). However, because the condition is rare, the predictive value of a positive test is low (see page 8); thus most positive screening tests are found not to be due to

PKU. This means that some children will be subjected to further investigation and subsequently shown not to have the disease, but this is acceptable if it ensures that genuine cases are not missed.

Comparison of PKU with another inherited metabolic disease, cystic fibrosis, provides an instructive contrast in relation to neonatal screening. Cystic fibrosis is a more common disease than PKU (incidence of one in 2500 live births in the United Kingdom). It is a generalized disorder of exocrine secretion in which these secretions have a greatly increased viscosity. Clinically, affected children develop recurrent respiratory infections, leading to bronchiectasis, and also pancreatic insufficiency, leading to malabsorption. The molecular basis is unknown. Although a screening test might seem an attractive notion, a suitable test has not yet been devised.

Sweat sodium secretion is increased in cystic fibrosis and the measurement of this provides the definitive test for the condition (a concentration of 80mmol/l being diagnostic). However, this is not a practical screening test. The tests suggested so far, for example, the detection of albumin in meconium, have been either insufficiently sensitive, insufficiently specific, or both. Further, there is no specific treatment for cystic fibrosis and although the respiratory and pancreatic complications can be treated, there is little evidence that treatment initiated before the development of symptoms is of any benefit.

PRENATAL DIAGNOSIS

When an inherited metabolic disease cannot be successfully treated, or the treatment imposes harsh restrictions on the patient, early prenatal diagnosis will allow parents the option of having the pregnancy terminated. The criteria for undertaking prenatal diagnosis are set out in Fig.18.6.

Satisfactory diagnostic tests are available for many inherited metabolic diseases but whether an attempt at prenatal diagnosis is justified depends upon the risk of the procedure. Most of these conditions have a recessive mode of inheritance, so that prenatal diagnosis should usually be considered only if there is an affected child from a previous pregnancy, if one parent is affected, or if there is a strong family history of the disease. Screening of selected populations may be justified if a disease has a high incidence in that population, for example, the lipid storage disorder Tay–Sachs disease in Ashkenazi Jews.

Maternal and fetal screening

Maternal screening is not diagnostic, but may point to the need to proceed to a more invasive, but definitive test. Although the diseases are not metabolic in origin, the method of screening for open neural tube defects (spina bifida and anencephaly) exemplifies this type of procedure. The screening test involves measurement of α-fetoprotein in maternal blood. If the concentration is raised, and no other cause is found, for example, wrong dates, twins, or spurious results, the woman is then offered amniocentesis for measurement of the protein in amniotic fluid, a more accurate predictor of the presence of a neural tube defect. Ultrasound examination is also a valuable diagnostic technique in this context.

An inherited metabolic defect may be reflected by the presence of an abnormally high concentration of a metabolite in maternal blood as, for example, in some organic acidaemias, but such metabolites, derived from the fetus, would normally be cleared by maternal enzymes. However, analysis of amniotic fluid, or cultures of amniotic cells obtained by amniocentesis, will give a more accurate reflection of fetal metabolism. Direct sampling of fetal blood by fetoscopy is a recently introduced technique and is being used in some centres for the prenatal diagnosis of haemophilia.

Measurement of fetal serum creatine kinase may also allow the prenatal diagnosis of Duchenne muscular dystrophy. Biopsy of chorionic villi is another new technique which provides direct access to fetal tissue at an early stage of pregnancy. This is a rapidly developing field and the introduction of gene probes to identify mutant genes, already in use for the diagnosis of α_1-antitrypsin deficiency and certain haemoglobinopathies in some centres, offers an exciting prospect for the prenatal diagnosis of inherited metabolic diseases in the future.

TREATMENT

Possible approaches to the treatment of inherited metabolic diseases are given in Fig.18.7.

Restriction of substrate intake

This is exemplified in the treatment of galactosaemia. If all foodstuffs containing galactose and lactose are removed from the diet, clinical symptoms regress. Similarly, hereditary fructose intolerance (see pages 187 & 188) is asymptomatic if fructose is avoided. The management is less straightforward, however, if the substrate is essential for life. In PKU, the metabolism of phenylalanine to tyrosine is blocked (see page 249), but phenylalanine is an essential amino acid and must therefore be provided in the diet to allow normal growth and development.

Supply of missing product

Congenital adrenal hyperplasia is managed by giving cortisol, production of which is impaired in this condition. In salt-losing types, a mineralocorticoid must also be given.

Addition of vitamin cofactors

If the defective enzyme has a vitamin cofactor, the supply of large amounts of the vitamin may, by a mass action effect, increase cofactor binding and thus enzyme activity. Many enzymes have separate catalytic and regulatory sites and amino acid substitution due to a mutant gene may affect either of such sites, or alter the way in which they interact. Homocystinuria is a condition in which the conversion of homocysteine to cystathionine is blocked. The enzyme involved, cystathionine β-

Inherited metabolic disease treatment strategies	
Treatment	**Example**
restriction of substrate intake	galactose in galactosaemia
supply of missing product	cortisol in congenital adrenal hyperplasia
supply of vitamin cofactors	pyridoxal phosphate in homocystinuria
increased excretion of toxic substances	copper in Wilson's disease
replacement of missing protein	lysosomal enzyme disorders
replacement of mutant gene	organ grafting

Fig.18.7 Treatment strategies for inherited metabolic disease.

synthase, requires pyridoxal phosphate as a cofactor and giving large amounts of this vitamin may be of therapeutic benefit in some cases. Some organic acidaemias may similarly respond to high-dose vitamin supplementation.

Increased excretion of toxic substances

This approach is used in the treatment of Wilson's disease (see page 86) to remove the excess copper which is responsible for the tissue damage in this condition. D-Penicillamine forms a soluble complex with copper which is then readily excreted in the urine. This drug is also used in the treatment of cystinuria, an inherited disorder characterized by defective renal tubular reabsorption of cystine and the dibasic amino acids, lysine, ornithine, and arginine. Cystine is relatively insoluble and there is a marked tendency to renal stone (renal calculus) formation. Cystine may be kept in solution if the urine is kept sufficiently dilute and alkaline. If calculi continue to form, penicillamine may be used; the drug complexes with cysteine (from which cystine is derived) and reduces the urinary excretion of cystine.

Replacement of missing protein

If the replacement of a missing protein were to be a feasible method of treatment it would need to be repeated at regular intervals as there is a continuous turnover of proteins in the body. Replacement of gammaglobulins is the mainstay of treatment of agammaglobulinaemia and, when necessary, factor VIII can be given in haemophilia. With the great majority of inherited metabolic diseases the defective protein is intracellular and thus replacement is not feasible. Attempts have been made to treat some disorders involving lysosomal enzymes by infusing liposomes (lipoprotein droplets) containing the missing enzyme into the blood stream. Unfortunately, the potential of this novel approach to treatment is limited.

Replacement of the defective gene

This should allow normal production of the product of that gene, such as an effective enzyme. Although the fields of gene cloning and genetic engineering are developing at astonishing speed, this is likely to remain only a theoretical possibility in the management of the inherited metabolic diseases for the foreseeable future.

Organ grafting may be appropriate in some conditions; liver transplantation has been used successfully in patients with Wilson's disease who have developed cirrhosis and patients with renal failure due to cystinuria have been treated by kidney transplantation. Organ grafting effectively replaces the missing or defective protein by providing the mechanism for its production in a genetically deficient patient.

SUMMARY

Inherited metabolic diseases are the result of gene mutations, which either prevent the synthesis of a protein or cause the production of an abnormal protein molecule. In the majority of these disorders, the protein is an enzyme and the result is a decrease in catalytic activity. There are several hundred known examples; most of them are rare. Their effects vary in severity from the completely benign (e.g., renal glycosuria) to the invariably fatal (e.g., Tay–Sach's disease). In some inherited metabolic diseases, the defective or missing protein is a receptor (e.g., familial hypercholesterolaemia) or a transport protein (e.g., cystinuria).

Most of these conditions have an autosomal recessive mode of inheritance; heterozygotes are usually phenotypically normal. However, some, notably most of the porphyrias (see *Chapter 19*), are unusual in having a dominant mode of inheritance.

A decrease in catalytic activity can have a number of consequences. In the case of an enzyme involved in a synthetic pathway, there could be decreased synthesis of the product of the enzyme or pathway, accumulation of the substrate and other precursor metabolites, or increased activity in a usually minor pathway which has, as its starting point, one of the intermediates which accumulates. Thus the clinical effects may relate to either decreased levels of a product of the pathway or increased levels of other metabolites (which may be toxic in excess), or to a combination of these.

The definitive diagnosis of an inherited metabolic disorder requires either measurement of the activity of the relevant enzyme, a procedure which may necessitate tissue biopsy unless the enzyme is present in blood cells, or detection of the defective gene by using a gene probe. The diagnosis can, however, often be inferred from the clinical features and measurements of the concentrations of metabolites or precursors of the enzyme, and may then be confirmed by the response to treatment.

Many inherited metabolic disorders can be screened for *in utero* when there is a significant risk of a fetus being affected, for instance when a previous child is known to have had the condition or when there is a strong family history of the disorder. Neonatal screening, though technically feasible for many inherited metabolic disorders, is widely practised in the general population only for phenylketonuria and congenital hypothyroidism. Screening tests must be highly sensitive and specific; the condition in question should have severe consequences which can be ameliorated by early treatment (or avoided by termination of the pregnancy in the case of antenatal screening); and the condition must occur sufficiently frequently in the population being screened for the exercise to be worthwhile.

Some inherited metabolic disorders can be treated relatively simply. Congenital adrenal hyperplasia, a group of conditions in each of which one of the enzymes involved in the synthesis of cortisol is defective, can be treated by replacing the missing product, cortisol. Others, such as galactosaemia and phenylketonuria, can be treated by dietary modifications which prevent the accumulation of toxic metabolites. A few metabolic disorders are due to decreased ability of the enzyme to bind a coenzyme; giving large amounts of the coenzyme may overcome this by a mass action effect and restore catalytic activity to normal.

The definitive treatment for an inherited metabolic disease would be replacement of the defective protein or gene. Attempts have been made to treat some lysosomal enzyme deficiencies by enzyme replacement but such treatment needs to be repeated frequently and has other disadvantages. Organ transplantation for the renal failure which can occur in cystinuria or the liver failure in Wilson's disease effectively replaces the defective gene but whether it will ever become possible to replace or modify specific genes is a matter for speculation.

FURTHER READING

Stanbury JB, Wyngaarden JB, Fredrickson DS, Goldstein J L & Brown MS (eds) (1983) *The Metabolic Basis of Inherited Disease*. 5th edition. New York: McGraw-Hill Book Company.

19. Haemoproteins, Porphyrins and Iron

INTRODUCTION

Haemoglobin, the oxygen-carrying pigment of blood, consists of a protein, globin, and four haem molecules (Fig.19.1). Globin comprises two pairs of polypeptide chains (the principal haemoglobin in adults, haemoglobin A, HbA, has two α- and two β-chains) and each polypeptide binds one haem molecule. Haem consists of a tetrapyrrole ring, protoporphyrin IXα, linked to an iron II ion (Fe^{2+}) to which oxygen becomes reversibly bound during oxygen transport. Other haemoproteins include myoglobin, which binds oxygen in skeletal muscle, and the cytochromes, enzymes responsible for catalyzing many oxidative processes in the body.

The major part of the body's iron is present in haemoglobin and the major product of porphyrin metabolism is haem. It is thus convenient to discuss the chemical pathology of the haemoproteins, the porphyrins and iron together, although disorders affecting any one of these do not necessarily (indeed do not often) involve the others.

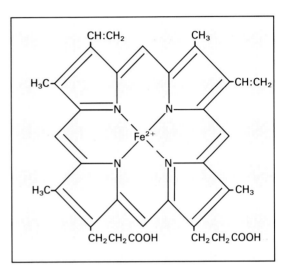

Fig.19.1 The structure of haem.

HAEMOPROTEINS

Haemoglobin and haemoglobinopathies

Haemoglobin is a very accessible protein and has been extensively studied. The haemoglobinopathies (genetically determined abnormalities of haemoglobin synthesis) fall into two groups: those involving amino acid substitutions; and the thalassaemias.

Amino acid substitutions

Here there is a single amino acid substitution in one of the polypeptide chains. Over 200 such variants have been described. Some of these involve amino acids which are not structurally or functionally vital, and are clinically silent; others have important consequences including effects on haemoglobin solubility, (for example, HbS, the haemoglobin of sickle cell disease), stability and oxygen-carrying capacity.

Thalassaemias

In the thalassaemias there is an inherited defect in the rate of synthesis of one of the globin chains. This can involve the α-chains (α-thalassaemia) or the β-chains (β-thalassaemia). The consequences include ineffective erythropoiesis, haemolysis and a variable degree of anaemia. The clinical severity varies between the different thalassaemias. Some are clinically silent except during periods of stress such as severe infection or pregnancy when anaemia may develop, while others cause severe, persistent anaemia. When α-chain synthesis is totally absent, affected infants are either stillborn or die shortly after birth.

Investigation of haemoglobinopathies

The investigation and management of the haemo-globinopathies are the province of the haematologist, and these disorders are not considered further in this book. It is, however, of relevance to point out that as a result of the amino acid substitution, some abnormal haemoglobins have a different electrophoretic mo-bility to normal adult haemoglobin (HbA). Indeed, this property is utilized in their identification. Because of this, however, they may co-migrate with HbA_1 (the glycosylated haemoglobin that is present in increased concentration in the blood of patients with diabetes mellitus), when an electrophoretic technique is used for its quantification, and thus give a falsely high result. This also applies to fetal haemoglobin (HbF) which, though normally present in only trace amounts in the blood from a few weeks after birth, is present in significant quantities in some haemoglobinopathies and thalassaemias, and in the benign condition, hereditary persistence of fetal haemoglobin (HPFH).

Abnormal derivatives of haemoglobin

Methaemoglobin

Methaemoglobin is oxidized haemoglobin, with iron in the Fe^{3+} form. It is incapable of carrying oxygen. A small amount is normally produced spontaneously in red blood cells but can be enzymatically reduced back to haemoglobin. Excessive methaemoglobin (met-haemoglobinaemia) can be congenital or acquired. It can occur in some haemoglobinopathies, with an inherited deficiency of the reductase enzyme, and also as a result of the ingestion of large amounts of certain drugs, such as sulphonamides. In toxic methaemo-globinaemia, the presence of methaemalbumin (formed as a result of haemolysis of red cells containing methaemoglobin) imparts a brown coloration to the serum, while the presence of free methaemoglobin may give the urine a similar colour.

The major clinical manifestation of congenital met-haemoglobinaemia is cyanosis. Acute toxic met-haemoglobinaemia causes symptoms of anaemia and may lead to vascular collapse and death. Methaemo-globinaemia, except when due to a haemoglobin-opathy, can be treated with methylene blue or ascorbic acid, agents which reduce the abnormal derivative back to haemoglobin.

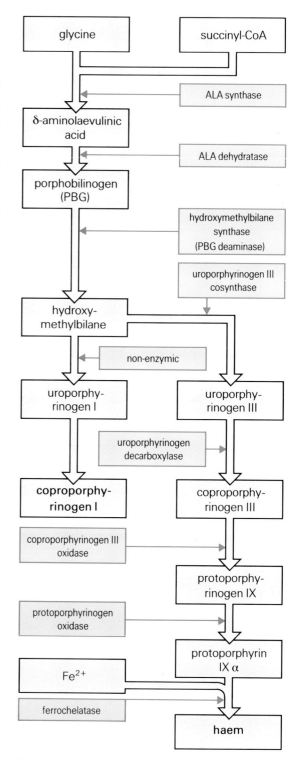

Fig.19.2 The biosynthesis of porphyrins.

Sulphaemoglobin

Sulphaemoglobin, a group of poorly characterized derivatives of haemoglobin, is often formed together with methaemoglobin. It is also incapable of carrying oxygen but cannot be converted back to haemoglobin.

Carboxyhaemoglobin

Carboxyhaemoglobin (COHb) is formed from haemoglobin in the presence of carbon monoxide, the affinity of the pigment for this gas being some 200 times greater than for oxygen. Because of this, only small quantities of carbon monoxide in the inspired air can result in the formation of large amounts of COHb and hence greatly reduce the oxygen-carrying capacity of the blood. Small amounts of COHb (less than 2%) are commonly present in the blood of urban dwellers and greater amounts (up to 10%) are found in the blood of tobacco smokers.

Haematin

Haematin is oxidized (Fe^{3+}) haem. It is released from methaemoglobin when red cells containing this pigment are haemolyzed but can be formed from free haem in severe intravascular haemolysis. Haematin combines with albumin in the blood stream to form methaemalbumin. Methaemalbuminaemia is sometimes a feature of acute haemorrhagic pancreatitis.

These various derivatives of haemoglobin can be detected by their spectral characteristics, and quantified when necessary.

PORPHYRINS

Protoporphyrin IXα, which combines with iron to form haem, is the end product of a complex series of reactions. The first step that is unique to this pathway is the combination of glycine and succinyl-CoA to form δ-aminolaevulinic acid (ALA), a reaction catalyzed by the enzyme ALA synthase (Fig.19.2). Two molecules of ALA then condense to form porphobilinogen (PBG), in a reaction catalyzed by ALA dehydratase.

The first porphyrins (strictly, porphyrinogens, see below) are formed when four molecules of PBG condense together. The initial product of this reaction, catalyzed by hydroxymethylbilane synthase (PBG deaminase) is hydroxymethylbilane. In the presence of uroporphyrinogen III cosynthase, this is converted to uroporphyrinogen III. In the absence of this enzyme, hydroxymethylbilane is converted non-enzymatically to uroporphyrinogen I. A series of enzyme-catalyzed reactions through isomers of the III series leads to the formation of protoporphyrin IXα. Haem is formed when iron is incorporated into the molecule in a reaction catalyzed by ferrochelatase.

The porphyrinogens are themselves unstable and become oxidized to their corresponding porphyrins when they are excreted in faeces or urine. Porphyrinogens and porphyrin precursors are colourless. Porphyrins are dark red in colour and intensely fluorescent. The major sites of porphyrin synthesis are the liver and the erythroid bone marrow.

The rate-limiting step in this sequence of reactions is the first, catalyzed by ALA synthase, which is susceptible to inhibition by the end-product, haem.

The porphyrias

These are a group of inherited diseases in which a partial deficiency of one of the enzymes of porphyrin synthesis leads to decreased formation of haem and thus, by releasing ALA synthase from inhibition, results in the formation of excessive quantities of porphyrin precursors (ALA and PBG) and porphyrins. When precursors are produced in excess, the clinical manifestations are primarily neurological (the precursors are neurotoxins). When porphyrins themselves are the major product, the predominant feature is photosensitivity; the porphyrins absorb light and become excited, inducing the formation of toxic free radicals. The porphyrias are diagnosed on the basis of their clinical features and the pattern of porphyrins and precursors present in blood and excreted in faeces and urine.

The porphyrias are classified as hepatic or erythropoietic, depending on the major site of abnormal metabolism, and as acute or non-acute, according to their clinical presentation (Fig.19.3). All the porphyrias are rare. Cutaneous hepatic porphyria is the most common but many cases are probably not inherited. Of the purely genetic types, acute intermittent porphyria is the most common, with a prevalence in the United Kingdom, where it occurs more frequently than in many countries, of only 1–2

Classification of the porphyrias		
acute	acute intermittent porphyria	hepatic
	hereditary coproporphyria	
	variegate porphyria	
chronic	cutaneous hepatic porphyria	
	congenital erythropoietic porphyria	erythropoietic
	erythropoietic protoporphyria	

Fig.19.3 Classification of the porphyrias.

per 100,000. Unusually for inherited metabolic diseases, their mode of inheritance is autosomal dominant, with the exception of congenital erythropoietic porphyria (autosomal recessive). The features of the porphyrias are summarized in Fig.19.4.

Acute porphyrias

Acute intermittent porphyria (AIP) is the commonest of these. Photosensitivity is never a feature of AIP although it may occur in patients with hereditary coproporphyria and variegate porphyria.

CLINICAL FEATURES

These conditions share the characteristics of acute attacks occurring separated by long periods of complete remission. The clinical features of acute attacks are summarized in Fig.19.5. They can be precipitated by various factors, including many drugs (see Fig.19.5); most commonly implicated are barbiturates, oral contraceptives and alcohol. These probably act by increasing the activity of ALA synthase, in many cases by increasing the synthesis of hepatic cytochrome P450 and hence the demand for haem, thereby decreasing intrahepatic haem concentration and releasing the enzyme from inhibition. Some drugs, notably the sulphonamides, inhibit PBG deaminase directly. Whatever the cause, the resulting increased activity of the metabolic pathway increases the formation of metabolites before the enzyme block.

Hormonal factors are also extremely important; symptoms rarely occur before puberty and may fluctuate in relation to menstruation or pregnancy.

Women are affected more commonly than men. In some ninety percent of individuals who inherit the defective gene for AIP, the disease remains clinically latent throughout adult life. In those in whom attacks do occur there is an additional enzyme deficiency, of steroid 5α-reductase, which may alter the metabolism of endogenous steroids in favour of the formation of epimers that induce ALA synthase.

DIAGNOSIS

In all acute porphyrias, excessive ALA and PBG are excreted in the urine during an acute attack. In suspicious circumstances, such as unexplained acute abdominal pain, peripheral neuritis or psychosis, the urine can be tested for PBG by a simple screening test, which if positive will indicate the need to carry out further investigations (see Fig.19.4) to establish the precise diagnosis. When an acute porphyria has been diagnosed, blood relatives should be screened for latent disease, and if necessary advised concerning the avoidance of precipitating factors. It is important to appreciate that the levels of porphyrins and their precursors in the blood, urine and faeces may be normal except during an attack, so that it may be necessary to measure the defective enzyme itself to establish who is at risk.

MANAGEMENT

Once the diagnosis of acute porphyria has been made, every effort must be made to prevent attacks by the avoidance of precipitating factors, which during an acute attack must be identified and treated appropriately. General supportive measures include maintenance of fluid and electrolyte balance,

Features of the porphyrias						
Condition	Enzyme defect	Photo-sensitivity	Neuro-logical	Abnormal levels of porphyrins:		
				red cell	urinary	faecal
acute intermittent porphyria	PBG deaminase	−	+	−	ALA **PBG**	−
hereditary coproporphyria	coproporphyrinogen III oxidase	+	+	−	ALA PBG **copro-**	copro-
variegate porphyria	protoporphyrinogen oxidase or ferrochelatase	+	+	−	ALA PBG **copro-**	copro- **proto-**
cutaneous hepatic porphyria	uroporphyrinogen decarboxylase	+	−	−	**uro-**	**isocopro-**
congenital erythropoietic porphyria	uroporphyrinogen III cosynthase	+	−	uro-* proto-	**uro-*** copro-	copro-*
erythropoietic protoporphyria	ferrochelatase	+	−	**proto-**	−	proto-
* I – isomers						

Fig.19.4 Features of the porphyrias. The most important abnormalities are in bold; the changes shown for the acute porphyrias may only be present during an attack.

Acute attacks of porphyrias		
Clinical features		**Factors involved**
Gastrointestinal	Central nervous system	drugs
abdominal pain	seizures	pregnancy
vomiting	depression	premenstrual
constipation	hysteria	infection
	psychosis	stress
Peripheral neuropathy		starvation
pain, stiffness and muscle weakness (limb and girdle muscles > trunk; upper limbs > lower; proximal > distal)	Cardiovascular	
paraesthesiae, numbness	sinus tachycardia systemic hypertension	

Fig.19.5 Clinical features and factors involved in acute attacks of porphyria; gastrointestinal, neuropsychiatric and cardiovascular derangements are all common.

adequate carbohydrate intake (intravenous fructose is often beneficial) and physiotherapy. Pain can be safely relieved with narcotic analgesics. Intravenous infusion of haematin, which decreases the activity of ALA synthase, has been used with success.

CASE HISTORY 19.1

A nineteen-year-old girl was admitted to hospital with colicky abdominal pain, which had started suddenly twelve hours before. She had vomited several times but had not opened her bowels since the pain started. Her abdomen was tender on examination but was otherwise normal. Her pulse was 140/min and her blood presure was 160/100mmHg. After she had been taken to the ward for observation, a nurse in the emergency room noticed that a specimen of the patient's urine, which had been collected for routine testing, had become a deep red colour although it had been normal when first passed. On being informed of this, the admitting doctor questioned the patient further and examined her more carefully. She said that she had also noticed cramping pains in her arms and was found to have bilateral partial wrist drop.

Investigations

screening test for urinary porphobilinogen:	strongly positive

quantitative analysis of urine:

porphobilinogen	very high
δ-aminolaevulinic acid	very high
uroporphyrin	slightly raised
coproporphyrin	slightly raised

Comment

Acute porphyrias may present as an acute abdomen; systemic hypertension and sinus tachycardia are often present. They are of course a very uncommon cause of abdominal pain and the diagnosis may be missed, at least initially. In this case, the nurse's observation of the changed colour of the urine was crucial and led to the presumptive diagnosis of an acute porphyria being made, supported clinically by the evidence of neuropathy and also by the positive screening test for PBG. There was no evidence of photosensitivity and the very high urinary excretion of porphyrin precursors, with only slightly increased excretion of intact porphyrins,

favours a diagnosis of acute intermittent porphyria rather than variegate or hereditary coproporphyria, both of which are anyway much less common.

The patient's symptoms and signs resolved rapidly with appropriate treatment. It transpired that she had started taking an oral contraceptive pill a few days before. The diagnosis was later confirmed by the demonstration of a reduced red cell PBG deaminase activity; she was advised to use an alternative method of contraception and told which drugs she should avoid. She remained well thereafter and no problems arose when she underwent elective surgery (cholecystectomy) three years later, nor during a subsequent pregnancy.

Non-acute porphyrias

Erythropoietic protoporphyria and congenital erythropoietic porphyria are both very rare. Photosensitivity occurs with both but is much more severe with the latter, causing extensive blistering and leading to tissue destruction and scarring.

Cutaneous hepatic porphyria (also known as porphyria cutanea tarda and symptomatic porphyria) also presents with photosensitivity. The initial lesion is just erythema but this progresses to the formation of vesicles and bullae, and eventually to scarring and pigmentation. In a very small number of cases the condition is inherited, but in the majority there is no family history although there is a deficiency of the enzyme uroporphyrinogen decarboxylase; it is not certain whether this is inherited or acquired in these cases. Precipitating factors can be identified in nearly all patients; the most important of these is alcohol and a history of excessive alcohol ingestion is present in more than ninety percent of cases. There is frequently evidence of liver disease. Other precipitating factors include various drugs and hepatotoxins.

MANAGEMENT
Management involves: the identification and removal of precipitating factors; venesection to remove excess iron from the liver (iron inhibits uroporphyrinogen III cosynthase and uroporphyrinogen decarboxylase); avoidance of direct sunlight; and the use of barrier creams to protect the skin.

Other causes of porphyrinuria

Increased urinary porphyrin excretion can occur in conditions other than the porphyrias. In patients with liver disease, particularly with cholestasis, the normal biliary excretion of porphyrins is impaired and there is increased urinary excretion — just as occurs with bilirubin. Porphyrinuria can also result from acquired defects in haem synthesis, as for example in lead poisoning. Lead inhibits ALA dehydratase (see Fig.19.2) and, to a lesser extent, coproporphyrinogen oxidase and ferrochelatase. As a result, the urinary excretion of ALA and coproporphyrin, and red cell protoporphyrin content may all be increased in lead poisoning. However, the measurement of blood lead concentration is to be preferred for the diagnosis of both lead poisoning and occupational overexposure to lead (see *Chapter 22*).

IRON

The total iron content of the adult body is approximately 4g (70mmol) of which some two-thirds is in haemoglobin. Iron stores (mainly spleen, liver and bone marrow) contain about one-quarter of the body's iron. Most of the remainder is in myoglobin and other haemoproteins; only 0.1% of the total body iron is in the plasma where it is almost all bound to a transport protein, transferrin.

Iron absorption and transport

The mean daily intake of iron is about 20mg, but less than ten percent of this is absorbed. The regulation of iron absorption is not fully understood. It is determined by the state of the body's iron stores, being increased when they are depleted and decreased when they are adequate. It is also increased when erythro-poiesis is increased (irrespective of the state of the iron stores).

The main site of iron absorption is the proximal small bowel. Iron is more readily absorbed in the Fe^{2+} form but dietary iron is mainly in the Fe^{3+} form. Gastric secretions are important in iron absorption in that they liberate iron from food (although haem can be absorbed intact) and promote the conversion of Fe^{3+} ions to Fe^{2+}. Ascorbic acid and other reducing substances facilitate iron absorption while phytic acid (in cereals), phosphates and oxalates form insoluble complexes with iron and prevent its absorption.

Once absorbed into intestinal mucosal cells, iron is either transported directly into the blood stream, or else combines with apoferritin, a complex iron-binding protein, to form ferritin. This iron is lost into the lumen of the gut when mucosal cells are shed. In iron deficiency, the apoferritin content of mucosal cells decreases and a greater proportion of absorbed iron reaches the blood stream.

In the blood, iron is transported bound mainly to transferrin, each molecule of which binds two Fe^{3+} ions. Transferrin is normally about one-third saturated with iron. In tissues, iron is bound in ferritin and haemosiderin. Free iron is very toxic and protein binding allows iron to be transported and stored in a non-toxic form.

Iron is lost from the body in faeces (non-absorbed and shed mucosal iron), by desquamation of skin and, in women, in menstrual blood loss. Very little iron is excreted in the urine.

Diagnostic tests for iron status

Iron status may require assessment when iron deficiency or overload is suspected or when the distribution or metabolism of iron is thought to be abnormal. Haematological tests used in this context include measurement of haemoglobin and red cell indices. Iron stores can be assessed directly by examination of the bone marrow, but measurement of serum ferritin concentration is the best non-invasive test for iron deficiency. It is rarely necessary to use biochemical tests merely to substantiate a diagnosis of iron deficiency, since this is by far the commonest cause of microcytic, hypochromic anaemia and the diagnosis is confirmed by a response to iron therapy.

Serum iron

The serum iron concentration is of little value in the investigation of iron metabolism, except in relation to haemochromatosis and in the diagnosis and management of iron poisoning. A fall in serum iron concentration is a late feature of iron deficiency, although a raised serum iron is usually present with iron overload. However, the concentration of iron in the serum of normal individuals fluctuates considerably; differences of more than twenty percent can occur within a few minutes, and of one hundred percent from one day to the next. Considerable catamenial variation occurs in women. Many conditions, including infection, trauma, chronic inflammatory disorders (especially rheumatoid arthritis) and neoplasia, are associated with low serum iron concentration (but normal iron stores), while others, for example hepatitis, cause an increase in concentration.

Serum total iron-binding capacity

Measurement of iron-binding capacity is effectively a functional measurement of transferrin concentration. Knowing the serum iron, the transferrin saturation can then be calculated; it is normally about thirty-three percent. Although serum total iron-binding capacity is increased in iron deficiency, many other factors can affect it while the saturation, dependent as it is upon the (highly variable) serum iron concentration, is itself highly variable. While it is true that low saturation is characteristic of iron deficiency, it also occurs in other conditions, such as pregnancy and chronic disease, in the absence of iron deficiency. The transferrin saturation is, however, a valuable screening test for idiopathic haemochromatosis (it is usually increased to a hundred percent in this condition) but is not useful otherwise.

Serum ferritin

Although serum ferritin concentration is more difficult to measure than iron or iron-binding capacity, it is by far superior to them for the assessment of body iron stores. The only known cause of a low serum ferritin concentration is a decrease in body iron stores; a concentration of less than $12\mu g/l$ suggests a complete absence of stored iron. In patients with anaemia and chronic disease, the serum ferritin concentration will indicate whether there is also iron deficiency and whether iron stores are adequate to meet the increased demand for incorporation into haemoglobin if the underlying condition can be treated successfully. Serum ferritin concentration is increased in iron overload, for example, in haemochromatosis, but may also be increased in some patients with liver disease and certain types of cancer, due to release of the protein from tissues. Raised concentrations should therefore be interpreted with caution, but a normal level militates against iron overload.

Iron deficiency

This may be due to inadequate intake, impaired absorption, excessive loss or a combination of these. The anaemia that develops is hypochromic and microcytic and if there is an obvious cause for iron deficiency, further investigation of the anaemia is not required. If the cause of an anaemia is in doubt, the finding of a low serum ferritin concentration will indicate iron deficiency.

Iron overload

This can occur with increased intestinal absorption of iron: either acutely, as in iron poisoning, or chronically, as is seen in peoples who traditionally cook their food in iron pots (Bantu siderosis). Increased parenteral iron administration occurs unavoidably in patients given repeated blood transfusions for the treatment of refractory anaemias and can also lead to overloading of the body's iron stores. The excess iron is deposited mainly as haemosiderin in reticuloendothelial cells in the liver and spleen where it is relatively innocuous, but with time parenchymal deposition may lead to hepatic fibrosis and myocardial damage.

Idiopathic haemochromatosis

The most severe iron overload is seen in patients with idiopathic haemochromatosis. This is an inherited metabolic disease, with an autosomal recessive mode of inheritance, characterized by excessive intestinal absorption of dietary iron; it shows a strong link with the histocompatibility gene HLA-A3. The molecular basis of the condition is unknown. The phenotypic expression in homozygotes depends upon the availability of dietary iron and overall iron turnover. Thus it is commoner in men than in women (because of menstrual iron loss), and when it does occur in women, does so on average at a later age. Even in men it is uncommon before the age of 40; although the defect is present from birth, it is only when the body becomes massively overloaded with iron that clinical features develop. Furthermore, the prevalence in homozygotes in countries with a high dietary content of available iron is higher than where it is low.

CASE HISTORY 19.2

A forty-five-year-old man presented with weight loss, lassitude and weakness. His skin was noticeably bronzed, although it was winter and he had not been out of the country. On examination, he was found to have hepato-splenomegaly, rather sparse body hair and small testes. On further questioning, he admitted that he had lost his libido and become impotent.

Investigations

urine	positive for glucose
blood glucose (fasting)	10mmol/l

serum:	iron	70μmol/l
	iron-binding capacity	67μmol/l
	ferritin	5000μg/l
	testosterone	9nmol/l
	luteinizing hormone	2mu/l

Comment

Skin pigmentation is virtually always present in idiopathic haemochromatosis, though it develops insidiously and may go unnoticed by the patient. It is a result of increased melanin deposition (and iron in advanced cases). Deposition of iron in the pancreas causes islet cell destruction and diabetes. Parenchymal iron deposition in the liver leads to cirrhosis which may be complicated by hepatoma formation; the liver disease is often exacerbated by excessive alcohol ingestion. Both primary and secondary hypogonadism can occur; in this case, the low luteinizing hormone level suggests that the hypogonadism is secondary to pituitary damage. The joints are often involved and deposition of iron in the myocardium can cause arrhythmias and cardiac failure.

The total iron-binding capacity is normal in this case but may be decreased as a result of the impaired ability of the liver to synthesize transferrin. It is, however, fully saturated with iron so that the serum iron concentration is always elevated and the serum ferritin concentration massively elevated. Values of several thousand micrograms per litre are typical of idiopathic haemochromatosis. The diagnosis of haemochromatosis can be confirmed by demonstration of the massive excess of parenchymal iron in a sample of liver obtained by percutaneous biopsy.

The families of patients with haemochromatosis should be screened to detect homozygotes for the defective gene, who are at risk of developing the condition and can be given prophylactic treatment. HLA typing and iron studies can be used for screening. Heterozygotes are not at risk; the results of iron studies in them may be normal but they can be detected by HLA typing.

MANAGEMENT AND PROGNOSIS

The mainstay of treatment of idiopathic haemochromatosis is repeated venesection; with each unit of blood, 200–250mg of iron are removed from the body. It is often possible to do this as often as once a week without rendering the patient anaemic. The serum iron and ferritin concentration are used to monitor treatment and once the excess iron has been removed, further accumulation can be prevented by less frequent (two to three monthly) venesection. Diabetes and heart failure are treated by conventional means, and hormonal deficiencies by appropriate replacement. Untreated, the prognosis is poor, but it is considerably improved by removal of the excess iron. There is often an improvement in cardiac and hepatic functions, but the diabetes, hypogonadism and joint disease are not affected.

Desferrioxamine, an iron-chelating agent, is valuable in patients receiving multiple blood transfusions for refractory anaemia and who are at risk of developing iron overload. Desferrioxamine has to be infused intravenously; unless this is performed daily, the rate of removal of iron is much slower than with venesection.

SUMMARY

Haemoproteins consist of a haem molecule (a tetra-pyrrole ring linked to an Fe^{2+} ion) bound to a protein. In haemoglobin, the protein consists of two pairs of identical polypeptide chains and four haem molecules; it is the latter which are responsible for binding oxygen. Haemoglobin contains two-thirds of the body's iron. Iron is essential for normal haemo-poiesis and iron deficiency is an important cause of anaemia. Many genetically determined variants of haemoglobin are known including haemoglobin S, responsible for sickle cell anaemia; some of these variants are of no clinical consequence but others, like HbS, can cause severe disease. Haemoglobin can undergo chemical changes in the blood, for example, binding carbon monoxide and forming carboxy-haemoglobin, which is incapable of transporting oxygen.

The synthesis of the tetrapyrrole ring of haem involves a complex metabolic pathway from glycine and succinyl-CoA through intermediates known as porphyrins. Inherited metabolic disorders are known affecting each of the enzymes of the haem synthetic pathway and are collectively called porphyrias. These conditions are classified into the acute porphyrias (whose effects are primarily neurological), for example, acute intermittent porphyria, and the chronic (with primarily cutaneous manifestations) such as cutaneous hepatic porphyria. The neuro-logical features are due to the accumulation of porphyrin precursors, whereas the accumulation of porphyrins themselves causes photosensitivity and hence leads to skin damage.

The porphyrias are also classified into hepatic and erythropoietic porphyrias, according to the major site of the enzyme abnormality. With the exception of congenital erythropoietic porphyria, which is inherited as an autosomal recessive, the porphyrias are unusual among the inherited metabolic disorders in having an autosomal dominant mode of inheritance. Although cutaneous hepatic porphyria may be inherited, it is often acquired, such as when it occurs with excessive alcohol intake. Each porphyria gives rise to a characteristic pattern of porphyrins and metabolites in blood, urine and faeces which can be used to make the diagnosis.

Dietary iron is absorbed in the proximal small intestine, more readily in the Fe^{2+} form. Almost all the iron in the plasma is protein-bound to transferrin. Iron deficiency can be due to inadequate intake or malabsorption, or excessive loss of iron, and gives rise to a microcytic, hypochromic anaemia. The serum iron concentration is an unreliable guide to the body's iron status; measurement of serum ferritin con-centration (ferritin is a primarily intracellular iron-binding protein) is the best biochemical test of iron status.

Free iron is highly toxic and iron poisoning, particularly in children, can be fatal. Chronic iron overload occurs in haemochromatosis, a genetically determined disorder characterized by excessive absorption of dietary iron which becomes deposited in many tissues of the body, including cardiac muscle, endocrine organs and parenchymal cells of the liver. Clinical features of haemochromatosis include skin pigmentation, cirrhosis, cardiomyopathy and im-paired endocrine function. The condition is best treated by repeated venesection to remove iron from the body. In haemosiderosis, in which iron accumulates, for example, as a result of repeated blood transfusion for refractory anaemia, the excess iron is deposited in reticuloendothelial cells and there is much less tissue damage.

FURTHER READING

Stanbury JS, Wyngaarden JB, Fredrickson DS, Goldstein J L & Brown MS (1983) *The Metabolic Basis of Inherited Disease*. 5th edition. New York: McGraw-Hill Book Company.

Cavill J, Jacobs A & Worwood M (1986) Diagnostic methods for iron status. *Annals of Clinical Biochemistry*, 23, 168–171.

20. Hyperuricaemia and Gout

INTRODUCTION

Uric acid is the end-product of purine nucleotide metabolism in humans. At physiological pH, uric acid is ninety-eight percent ionized and is therefore present mainly as the urate ion. In the extracellular fluid (ECF), where sodium is the predominant cation, uric acid effectively exists as a solution of its sodium salt, monosodium urate. This salt has a low solubility, the ECF becoming saturated at urate concentrations little above the upper limit of the normal range. In consequence, there is a tendency for crystalline monosodium urate to form in subjects with hyperuricaemia.

The most clinically obvious manifestation of this process is gout in which crystals form in the cartilage, synovium and synovial fluid of the joints. Other manifestations are deposition in the interstitial tissues of the kidney, which may cause renal damage; tophi, which are accretions of monosodium urate crystals in soft tissues; and precipitation in the urine which may lead to calculus formation.

URIC ACID METABOLISM

Purine nucleotides are essential components of nucleic acids; they are intimately involved in energy transformation and phosphorylation reactions and act as intracellular messengers. There are three sources of purines in man: the diet, degradation of endogenous nucleotides and *de novo* synthesis (Fig.20.1). Since purines are metabolized to uric acid, the body urate pool (and hence serum concentration) depends on the relative rates of both urate formation from these sources and urate excretion. Urate is excreted by both kidneys and gut, renal excretion accounting for approximately two-thirds of the total. Urate secreted into the gut is metabolized to carbon dioxide and ammonia by bacterial action (uricolysis).

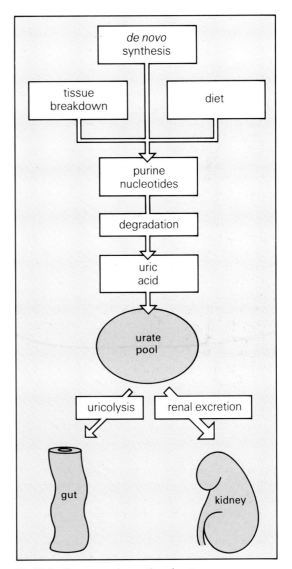

Fig.20.1 Sources and excretion of urate.

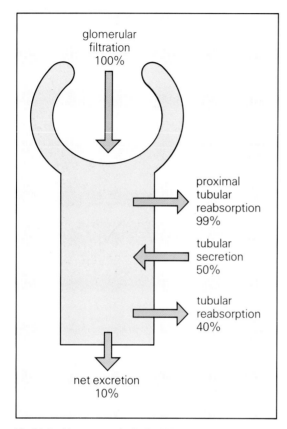

Fig.20.2 Urate excretion in the kidney.

Urate handling by the kidney is complex (Fig.20.2). It is filtered at the glomerulus and is almost totally absorbed in the proximal convoluted tubule; distally, both secretion and reabsorption occur. Normal urate clearance is about ten percent of the filtered load. In normal subjects, urate excretion is increased if the filtered load is increased. In chronic renal failure, the serum concentration rises only when the glomerular filtration rate falls below about 20ml/min.

Dietary purines account for only about thirty percent of excreted urate. The introduction of a purine-free diet reduces serum urate levels by only ten to twenty percent.

The metabolic pathways leading to uric acid synthesis are shown in outline in Fig.20.3. *De novo* synthesis (see page 267) leads to the formation of inosine monophosphate (IMP), which can be converted to the nucleotides adenosine monophosphate (AMP) and guanosine monophosphate (GMP). Nucleotide degradation involves the formation of the respective nucleosides (inosine, adenosine and guanosine); these are then metabolized to purines. The purine derived from IMP is hypoxanthine, which is converted by the enzyme xanthine oxidase first to xanthine and then to uric acid. Guanine can be metabolized to xanthine (and so to uric acid) directly, but adenine cannot. However, AMP can be converted to IMP by the enzyme AMP deaminase and, at the nucleoside level, adenosine can be converted to inosine. Thus, surplus GMP and AMP can be converted to uric acid and excreted.

However, the excretion of uric acid represents the waste of a metabolic investment since purine synthesis requires considerable energy expenditure. Pathways exist whereby purines may be salvaged and converted back to their parent nucleotides. For guanine and hypoxanthine, this is accomplished by the enzyme, hypoxanthine–guanine phosphoribosyl transferase (HGPRT), and for adenine by adenine phosphoribosyl transferase (APRT).

URIC ACID IN SERUM

Serum urate concentrations are, in general, higher in men than in women, tend to rise with age (Fig.20.4), and be elevated in people in the higher socioeconomic groups and in the obese. There is considerable variation in serum urate concentrations between different ethnic groups.

In adult males in the United Kingdom, the upper limit of normal is usually taken as 0.42mmol/l. In an aqueous solution of pH 7.4, at 37°C, and with an ionic strength similar to that of plasma, the solubility of monosodium urate is 0.57mmol/l; in serum, the presence of protein appears to reduce this somewhat. The risk of gout increases with increasing serum urate levels (Fig.20.5). Undoubtedly, many factors are involved in the precipitation of monosodium urate crystals in connective tissue; gout may not necessarily occur with hyperuricaemia but the presence of hyperuricaemia is a prerequisite for the development of gout. The solubility of monosodium urate declines rapidly with decreasing temperature and this may, to some extent, explain the tendency for the more peripheral joints, which have lower intra-articular temperatures, to be more frequently affected.

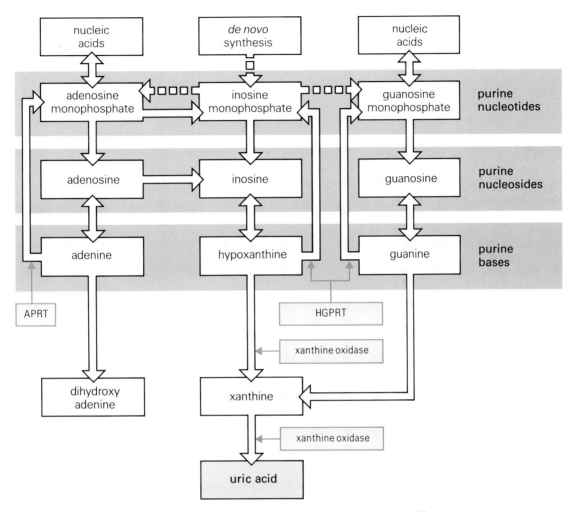

Fig.20.3 Simplified diagram of the pathways of purine nucleotide metabolism and uric acid synthesis in man. APRT = adenine phosphoribosyl transferase; HGPRT = hypoxanthine–guanine phosphoribosyl transferase.

HYPERURICAEMIA

Hyperuricaemia may occur due to increased formation of uric acid, decreased excretion, or a combination of both. Some causes of increased formation are given in Fig.20.6.

When hyperuricaemia is due to decreased excretion, it is renal excretion of urate that is usually decreased. Indeed, in hyperuricaemia, the total amount of urate removed by uricolysis in the gut is increased. Reference to Fig.20.2 will show that decreased renal urate excretion could result from a decrease in either filtration or tubular secretion. Since almost all filtered urate is reabsorbed, the glomerular filtration rate (GFR) itself is only important in chronic renal failure in that it reflects the total number of functioning nephrons remaining. Serum urate only rises late in chronic renal failure, but many factors can affect tubular function and thereby cause hyperuricaemia; the more important of these are given in Fig.20.6. Excessive alcohol ingestion probably increases *de novo* purine synthesis, but any increase in lactate production due to alcohol may also impair urate excretion.

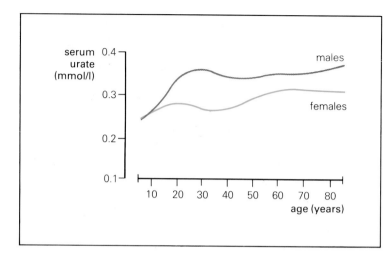

Fig.20.4 Mean serum urate concentrations in men and women.

Serum urate (mmol/l)	Risk of developing gout (%)	
	males	females
<0.41	2	3
0.42–0.47	17	17
0.48–0.53	25	*
>0.54	90	*
*insufficient data; such levels are exceptional in females		

Fig.20.5 Risk of developing gout in relation to serum urate concentration.

Causes of hyperuricaemia	
Increased formation	**Reduced excretion by the kidneys**

high dietary intake of uric acid
 	
increased purine synthesis:
  idiopathic
  inherited metabolic disease
 	
increased nucleic acid turnover:
  myeloproliferative disorders
  psoriasis
  secondary polycythaemia
  carcinomatosis
  chronic haemolytic anaemias
  cytotoxic drugs
	chronic renal disease
	
	drugs:
	  diuretics (especially thiazides,
	    not spironolactone)
	  salicylates
	
	poisons:
	  lead
	
	organic acids,
	  e.g. lactic, acetoacetic and
	    β-hydroxybutyric acids
	
	hyperparathyroidism

Fig.20.6 Causes of increased formation of uric acid and reduced uric acid excretion by the kidneys. Note that salicylates reduce uric acid excretion at low doses only; at high doses (>4g/day) aspirin is uricosuric as it blocks the tubular reabsorption or uric acid.

Gout

Gout is customarily classified as primary (idiopathic) or secondary (when a condition known to cause hyperuricaemia is present). However, gout is uncommon when hyperuricaemia develops secondarily to other conditions. The tendency for hyperuricaemia and gout to be familial has led to investigation for a causal inherited metabolic defect. Although there are a few rare conditions in which such a defect does lead to hyperuricaemia, none has been found in the great majority of cases of primary gout. Some seventy percent of patients appear to excrete urate at a rate inappropriately low for the serum concentration, while about thirty percent have excessive urate production. Dietary factors and alcohol ingestion exacerbate hyperuricaemia in about half the cases, but while their amelioration may reduce the serum urate levels somewhat, these still remain elevated. Gout is rare in premenstrual women, in whom mean serum levels of urate are much lower than in men of corresponding ages (see Fig.20.4), but the incidence increases markedly after the menopause.

CASE HISTORY 20.1

An obese fifty-five-year-old male was awoken from sleep, after spending the evening at a business dinner, by excruciating pain in his left first metatarsophalangeal joint. He was unable to put his foot to the floor. The affected joint was hot, swollen, red and extremely tender. He was treated with indomethacin and the symptoms resolved rapidly. One year previously he had had an episode of renal colic but had declared himself to be too busy to be investigated in connection with this.

Investigations

serum urate   0.78mmol/l

Comment

This is the classic presentation of gout. The onset is often sudden and nocturnal. In seventy percent of cases the metatarsophalangeal joint of the great toe is the first to be affected. The fundamental signs of inflammation are present and hyperuricaemia was confirmed. In this case, the previous episode of renal colic may well have been due to a renal urate stone. Gout is more common in men and is associated with higher social class, driving (type A) personality, obesity, hypertriglyceridaemia, hypertension and excessive food and alcohol intake.

Differential diagnosis

The differential diagnosis includes other crystalline arthropathies and septic arthritis. In gout, examination of aspirated joint fluid will reveal typical $2-10\mu m$ long needle-shaped crystals which show strong negative birefringence when viewed with polarized light.

Pathogenesis

Monosodium urate crystals forming in joints are engulfed by neutrophil leucocytes but damage the lysosomal membranes of these cells, so causing cellular disruption. Lysosomal enzymes are thus released into the joint, precipitating an acute inflammatory reaction.

Management

Anti-inflammatory drugs (of which indomethacin is the most efficacious) are used to treat acute gout, but have no effect on the hyperuricaemia. Allopurinol is used in the long-term treatment of gout when there have been recurrent acute attacks, when there is renal damage or renal calculi with hyperuricaemia, in tophaceous gout and if serum urate concentrations persistently exceed 0.6mmol/l. Allopurinol is a competitive inhibitor of xanthine oxidase; it is thus hypouricaemic. Urinary urate excretion is decreased and there is an increase in the excretion of xanthine, which is more water-soluble than uric acid.

Purine-rich foods, for example, offal and fish roes, should be avoided and alcohol intake reduced.

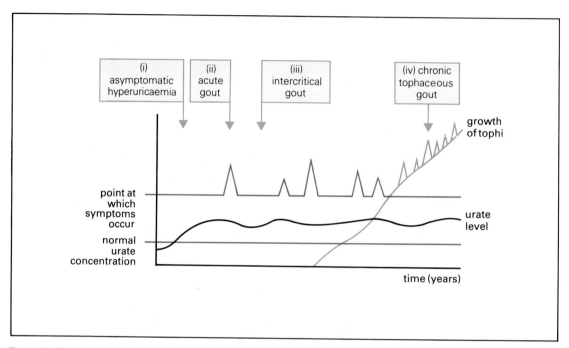

Fig.20.7   The natural history of gout. Modified from Dieppe & Calvert (1983).

Stages of gout

Four stages in the natural history of gout have been described (Fig.20.7). Asymptomatic hyperuricaemia (i) can be present for years before an acute attack (ii) is precipitated, for example, by trauma or dietary indiscretion. Symptom-free periods of months or years follow ('intercritical gout'; iii) punctuated by acute attacks leading, if untreated, to chronic tophaceous gout (iv). Since the introduction of allopurinol, tophaceous gout, once common, is now rarely seen. It tends to occur in elderly women treated with diuretics for many years (in particular thiazides, which inhibit renal tubular secretion of urate) rather than as a sequel to recurrent attacks of acute gout.

Rare causes of hyperuricaemia

There are a number of rare, inherited metabolic diseases associated with hyperuricaemia and gout (Fig.20.8). In all of them, hyperuricaemia results from increased uric acid synthesis.

HYPOURICAEMIA

This is uncommon and clinically inconsequential. It may be due to either decreased urate synthesis or to increased excretion and so is seen in congenital xanthine oxidase deficiency (xanthinuria), severe liver disease and renal tubular disorders such as the Fanconi syndrome. It may also result from excessive medication with allopurinol and the use of uricosuric drugs such as probenecid.

OTHER CRYSTALLINE ARTHROPATHIES

Gout is not the only crystalline arthropathy. The deposition of calcium pyrophosphate in joints may mimic gout clinically (pseudogout) and chondro-calcinosis (calcium deposition in joint cartilage) may be present. The cause of this condition is unknown. The rhomboid-shaped crystals formed show weak positive birefringence when viewed with polarized light. The deposition of hydroxyapatite in joints has also been described.

Major inherited metabolic diseases associated with hyperuricaemia	
Enzyme abnormality	**Consequence**
hypoxanthine–guanine phospho-ribosyltransferase deficiency (Lesch–Nyhan syndrome and less severe variants)	decreased activity of salvage pathway decreases purine reutilization and thus increases uric acid synthesis
glucose-6-phosphatase deficiency (glycogen storage disease type 1)	(i) increased metabolism of glucose-6-phosphate through pentose phosphate pathway increases formation of ribose-5-phosphate, a substrate for purine nucleotide synthesis (ii) hyperlactataemia decreases uric acid secretion in renal tubules
phosphoribosylpyrophosphate synthetase (PRPP synthetase) variant (with increased activity)	PRPP is a substrate for purine nucleotide synthesis and also activates the rate limiting enzyme

Fig.20.8 Inherited metabolic diseases associated with hyperuricaemia.

SUMMARY

Uric acid is the end product of the metabolism of purines. These substances are components of nucleotides which in turn are components of nucleic acids. They also have an essential role in energy transformations; act as intracellular messengers; and are involved in neurotransmission.

The pathological significance of uric acid is related primarily to its low solubility. At concentrations little above those in which it is normally present in body fluids, uric acid can precipitate in the form of monosodium urate crystals. In joints this causes gout, an acute inflammatory arthritis and can lead to chronic, destructive joint disease; in the kidneys, renal damage can occur and uric acid can precipitate from the urine to form calculi; and with long-standing hyperuricaemia, accretions of urate known as tophi can form in soft tissues.

The concentration of uric acid in the serum depends upon the relative rates of its formation and excretion. Formation in turn depends upon the rate of purine turnover which is dependent upon the balance between dietary purine intake, *de novo* synthesis and degradation. Excretion takes place through the kidneys (about two-thirds) and gut (where uric acid is broken down by bacterial action).

Hyperuricaemia can occur secondarily to many conditions, in some of which there is increased urate formation (e.g., malignant disease, especially haematological malignancies) and in others, decreased excretion (e.g., renal failure). Gout, however, is more frequently seen in patients who do not suffer from such conditions and in these cases is known as primary or idiopathic gout. There is frequently a family history of the condition but although a small number of rare inherited metabolic disorders are known in which the presence of an abnormal enzyme leads to excessive urate synthesis, in the majority of patients, no such defect is demonstrable. About thirty percent of patients with idiopathic gout appear to have excessive urate production; in the remainder, the rate of urate excretion appears to be inappropriately low for the rate of production. Dietary factors and alcohol often exacerbate hyperuricaemia.

Patients with acute episodes of gout require treatment with analgesics and anti-inflammatory drugs; in the long term, serum urate levels can be reduced by dietary modification and the use of allopurinol, an inhibitor of xanthine oxidase, the enzyme responsible for the synthesis of uric acid. This drug thus reduces the formation of uric acid and instead xanthine, which is far more soluble, becomes the principal end product of purine metabolism.

Pseudogout can resemble gout clinically. This condition is due to the deposition of crystals of calcium pyrophosphate in joints. These can be distinguished from monosodium urate crystals by using polarizing microscopy.

FURTHER READING

Dieppe P & Calvert P (1983) *Crystals and Joint Disease*. London: Chapman and Hall.

Stanbury JB, Wyngaarden JB, Fredrickson DS, Goldstein JC & Brown MS (1983) *The Metabolic Basis of Inherited Disease*. 5th edition. New York: McGraw-Hill Book Company.

21. Metabolic Aspects of Malignant Disease

INTRODUCTION

The clinical signs and symptoms in patients suffering from cancer are often directly related to the physical presence of the tumour. For example, the tumour may destroy essential normal tissue, cause obstruction of ducts, or exert pressure on nerves. Systemic manifestations, including cachexia and pyrexia, are also commonly present and indeed may be the only evidence of the presence of a tumour; in some patients, they may resemble those of a known endocrine syndrome. This would be expected with a tumour of endocrine origin, such as a malignant insulinoma producing hypoglycaemia or an adrenal carcinoma producing Cushing's syndrome. However, the tumour may commonly be sited in tissue, for example, the bronchus, that is not of obvious endocrine origin.

In some cases, there is good evidence that such syndromes are due to the secretion of a hormone by the non-endocrine tumour. This has been termed *ectopic* hormone secretion since the hormone is not secreted from its normal site, while *eutopic* hormone secretion describes secretion from the endocrine gland. The term ectopic hormone secretion is often used loosely; in some cases when an endocrine syndrome is present in a patient with a non-endocrine tumour there is little direct evidence that a known hormone is being secreted. For this reason, the term 'paraneoplastic syndrome' is preferable; this also has the advantage that it embraces all the systemic manifestations of cancer not directly related to the physical presence of the primary tumour, whether or not they are due to a hormone.

This chapter discusses paraneoplastic syndromes, certain familial endocrine syndromes and tumour markers, substances whose presence is a reflection of tumour activity and whose concentration can be determined as an aid to the diagnosis or monitoring of malignant disease.

PARANEOPLASTIC SYNDROMES

Origins and classification

The mechanism whereby a non-endocrine tumour can produce a known hormone or other humoral agent remains unknown. All somatic cells are believed to contain a full complement of genes and presumably certain of these, related to hormone production, become derepressed in the paraneoplastic syndromes. However, these syndromes are more common with certain tumours, notably small-cell carcinoma of the bronchus, and particular tumours tend to be associated with particular syndromes. Renal cell carcinoma, for example, is associated with a form of hypercalcaemia, similar in many ways to that seen in hyperparathyroidism, but not with other endocrine syndromes. It therefore seems unlikely that these syndromes arise as a result of random derepression of genes.

Paraneoplastic syndromes are frequently associated with tumours comprising cells of the amine precursor uptake and decarboxylation (APUD) system. This term was originally used to describe tumours of neuroectodermal origin sharing similar amine-handling characteristics. In fact, most of these tumours have low molecular weight polypeptides, including hormones, as their major product. Some of the cells of the APUD series are shown in Fig.21.1. APUD cells are also present in other tissues, for example, the bronchi, although their function in this location is unknown. Small-cell carcinoma of the bronchus, which is frequently implicated in paraneoplastic syndromes, is a tumour of APUD origin. However, paraneoplastic syndromes also occur in association with tumours that do not appear to be derived from APUD cells. No single distinctive feature has been shown to be common to all non-endocrine tumours which secrete hormones.

Endocrine cells of the APUD series	
Cell/tissue	**Hormone**
hypothalamus	oxytocin, vasopressin; releasing/inhibitory hormones
anterior pituitary	TSH, ACTH, FSH, LH, prolactin and growth hormone
pancreatic islets	insulin and glucagon
pancreatic islet and non-islet	somatostatin, VIP and pancreatic polypeptide
gastric endocrine	gastrin and glucagon
intestinal endocrine (argentaffin cells)	e.g. secretin, cholecystokinin and enteroglucagon
thyroid C cells	calcitonin
parathyroid	parathyroid hormone

Fig.21.1 Endocrine cells of the APUD series and the hormones they produce.

Tumours associated with endocrine syndromes are shown in Fig.21.2. The commonest paraneoplastic endocrine syndromes encountered clinically are Cushing's syndrome, ectopic ADH secretion and hypercalcaemia. Calcitonin secretion is common, but clinically silent.

Cushing's syndrome

Cushing's syndrome is the condition which results when tissues are exposed to supraphysiological concentrations of glucocorticoids. It is discussed in detail in *Chapter 9*.

CASE HISTORY 21.1

A retired warehouseman presented with muscle weakness and back pain. He had also lost 5kg in weight in the previous two months and had recently been passing more urine than usual. He had smoked twenty-five to thirty cigarettes a day for many years but had generally enjoyed good health. On examination, in addition to the weakness and signs of weight loss, he was found to have glycosuria and was hypertensive, but his appearance was otherwise normal and no abnormal physical signs were elicited.

Investigations

serum:	sodium	144mmol/l
	potassium	2.2mmol/l
	bicarbonate	39mmol/l

blood glucose   10.2mmol/l

High-dose dexamethasone suppression test:

plasma cortisol
  (0900h)   1520nmol/l
  (0900h) after dexamethasone
  2mg 4 times daily for 2 days   1500nmol/l
plasma ACTH   460ng/l
  (normal <80ng/l)

A discrete mass was present in the left lower zone on chest radiography.

Comment

The greatly elevated plasma cortisol and ACTH levels are typical of ectopic ACTH secretion. Plasma ACTH levels are generally much higher than those seen in Cushing's disease, except when

Some non-endocrine tumours associated with paraneoplastic hormone secretion		
Tumour	Hormone	Syndrome
small cell carcinoma of bronchus*	ACTH (and precursors) vasopressin hCG	Cushing's syndrome dilutional hyponatraemia gynaecomastia
squamous cell carcinoma of bronchus	?	hypercalcaemia
breast carcinoma	calcitonin	none
carcinoid tumours*	ACTH	Cushing's syndrome
renal cell carcinoma	? erythropoietin	hypercalcaemia polycythaemia
hepatocellular carcinoma	? NSILA	hypercalcaemia hypoglycaemia
mesenchymal tumours	NSILA	hypoglycaemia
*known or possible APUD tumours		

Fig.21.2 Non-endocrine tumours associated with paraneoplastic hormone secretion. Only the common associations of hormone production and non-endocrine tumours are included. NSILA=non-suppressible insulin-like activity.

a carcinoid or thymic tumour is responsible. Since the ACTH secretion is not under normal feedback control, the hypercortisolaemia is not suppressed by dexamethasone.

The clinical picture of Cushing's syndrome is typically dominated by the metabolic sequelae of excessive cortisol secretion, as in this case. These include hypokalaemia with alkalosis, which exacerbates the physical weakness due to steroid-induced myopathy; glucose intolerance, sometimes sufficient to cause frank diabetes; and hypertension. Osteoporosis predisposes to crush fractures of the vertebrae and the presence of secondary tumour deposits may also give rise to back pain. On the other hand, the classic somatic manifestations of Cushing's syndrome are often absent, a reflection of the very rapid progression of the condition in most cases. ACTH-secreting carcinoid and thymic tumours are an exception; the clinical syndrome in these cases may closely resemble Cushing's disease even to the extent that ACTH secretion, and hence that of cortisol, is suppressed by dexamethasone.

Ectopic ACTH secretion by non-endocrine tumours is common. Evidence of it has been found in upto fifty percent of patients with small-cell bronchial carcinomas, though massive secretion, giving rise to the typical features as shown by this case, is uncommon. Precursors of ACTH are secreted in some cases.

With bronchial carcinomas, the prognosis is usually very poor unless the tumour is suitable for surgical excision. As discussed on page 134, drug treatment may provide symptomatic relief.

Ectopic antidiuretic hormone secretion

A case of this syndrome is described in *Case history 2.2.* The secretion of antidiuretic hormone (vasopressin) by the tumour is uncontrolled and thus likely to be greater than the body's normal requirements, resulting in water retention with dilutional hyponatraemia. When this is mild and develops slowly, it is asymptomatic; however, severe hyponatraemia in this condition indicates water intoxication which can be fatal. The symptoms (drowsiness, confusion, fits and coma) may mimic cerebral metastases. Ectopic antidiuretic hormone secretion is most commonly seen with small-cell carcinomas of the bronchus but other tumours may be responsible. A similar syndrome results from inappropriate secretion of antidiuretic hormone and occurs in a variety of non-malignant diseases (see page 23).

Tumour-associated hypercalcaemia

Hypercalcaemia is common in malignant disease. When bony metastases are present, dissolution of calcium from bone by the secondary tumour is an important factor. However, hypercalcaemia can occur in the absence of metastases and the clinical syndrome is commonly suggestive of the presence of a tumour secreting parathyroid hormone.

CASE HISTORY 21.2

An elderly man presented with loin pain and increasing thirst. Examination of the urine showed haematuria but no glycosuria.

Investigations

serum:	calcium	3.2mmol/l
	phosphate	0.7mmol/l
	alkaline phosphatase	80iu/l
	parathyroid hormone	below detection limit of the assay

An intravenous urogram showed an irregularly enlarged left kidney with a distorted pelvicalyceal system.

Arteriography demonstrated an abnormal circulation in the left kidney, strongly suggestive of a tumour.

Skeletal survey and isotopic bone scan showed no evidence of metastatic disease and the chest radiograph was normal.

At operation, a tumour was seen in the upper part of the left kidney. The patient underwent nephrectomy and made an uneventful recovery. Following the operation his serum calcium concentration decreased and subsequently remained normal.

Comment

The combination of hypercalcaemia and hypophosphataemia are compatible with excessive parathyroid hormone secretion, although the serum phosphate may be normal or even raised if there is renal impairment. The absence of parathyroid hormone suggests that secretion of the hormone by the parathyroid glands is suppressed. Parathyroid hormone assays suffer from a number of problems that affect their reliability and clinical usefulness (see page 192), but in the majority of cases of tumour-associated hypercalcaemia that have been fully investigated there has been no direct evidence of parathyroid hormone secretion. The nature of the clinical syndrome and its resolution in cases where the tumour can be removed suggest that a humoral factor, which has effects on calcium and phosphate metabolism similar to those of parathyroid hormone, is involved and a tumour-derived polypeptide, showing some sequence homology with parathyroid hormone has recently been described.

Even if bony metastases are present in a patient with a malignant disease and hypercalcaemia, it seems likely that a humoral agent is secreted which directly affects calcium metabolism. Hypercalcaemia itself can affect renal function, but if this is normal the suppression of endogenous parathyroid hormone secretion should decrease renal tubular reabsorption of calcium, allowing the excess calcium mobilized from bone to be excreted. The persistence of hypercalcaemia suggests that renal excretion of calcium is decreased in such cases and the inference is that a humoral agent, secreted by the tumour, is responsible. In haematological malignancies, particularly myeloma, hypercalcaemia is due to the release of lymphokines, termed osteoclast-activating factors, by the tumour cells which activate the osteoclasts. Osteoclasts may also be activated by prostaglandins produced by tumour metastases in bone, for example, metastases from breast carcinoma.

Tumour-associated hypoglycaemia

This condition is discussed in detail in *Chapter 13*. It is only rarely due to ectopic insulin secretion by non-B

Cancer cachexia	
Characteristics	**Pathogenesis**
anorexia and early satiety	inadequate food intake
weight loss	excessive loss of body protein
muscle weakness	malabsorption
non-specific anaemia	increased metabolic rate
altered host metabolism often with increased metabolic rate	abnormal host metabolism e.g. related to hormone secretion or other humoral factors
	trapping of nutrients by tumour

Fig.21.3 Characteristics and pathogenesis of cancer cachexia.

cell tumours. Other humoral factors, for example, somatomedins and growth factors, and the metabolic activity of the tumours themselves are thought to be responsible. Much of the hypoglycaemic activity of normal serum is associated with such substances. This activity persists when insulin secretion is suppressed and is termed non-suppressible insulin-like activity (NSILA).

Other paraneoplastic endocrine syndromes

Gynaecomastia may occur in patients with bronchial carcinomas, due to secretion of human chorionic gonadotrophin (hCG). Precocious puberty may develop in male children with hepatic tumours secreting hCG, although this is very rare. Secretion of erythropoietin is responsible for the polycythaemia which is sometimes a feature of renal cell tumours, uterine fibromyomata and the rare cerebellar haemangioblastoma.

Paraneoplastic syndromes are common, but it must be remembered that an endocrine syndrome in a patient with a tumour may be due to coexistent endocrine disease and not necessarily to the secretion of a hormone or other factor by the tumour.

CANCER CACHEXIA

The characteristics of cancer cachexia are summarized in Fig.21.3. It is a common feature of malignant disease although its causes are imperfectly understood. Deficient food intake, due either to mechanical obstruction of the alimentary tract or to the anorexia that is often present in malignant disease, may be partly responsible, and there may also be loss of body protein from ulcerated mucosa or blood loss. The basal metabolic rate is increased in some patients with cancer, the tumour itself requiring energy and nitrogen for growth which are met from body stores if intake is insufficient.

The metabolism of many tumours is primarily anaerobic. The effect of this is that lactate is produced which is converted back to glucose by gluconeogenesis in the liver and kidney. While glycolysis results in the net liberation of only two molecules of ATP per molecule of glucose, gluconeogenesis, the reverse process, requires six molecules of ATP. The presence of an anaerobically metabolizing tumour thus represents a considerable net drain on the energy stores of the host. Tumours can adversely affect many of the metabolic processes of the body and humoral factors elaborated by the tumour may be mediators of these effects.

In the majority of cases, the pathogenesis of cancer cachexia is probably multifactorial. The possible causes are summarized in Fig.21.3. Management is a difficult problem and nutritional support, though an apparently obvious solution, could in theory accelerate tumour growth although there is little evidence that this does occur to any significant extent in man.

CARCINOID TUMOURS

Carcinoid tumours arise from the argentaffin cells of the gut, cells of the APUD series; ninety per cent of these tumours are found in the appendix and ileocaecal region but they also occur elsewhere in the gut, gallbladder, biliary and pancreatic ducts, and in the bronchi. They are of low-grade malignancy; while they frequently invade local tissue, distant metastases are rare.

The carcinoid syndrome is a result of the liberation of vasoactive amines, such as serotonin and peptides (including substance P), from the tumour into the circulation. It is usually only seen with bronchial tumours, which liberate their products directly into the systemic circulation, or when tumours in the gut have metastasized to the liver. Since the greater part of the gut is drained by the portal circulation, the secreted products of tumours in the gut pass to the liver and are inactivated there. However, the secreted products of hepatic metastases reach the systemic circulation via the hepatic veins.

Serotonin (5-hydroxytryptamine, 5-HT) is synthesized from tryptophan (Fig.21.4). In patients with carcinoid syndrome, fifty percent of dietary tryptophan (rather than the usual one percent) may be metabolized by this pathway, diverting tryptophan away from protein and nicotinic acid synthesis. (Pellagra-like skin lesions due to nicotinic acid deficiency are an occasional feature of the carcinoid syndrome.) The major amine secreted by intestinal carcinoid tumours (derived from embryonic midgut) is 5-hydroxytryptamine. Bronchial carcinoids (derived from foregut) tend to produce 5-hydroxytryptophan since they often lack the decarboxylase enzyme. All carcinoid tumours may also produce histamine, kinins and substance P which are important in the symptomatology of the carcinoid syndrome. Further, the secretion of peptide hormones, including ACTH, calcitonin and other products of cells of the APUD series, is often demonstrable and may contribute to the clinical presentation.

CASE HISTORY 21.3

A fifty-year-old woman presented with a history of episodic facial flushing and dizziness, sometimes accompanied by wheezing respiration.

These attacks could occur at any time but she was frequently embarrassed by them at meal times.

Investigations

urinary 5-hydroxyindoleacetic acid excretion
270µmol/24h (normal 10–50µmol/24h)

An isotopic scan of the liver revealed multiple filling defects suggestive of tumour deposits.

A distorted hepatic vasculature with evidence of tumour circulation was demonstrated on arteriography, but the primary tumour could not be located.

Comment

Facial flushing is the commonest clinical feature of carcinoid syndrome and may be provoked by the ingestion of food or alcohol, or by emotional stimuli. It may become continuous and spread to other parts of the body. The vasodilatation causes transient hypotension and patients may complain of dizziness. Other clinical features are listed in Fig.21.5 and include intermittent abdominal discomfort, diarrhoea and bronchospasm with wheezing. Right-sided valvular lesions of the heart, particularly pulmonary stenosis, may lead to cardiac failure.

The diagnosis is confirmed by demonstrating an increase in the urinary excretion of 5-hydroxyindoleacetic acid. This is usually more than twice the upper limit of normal and may be much greater. Foodstuffs containing serotonin (bananas, tomatoes) or drugs such as reserpine, which stimulate endogenous serotonin release, must be avoided during the urine collection.

Management

Carcinoid tumours are difficult to manage. Once metastases have developed, surgical removal is usually not possible though partial resection may be palliative since the tumours are usually slow-growing. Medical treatment, for example, with serotonin antagonists such as methysergide, is rarely fully effective at controlling symptoms. *p*-Chlorophenylalanine, an

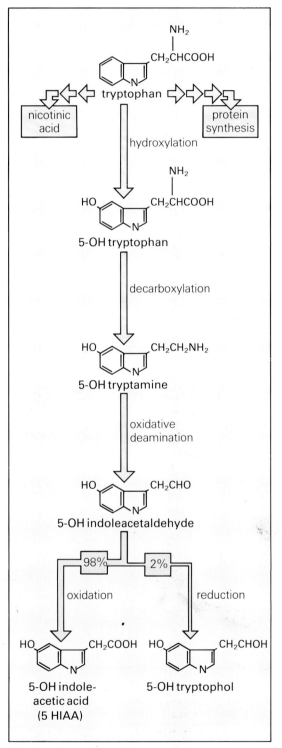

Fig.21.4 Metabolism of 5-hydroxyindoles.

Clinical features of the carcinoid syndrome

Gastrointestinal

discomfort, hyperperistalsis
    and borborygmi
diarrhoea
nausea and vomiting
colicky pain

Cardiovascular

flushing
pulmonary stenosis (may lead to
    right heart failure and occasionally
    mitral stenosis)

Respiratory

bronchospasm
variable rate and depth of breathing

Other

pellagra
manifestations of secretion of other
    hormones

Fig.21.5 Clinical features of the carcinoid syndrome.

inhibitor of tryptophan 5-hydroxylase, has been used, but the best results are seen with cytotoxic drugs or, in selected cases, with destruction of the blood supply to hepatic metastases by embolization of the hepatic artery.

PLURIGLANDULAR SYNDROMES

The pluriglandular syndromes, also known as syndromes of multiple endocrine adenopathy (MEA) and syndromes of multiple endocrine neoplasia (MEN), are familial disorders with an autosomal dominant inheritance, in which tumours (benign or malignant) or hyperplasia develop in two or more endocrine glands. These syndromes are uncommon, but it is important to recognize that a patient presenting with certain endocrinopathies could have one of these syndromes. The glands affected in the two major types of pluriglandular syndrome are shown in Fig.21.6. Although the syndromes are inherited, the predominant features vary in different members of the same family; thus one person may present with recurrent peptic ulceration due to a gastrinoma (Zollinger–Ellison syndrome), while a sibling may have renal calculi as a result of hyperparathyroidism.

TUMOUR MARKERS

Tumour markers are substances which can be related to the presence or progress of a tumour. They include substances secreted into body fluids by tumours and antigens expressed on cell surfaces. Clinical chemistry laboratories are usually only involved in the measurement of tumour markers falling into the first category.

The concept that a tumour can be diagnosed and its progress monitored by measurement of a tumour marker is an attractive one, but at present, in spite of the expenditure of much effort, the number of clinically useful markers is still very small. Some of the problems associated with them are discussed below; poor specificity and inadequate sensitivity are particularly important.

α-Fetoprotein

α-Fetoprotein is a glycoprotein of molecular weight 67,000 daltons. It is synthesized by the yolk sac and the fetal liver and gut. In the fetus, it is a major serum protein; in adults, the normal concentration is less than 16ug/l.

Increased serum levels of α-fetoprotein are seen in normal pregnancy (its use in the diagnosis of neural tube defects is discussed in *Chapter 18*) and in hepatic regeneration as may occur, for example, following hepatitis. However, elevated levels are also found in some seventy percent of patients with hepatocellular carcinomas and in up to ninety percent with malignant testicular teratomas. Liver cancer is becoming more frequent in the United Kingdom, commonly developing in the cirrhotic liver. While the lack of sensitivity limits the use of α-fetoprotein in the diagnosis of hepatic cancer, in cases where it is raised

---

Glands affected in multiple endocrine adenomatosis
MEA type I
parathyroids
pancreatic islets
anterior pituitary
adrenal cortex
thyroid (follicular cells)
MEA type IIa
thyroid (medullary cell carcinoma)
adrenal medulla (phaeochromocytoma)
MEA type IIb
thyroid (medullary cell carcinoma)
adrenal cortex (phaeochromocytoma)
parathyroids
various somatic abnormalities,
Marfanoid habitus
mucosal neuromata
pigmentation

Fig.21.6 Glands affected in multiple endocrine adenomatosis (MEA).

and which are amenable to treatment, repeated serum α-fetoprotein measurements are a useful guide to the efficacy of treatment.

CASE HISTORY 21.4

A two-year-old boy presented with progressive abdominal swelling. On examination, the liver was found to be massively enlarged. Ultrasound and radiological examinations suggested the presence of a tumour and histological examination of tissue obtained by percutaneous needle biopsy showed this to be a hepatoblastoma.

Investigations

serum α-fetoprotein   33,000 ku/l

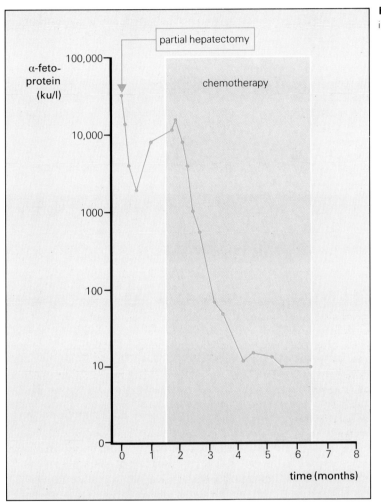

Fig.21.7 α-Fetoprotein concentration in a patient with hepatoblastoma.

A partial hepatectomy was performed but complete removal of the tumour was not possible because of its extent. He was therefore started on a course of cytotoxic therapy.

Comment

The change in serum α-fetoprotein concentration is shown in Fig.21.7. Partial hepatectomy produced a temporary fall in α-fetoprotein but continued growth of the tumour resulted in a further increase. Cytotoxic treatment produced a sustained decline in α-fetoprotein levels, and this corresponded to clinical remission.

In patients with testicular teratomas, α-fetoprotein measurements are valuable in assessing prognosis, in staging and in monitoring therapy. A very high concentration indicates a massive tumour load and a poor prognosis (a mortality rate greater than forty percent if α-fetoprotein concentration is greater than 1000ku/l). A rapid fall to normal after orchidectomy would imply that the disease was limited to the testis. Remission is achieved in eighty percent of patients with metastatic teratoma of the testis, using a combination of surgery and chemotherapy.

The efficacy of treatment can be assessed from the decline in serum α-fetoprotein which reflects the decrease in tumour mass. Once a patient is in remission, repeated measurements are essential; a rise in concentration will be due to recurrence of the tumour and indicates the need for further treatment or a change in the chemotherapeutic regimen. It is important to appreciate, however, that a tumour may lose the ability to secrete α-fetoprotein so that vigorous clinical assessment remains an important part of the follow-up of these patients.

Carcinoembryonic antigen (CEA)

This tumour marker is present in elevated concentrations in the serum of sixty percent of patients with colorectal cancer, more commonly so with advanced disease (eighty to one hundred percent if hepatic metastases are present) than with tumours confined to the colon. However, elevated levels are also found in a variety of non-malignant conditions, including liver disease of various types, pancreatitis and inflammatory bowel disease, and in some people who smoke heavily.

CEA is neither sufficiently specific nor sensitive to be used in screening for colorectal carcinoma. Its place in the follow-up of patients with established disease has yet to be determined. It does not appear to be a reliable indicator of early (and possibly curable) recurrence, and the fact that serum CEA levels do not relate directly to the mass of the tumour militates against its use in monitoring treatment response.

Paraproteins

Paraproteins (see pages 215–217) are detectable in either serum or urine in ninety-eight to ninety-nine percent of patients with myeloma. Not only is their detection valuable in the diagnosis of this condition, but paraprotein levels correlate well with tumour bulk so that the reduction in the amount of paraprotein is a good indicator of the efficacy of treatment. There is, however, no correlation between the initial paraprotein level and long-term survival.

Human chorionic gonadotrophin (hCG)

hCG is a hormone produced by the normal placenta, reaching a maximum concentration in serum by the eighth week of pregnancy. hCG is composed of an α- and β-subunit: the α-subunit is identical to that of luteinizing hormone (LH), follicle stimulating hormone (FSH) and thyroid stimulating hormone (TSH); the β-subunit, however, is specific to hCG and is therefore measured in assays for the hormone. The presence of hCG in the serum at other times indicates the presence of abnormal trophoblastic tissue or a tumour secreting the hormone ectopically.

β-hCG is an almost ideal tumour marker for choriocarcinoma, a malignant proliferation of chorionic villi which may develop from a hydatidiform mole, itself a potentially malignant proliferation of this tissue which occurs in approximately one in 2000 pregnancies in the United Kingdom. Hydatidiform mole is treated by uterine curettage, but the patient is at risk of developing choriocarcinoma if removal is incomplete. β-hCG is an extremely sensitive tumour marker; tumours weighing only 1mg (corresponding to 10^5 cells) may be detectable. All patients who have had hydatidiform moles must be followed up with regular checks of serum β-hCG concentration. Should a tumour develop, the marker can be used as an indicator of the response to treatment.

hCG is also secreted by approximately fifty percent of testicular teratomas and is therefore usually measured together with α-fetoprotein in the follow-up of patients after treatment of the tumour. Since LH concentrations rise after orchidectomy, it is important that an assay specific to the β-chains of hCG is employed to avoid cross-reactivity.

Other hormones

Hormones secreted both eutopically and ectopically can provide useful tumour markers. The measurement, for example, of metabolites of serotonin in the diagnosis of carcinoid syndrome and of catecholamines in phaeochromocytomas, has been discussed elsewhere (see page 136). Calcitonin is a valuable marker, not only for medullary cell carcinoma of the thyroid (eutopic secretion) but in some cases of carcinoma of the breast (ectopic secretion). Medullary cell carcinoma of thyroid is commonly familial and may be part of a pluriglandular syndrome. Calcitonin measurements may be used to screen for this tumour in the families of affected patients. Although basal levels of calcitonin may be normal, an excessive rise in calcitonin following provocation, for example, with alcohol or a calcium infusion, is seen in patients with medullary carcinoma.

Ectopic hormonal markers of other tumours, for example, bronchial carcinomas, are of little practical use in the management of patients. They are not present sufficiently frequently to be of use in screening and the response to treatment is, in general, so poor that their measurement provides no practical support to the clinician.

Enzymes

Prostatic acid phosphatase is used as a marker to monitor patients with disseminated prostatic carcinoma being treated with oestrogens. Although elevated levels of the enzyme are seen in more than ninety percent of patients with disseminated disease, they are seen in only about one-third of patients with a tumour confined to the prostate. This, combined with the comparative rarity of prostatic carcinoma, negates its use as a screening test.

CONCLUSION

The clinical usefulness of secreted tumour markers for screening, diagnosis and assessing the response to treatment is limited to a small number of markers and relatively rare tumours. However, antibodies that have been developed against surface antigens on tumour cells have already proved of great value in the differential diagnosis of lymphomas and leukaemia, and this field will undoubtedly develop further. The use of such antibodies to carry a cytotoxic drug or radioactive isotope directly to the tumour, currently in its infancy, offers exciting prospects for the treatment of cancer in the future.

SUMMARY

Patients with malignant disease frequently suffer from disorders not directly attributable to the physical presence of the tumour. These 'paraneoplastic' syndromes include metabolic disorders, notably those in which ectopic hormone secretion occurs. This term refers to the secretion of a known hormone (or a substance with hormone-like activity) by a non-endocrine tumour.

The clinical features produced can closely resemble those seen when a hormone is secreted in excess by its normal tissue of origin (eutopic secretion). Examples include: Cushing's syndrome and the syndrome of inappropriate antidiuretic hormone secretion caused by the production of adrenocorticotrophin and anti-diuretic hormone (vasopressin), respectively, by small cell carcinomas of bronchus and various other tumours; a syndrome resembling hyperparathyroidism (but in fact rarely due to parathyroid hormone itself) occurring in some patients with squamous cell carcinomas of bronchus and renal cell carcinomas; and hypoglycaemia which may occur in patients with large mesenchymal tumours, though it is not due to insulin. Other hormones that have been shown to be secreted by non-endocrine tumours include chorionic gonadotrophin, growth hormone releasing hormone, calcitonin and erythropoietin. The mechanism of ectopic hormone secretion is unclear but it is presumed that selective derepression of the appropriate genes occurs in the tumour cells.

Cancer cachexia, a non-specific syndrome of weight loss, anorexia and weakness, is common in patients with cancer. It is probably multifactorial in origin but the secretion of a humoral substance by the tumour may be partly responsible.

Many tumours capable of secreting hormones are derived from cells of the APUD system (cells of neuro-ectodermal origin characterized by their ability to take up and decarboxylate amines). Carcinoid tumours are derived from argentaffin cells, themselves members of the APUD family. These tumours are mainly found in the gut; they are of low grade malignancy and may go unnoticed unless metastasis occurs. They tend to secrete 5-hydroxytryptamine (5-HT); this is released into the portal circulation and usually metabolized by the liver but when hepatic metastases are present, 5-HT reaches the systemic circulation and may produce the carcinoid syndrome. The diagnosis is made by demonstrating an increased urinary excretion of the 5-HT metabolite, 5-hydroxyindoleacetic acid.

Some tumours occur in association and this tendency is often inherited. There are several syndromes of multiple endocrine adenomatosis or multiple endocrine neoplasia. Tumours that may be present in these syndromes include parathyroid adenomas, medullary cell carcinomas, phaeochromocytomas and pancreatic endocrine tumours.

Hormones secreted by tumours are occasionally useful as markers for the presence of the tumour but in general the fact that the hormone is also being produced by its normal source and is usually not consistently secreted by a particular tumour vitiates their measurement for this purpose. Some tumours, however, regularly secrete substances which are not usually detectable in the serum and these may be useful markers both for diagnosis and for following the progress of a malignancy. The best established examples of such tumour markers are α-fetoprotein (for testicular teratoma and hepatocellular carcinoma), β-human chorionic gonadotrophin (for choriocarcinoma), acid phosphatase (for carcinoma of prostate) and paraproteins (for myeloma). The place of carcinoembryonic antigen in the management of cancer of the large bowel has not yet been established.

FURTHER READING

Abe K (ed) (1980) Endocrinology and cancer. *Clinics in Endocrinology and Metabolism*, 9, 209–433.

Buckman R (1982) Tumour markers in clinical practice. *British Journal of Hospital Medicine*, 27, 9–20.

Calman KC (1982) Cancer cachexia. *British Journal of Hospital Medicine*, 27, 28–34.

Coombes RC (1982) Metabolic manifestations of cancer. *British Journal of Hospital Medicine*, 27, 21–27.

Wilson JD & Foster DW (eds) (1985) *Williams – Textbook of Endocrinology*. 7th edition. Philadelphia: WB Saunders Company.

22. Therapeutic Drug Monitoring and Chemical Aspects of Toxicology

INTRODUCTION

Clinical chemistry laboratories are called upon to measure drugs in body fluids for two main purposes: (i) to provide information relevant to the diagnosis and management of patients suspected to have taken drug overdoses; and (ii) to provide such information in patients taking drugs therapeutically. This chapter covers these topics and discusses the metabolic sequelae of some common poisonings.

Few laboratories in the United Kingdom are at present involved in screening for drug abuse. This may become a more widespread activity in the future but is not considered further here.

THERAPEUTIC DRUG MONITORING

The questions that should be addressed when prescribing a drug are summarized in Fig.22.1. It is obvious that drug treatment must be monitored if the efficacy of the therapy is to be accurately assessed. Clinical monitoring must always be carried out, but quantitative laboratory investigations also play an important role.

Clinical monitoring includes assessment of the therapeutic response and recognition of any side-effects or signs of toxicity. The laboratory may be called upon to help in these assessments by measuring a particular index of therapeutic response: for example, the blood glucose in a patient with diabetes treated with insulin; or thyroid function tests in a patient with thyrotoxicosis treated with carbimazole. The laboratory may also be asked to monitor for possible toxic effects: for example, proteinuria in patients treated with penicillamine; and abnormalities of thyroid function in patients treated with the iodine-containing antiarrhythmic drug, amiodarone.

The individual's response to a particular drug is dependent upon many factors, for example, age, sex, renal function and the concurrent administration of other drugs. These factors must be borne in mind when deciding what drug dosage to prescribe, but in many cases the optimum dosage can be arrived at by commencing treatment with a standard dose and

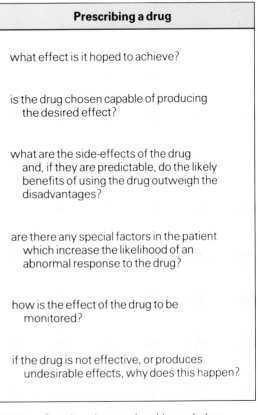

Prescribing a drug

what effect is it hoped to achieve?

is the drug chosen capable of producing the desired effect?

what are the side-effects of the drug and, if they are predictable, do the likely benefits of using the drug outweigh the disadvantages?

are there any special factors in the patient which increase the likelihood of an abnormal response to the drug?

how is the effect of the drug to be monitored?

if the drug is not effective, or produces undesirable effects, why does this happen?

Fig.22.1 Questions that must be addressed when prescribing a drug.

modifying this as necessary in the light of the clinical response.

This approach is suitable for the many drugs whose effects can be assessed reliably, such as hypotensive and hypoglycaemic agents, and anticoagulants, but it is not universally applicable. Obviously, optimization of drug dosage in this way is impossible when the effect of treatment is not easily ascertainable. An example is the use of anticonvulsants as prophylaxis in epilepsy. Many patients, if untreated, would have an unpredictable incidence of seizures, making it difficult to assess the effect of the drug in preventing them. It is also difficult to adjust dosage on the basis of the therapeutic effect when a drug has a narrow therapeutic ratio (that is, the dose required to produce a therapeutic effect is close to that at which features of toxicity are seen) especially if the adverse effects are hard to recognize. In such cases, the measurement of the concentration of the drug in the serum may provide valuable objective information.

It is outside the scope of this chapter to discuss in detail the many factors that can influence the relationship between the dose of a drug and the intensity of its effects. Some of these are listed in Fig.22.2. It is reasonable to assume that there will be a greater correlation between the serum concentration of a drug and the intensity of its effects than between the dose prescribed and this effect. Despite this, serum levels and tissue effects may correlate poorly since the drug must first travel from the serum to its site of action, and once there the responsiveness of the tissues may not be constant or predictable. Additionally there may be no correlation at all when a drug is itself inactive (but is metabolized to an active substance in the body) or when a drug acts irreversibly.

Nevertheless, the correlation between the serum concentration and pharmacological effect is surprisingly strong for many drugs and provides the rationale for the use of serum or plasma concentrations in therapeutic drug monitoring. It is important that any experimentally determined relationship between serum drug concentration and the effect of a drug is confirmed in a clinical setting, and serum drug levels must be interpreted in the clinical context. The time of sampling in relation to the time of dosage may be critical; and sensitivity of the target organ may vary, being influenced, for example, by genetic factors, nutritional status, other drugs and the health of the patient.

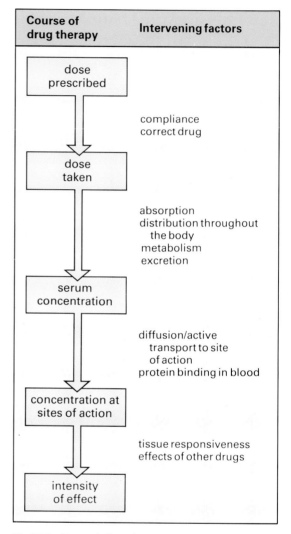

Fig.22.2 Factors influencing the relationship between drug usage and the intensity of its effect; the latter is not necessarily directly related to the serum concentration but may be more closely related to it than the dose prescribed.

Measuring serum concentration

The most frequently used assays measure total serum concentration of a drug. With drugs that are protein bound, changes in serum protein concentration may have a disproportionate effect on the total drug concentration relative to the amount free in the serum and thus available to tissues. The assay chosen must be specific for the drug itself (or its active metabolite where appropriate) and should not measure inactive metabolites or be affected by other drugs that the patient may be taking.

In the following section the use of serum measurements of a few representative and commonly used drugs is discussed to illustrate the general principles of therapeutic drug monitoring. Although 'serum' is used throughout, some assays may equally well be performed on plasma.

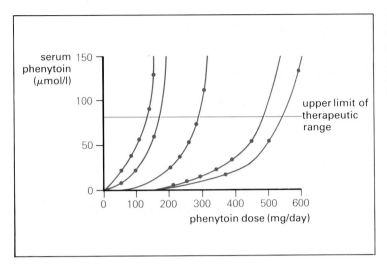

Fig.22.3 Relationship between the steady state serum concentration of phenytoin and dose; 80μmol/l is the upper limit of therapeutic dose. Data for five patients are shown. Redrawn from Richens and Dunlop (1975).

MONITORING OF SPECIFIC DRUGS

Phenytoin

The therapeutic effectiveness of this popular anti-convulsant drug is difficult to assess without monitoring. It has a low therapeutic ratio and the signs of toxicity may mimic the neurological diseases associated with epilepsy. Further, phenytoin has unusual pharmacokinetic properties; the enzyme responsible for the elimination of the drug becomes saturated within the therapeutic range of serum concentrations. This phenomenon has several important implications. In particular, the relationship between serum concentration and dose is non-linear (Fig.22.3); thus small increments in dose may lead to disproportionate increases in steady state serum levels. On the other hand, if the dose is unchanged, a small decrease in drug-metabolizing enzyme activity, or the presence of other drugs that inhibit phenytoin metabolism, could transform a therapeutic serum level to a toxic level. Fig.22.3 also indicates the wide variation of doses required to achieve therapeutic plasma levels in individuals.

CASE HISTORY 22.1

A young woman developed idiopathic epilepsy at the age of nineteen and had three generalized convulsions in ten days before being started on phenytoin, 150mg/day. She had a further fit two days after the first dose but thereafter remained fit-free.

Investigations

serum phenytoin (four weeks after starting treatment)   30μmol/l

Comment

Steady state serum concentrations of phenytoin are not reached for three to four weeks. The upper limit of the therapeutic range is 80μmol/l. The usual procedure when commencing treatment is to give a standard dose of 150–200mg/day (in adults) and to measure the serum concentration after three to four weeks. If the patient is well controlled and there are no features of toxicity the same dose may be continued even if, as in this case, the serum concentration is low in the therapeutic range. A dose increment is not indicated if the patient is fit-free just on the basis of the serum concentration of the drug. In the well-controlled patient, this initial serum level may be useful later to help ascertain the cause (for example, poor compliance, drug interaction) should seizures recur.

Therapeutic monitoring for anticonvulsant drugs		
Drug	**Therapeutic range**	**Monitoring**
phenytoin	0–80 μmol/l	essential
carbamazepine	0–42 μmol/l	useful but not essential
ethosuximide	0–710 μmol/l	useful not not essential
phenobarbitone	0–172 μmol/l	tolerance makes upper limit imprecise
primidone	0–55 μmol/l	metabolized to phenobarbitone (which should be monitored) primidone levels not useful
sodium valproate	0–700 μmol/l	not known to be useful
clonazepam	0–285 μmol/l	not known to be useful

Fig.22.4 Therapeutic monitoring of anticonvulsant drugs.

If the patient is not well controlled, increments in dose can be made, guided by measurement of serum levels, to produce a steady state concentration in the therapeutic range. Phenytoin has a long half-life when used chronically and serum levels remain relatively constant throughout the day. For this reason (unusual in therapeutic drug monitoring) the time of sampling in relation to the time the drug was taken is not critical.

The measurement of serum levels of phenytoin is also useful if adverse effects occur, if there is an unexplained deterioration in the patient's control, and if a drug known to interact with phenytoin has to be prescribed. It is of particular value in children and during pregnancy, when dramatic fluctuations in serum levels and in epileptic control may be seen.

The value of measuring the serum concentration of other anticonvulsants is indicated in Fig.22.4. Although toxicity is less of a problem than with phenytoin, monitoring serum drug levels is valuable in assessing compliance in what is often a difficult group of patients in this respect.

Digoxin

Digoxin is frequently used in the management of cardiac failure with atrial fibrillation, a common problem in the elderly. Serum digoxin measurements are valuable not only in the assessment of the appropriate dose to prescribe, but also in the diagnosis of digoxin toxicity and in assessing patient compliance. Failure to take a prescribed medication (non-compliance) is a common cause of failure to achieve a therapeutic response.

The therapeutic range for serum digoxin concentration is generally taken as 1.0–2.6μmol/l. There is a significant increase in serum levels after taking a dose of the drug and a minimum period of six hours should elapse before blood is drawn for assessment of the mean steady state concentration. In practice it is often simplest, and satisfactory for diagnostic purposes, if a blood sample is taken shortly before a dose is due.

While the therapeutic effect is minimal when serum concentration is below 1μmol/l and toxicity becomes more common at concentrations above 2.6μmol/l, there is in general a rather poor correlation between serum concentration and therapeutic effect. Furthermore, evidence of toxicity may sometimes be seen in patients whose serum concentration is below 2.6μmol/l while others may tolerate levels of fifty percent higher than this without ill effect.

Sensitivity to digoxin
Stimulatory factors
hypokalaemia hypercalcaemia hypomagnesaemia hypoxia hypothyroidism
Inhibitory factors
hypocalcaemia hyperthyroidism

Fig.22.5 Factors affecting sensitivity to digoxin. In addition, renal impairment and hypothyroidism may increase the serum level of digoxin in relation to the dose taken; hyperthyroidism may decrease the level.

This phenomenon is partly a result of the existence of various factors which alter either the therapeutic response to a given serum concentration of digoxin or the serum concentration achieved on a particular dose (Fig.22.5). Hypokalaemia is a particular problem since many patients treated with digoxin are also receiving diuretics, which may cause this (see *Case history 24.2*); also, renal impairment may be a consequence of congestive cardiac failure. It is thus very important to consider the clinical setting when assessing the significance of serum digoxin levels.

Digoxin levels are also useful in the diagnosis of digoxin toxicity. This is important because some of the features of toxicity are relatively non-specific (for example, nausea and vomiting), while others include dysrhythmias which could possibly be a complication of the underlying heart disease. It is important that the possible influence of pathological and physiological factors is considered (see Fig.22.5).

Antidysrhythmics

Methods are available for the measurement of many other drugs used in patients with heart disease, in particular antidysrhythmics. The arguments relating to the value of serum levels in monitoring treatment are complex and the place of therapeutic drug monitoring is arguable. It is probably useful for procainamide, lignocaine and quinidine, but its role with the newer antidysrhythmics such as amiodarone and verapamil has yet to be established.

Lithium

Lithium is widely used in the management of acute mania and chronically in manic-depressive psychosis. The optimum therapeutic serum level varies from patient to patient with an overall range of 0.3–1.3mmol/l, twelve hours after the last dose. Lithium has a low therapeutic ratio and there are wide inter-individual differences in dose requirements, so that the monitoring of serum levels has become vital to clinical assessment in patients on lithium therapy.

Serum levels are also vital in the management of lithium toxicity. Lithium is excreted by the kidneys, but lithium toxicity impairs the ability of the kidney to concentrate the urine leading to dehydration with a fall in glomerular filtration rate and consequent decrease in lithium excretion. Patients with lithium toxicity may require dialysis to remove the drug; dialysis should usually be undertaken if the serum level is more than 3.5mmol/l but may be necessary with concentrations lower than this in sick patients. The serum level can be used to monitor the efficacy of dialysis.

Other drugs

Therapeutic drug monitoring is firmly established as a tool in the management of patients treated with phenytoin, digoxin and lithium. It is also valuable in the management of patients treated with aminophylline, a bronchodilator used in the treatment of asthma, cyclosporin A, an immunosuppressive drug, and several antibiotics, the measurement of which is usually the responsibility of departments of medical microbiology.

A case can also be made for the measurement of serum levels of antidepressant drugs and undoubtedly the list will increase as more drugs are introduced. While at present most therapeutic drug monitoring is based on serum or plasma measurements, there is increasing interest in salivary assays. These tend to reflect the serum concentration of the non-protein bound, that is, free, drug which is directly available to the tissues and the advantage of this technique is that venepuncture is not required.

CHEMICAL TOXICOLOGY

Poisoning is a common reason for hospital admission. In most cases, the patient has taken an overdose of a prescribed or over-the-counter drug, but poisoning may also be accidental (common in children), suicidal, or homicidal and the range of toxic substances is vast, including industrial and domestic chemicals, plants and fungi as well as drugs.

Management

There are no specific antidotes for most poisons. Management is therefore primarily directed towards the support of vital functions. This may be supplemented by measures to remove the drug from the body if this is possible. In cases of this sort, the chemical pathology laboratory has an important role in monitoring vital functions, for example, by measuring arterial blood gases. Measurement of the serum concentration of the poison may indicate the need to take steps to increase the elimination of the drug and is valuable in monitoring such treatment. For some poisons, specific treatments are available but these may themselves not be without risk to the patient. Serum levels can then be used to indicate whether such treatment is likely to be of value.

Few poisons produce specific physical signs: patients' histories, if available, may not be reliable and mixed drug overdoses are common. There is therefore a need for an analytical service to identify what poisons may have been ingested, particularly if a patient does not respond to conventional management. This presents an entirely different problem for the laboratory since what is required is a screening service capable of identifying any of a large number of toxins, rather than providing quantitative data on a small number (see page 295).

POISONING WITH SPECIFIC AGENTS

Paracetamol

A specific antidote is available for paracetamol, the metabolism of which is summarized in Fig.22.6. The major products of its metabolism are harmless glucuronide and sulphate conjugates, which are excreted in the urine together with a small amount of the unchanged drug. Small quantities of a highly hepatotoxic metabolite (probably N-acetyl p-benzo-

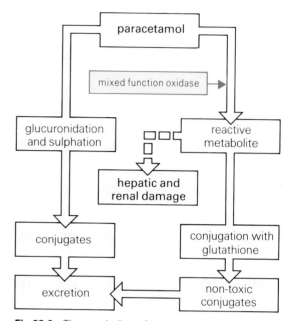

Fig.22.6 The metabolism of paracetamol. When the drug is taken in therapeutic doses, all the reactive metabolite formed is detoxified by conjugation with glutathione; when taken in overdose, glutathione supplies are rapidly exhausted and the metabolite accumulates causing cell damage.

quinone-imine) are also formed through the action of the mixed function oxidase (cytochrome P450) enzyme system; this is normally detoxified by conjugation with glutathione. However, the glucuronidation and sulphation pathways are saturable so that when an overdose of the drug is taken, a greater proportion is converted to the toxic metabolite. Glutathione supplies are limited, and if they are insufficient to detoxify this metabolite, liver damage will result. The metabolite is also nephrotoxic so renal failure may also develop.

Clinical features

Paracetamol is an insidious poison since there may be no clinical disturbance in the first twenty-four hours after taking an overdose, except anorexia, nausea and vomiting (Fig.22.7). The conscious state is normal unless a sedative drug has been taken concurrently (compound preparations containing paracetamol

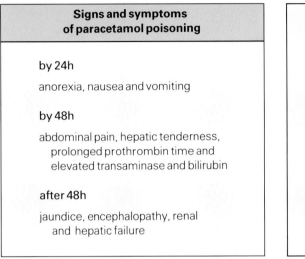

Signs and symptoms of paracetamol poisoning

by 24h

anorexia, nausea and vomiting

by 48h

abdominal pain, hepatic tenderness, prolonged prothrombin time and elevated transaminase and bilirubin

after 48h

jaundice, encephalopathy, renal and hepatic failure

Fig.22.7 Signs and symptoms of paracetamol poisoning.

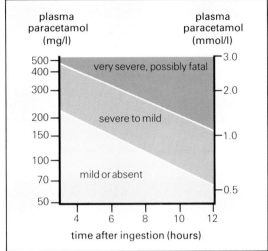

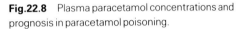

Fig.22.8 Plasma paracetamol concentrations and prognosis in paracetamol poisoning.

and a sedative, such as dextropropoxyphene, are common). If liver damage occurs, abdominal pain with hepatic tenderness will develop and liver function tests become abnormal (prolonged prothrombin time, elevated transaminase, elevated bilirubin). Features of renal impairment may also be present but the serum creatinine concentration will provide a better guide to its severity than the serum urea since hepatic urea synthesis may be decreased. With massive overdoses patients may develop fulminant hepatic failure with hepatic encephalopathy which is commonly fatal.

It is possible to predict the likelihood of liver damage from the plasma level of paracetamol. The blood sample must be taken at least four hours after ingestion of the drug. The plasma level can be used as a guide to patient management, that is, whether to treat the patient with an antidote (Fig.22.8). Unfortunately, the time at which the drug was taken may not be known and it is then wisest to treat the patient actively if the plasma level falls within a treatment zone. There is no rational basis for the use of an antidote once liver damage has occurred, therefore the measurement of plasma paracetamol levels more than twelve to fourteen hours after an overdose is of little value. Indeed, the antidotes themselves may be positively harmful if there is liver damage.

Management

The most widely used antidotes are N-acetyl cysteine (which must be given parenterally) and methionine (given orally). Although it is more expensive, N-acetyl cysteine is usually preferred as there will be no doubt about absorption whereas the absorption of methionine will be more variable, especially if the patient is vomiting. N-Acetyl cysteine is given by intravenous infusion in five percent dextrose, initially at a high dose and then at a much lower dose over a period of twenty hours. Both N-acetyl cysteine and methionine act by promoting hepatic glutathione synthesis, thereby increasing the capacity of the liver to detoxify the active metabolite.

In treating paracetamol poisoning, general emergency measures must not be forgotten. Gastric lavage or administration of an emetic is only of use in the first six hours after an overdose. The patient must be kept hydrated, preferably using five percent dextrose since there may be a tendency to hypoglycaemia with hepatic damage. Vitamin K should be given prophylactically. Should liver failure develop, close clinical and laboratory monitoring are vital.

Salicylates

Salicylate poisoning, usually with aspirin (acetyl-salicylic acid), is common. It can produce profound metabolic disturbances and though there is no specific antidote, measures can be taken to increase the excretion of the drug which, though effective, are not without hazard in themselves. The upper limit of the therapeutic range of serum salicylate concentration is approximately 2.5mmol/l (35mg/100ml), but tinnitus, an early symptom of toxicity, may become apparent at levels lower than this.

The effects of salicylates which lead to metabolic disturbances are summarized in Fig.22.9, and include stimulation of the respiratory centre, a non-respiratory acidosis, uncoupling of oxidative phosphorylation and a central emetic effect.

CASE HISTORY 22.2

A twenty-year-old male student was brought into hospital in a confused state, having been found at home by his flatmate with an empty bottle of aspirin tablets on his desk.

On admission, he was hyperventilating and sweating profusely. He was pale but not anaemic. He was not grossly dehydrated but the inside of his mouth was dry and there was a smell of ketones on his breath. His pulse was 112/min, blood pressure 110/60mmHg and temperature 39.5°C.

Investigations

serum:	sodium	131mmol/l
	potassium	3.2mmol/l
	bicarbonate	10mmol/l
	urea	10mmol/l
	glucose	3.2mmol/l
	salicylate	3.9mmol/l

arterial blood:	
hydrogen ion	62nmol/l (pH 7.20)
P_{CO_2}	3.5kPa (26mmHg)

prothrombin time	18s (control 14s)

Comment

The results are consistent with the metabolic effects of salicylates described above. There is an acidosis, compensated to some extent by hyperventilation (see *Chapter 4*). The initial acid–base disturbance (in adults, but usually not in children) is a respiratory alkalosis due to direct stimulation of the respiratory centre. This is usually overwhelmed by the developing acidosis, but during the alkalotic phase any compensatory renal excretion of bicarbonate will deplete the capacity of the body to buffer excess hydrogen ions, thus making the acidosis more dangerous.

Patients who have taken overdoses of salicylates are rarely comatose; irritability is an early feature and later hallucination and delirium may occur. Tinnitus may be a prominent feature.

The prothrombin time is commonly elevated, as in this case, due to decreased hepatic synthesis of clotting factors. Salicylates also inhibit platelet aggregation. However, although gastric erosions may occur, due directly to the action of salicylate on the gastric mucosa, severe bleeding is uncommon in aspirin overdosage. Nevertheless, prophylactic vitamin K is usually administered.

Management

There is no specific antidote to aspirin. It is metabolized by hydrolysis to salicylic acid, the active form of the drug, which is excreted unchanged in the urine; other metabolites include various inactive conjugates. The conjugation pathways are saturable, so once they are saturated urinary excretion becomes the major route for elimination of the drug. If the pH of the urine is acidic, salicylic acid is un-ionized and, though filtered by the glomeruli, is reabsorbed by the tubules. If the urine pH is alkaline, however, salicylic acid is ionized. Its tubular reabsorption is then reduced while urinary excretion is enhanced. This is the rationale for alkalinization using sodium bicarbonate infusions in the treatment of salicylate poisoning. However, this process is in itself potentially dangerous and requires careful monitoring. It should not be attempted if the patient already has a systemic alkalosis or if the urine pH exceeds 8, the aim being to maintain a urine pH greater than 7.5 during treatment. Potassium supplements are required (hypokalaemia may hinder

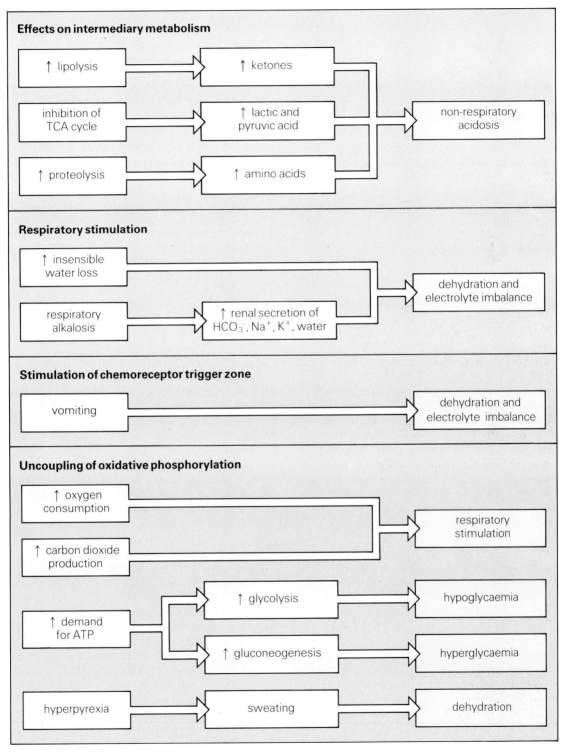

Fig.22.9 Pathophysiology of salicylate poisoning.

effective alkalinization of the urine); dehydration and hypoglycaemia must be corrected, and fluid balance, blood glucose, arterial hydrogen ion concentration and urine pH must be monitored. It is commonly advocated that a high intravenous fluid input is maintained to promote a diuresis, but fluid overload must be avoided. It is far more important to ensure adequate alkalinization of the urine.

Aspirin is absorbed only slowly from the gut so gastric lavage is always worth attempting in overdose unless specifically contra-indicated. The decision whether to embark on active treatment should be based on clinical grounds but guided by laboratory data. Maintenance of adequate hydration and general supportive measures are important for all patients. Alkalinization should be undertaken if serum salicylate concentration exceeds 3.6mmol/l (50mg/100ml) in adults and 2.2mmol/l (30mg/100ml) in children more than six hours after the overdose. If the initial concentration exceeds 6.5mmol/l (90mg/100ml) and if there is renal impairment or if other therapeutic measures fail, haemoperfusion or haemodialysis will usually be necessary. Serum salicylate levels should be measured during treatment as an indication of its efficacy.

Iron

Iron poisoning, although much less common now than in the past, can still cause severe illness especially in young children. Iron causes necrosis of the gastrointestinal mucosa with resultant haemorrhage and fluid and electrolyte loss. Patients may develop encephalopathy and renal failure with circulatory collapse and acute liver necrosis may develop in those who survive these complications.

Severe poisoning is indicated by serum iron levels in excess of 90μmol/l in a child, or 145μmol/l in adults. Management involves the use of desferrioxamine, an iron-chelating agent, to promote iron excretion, together with appropriate supportive measures.

Lead

Acute lead poisoning is very uncommon but chronic poisoning is frequently diagnosed. In children the source may be old paint or toys, cosmetics and patent medicines imported from the Indian subcontinent. In adults most cases are associated with occupational exposure (battery manufacture, smelting, ship-breaking) and lead poisoning is a notifiable industrial disease. Lead is concentrated in erythrocytes and in persons who are not occupationally exposed to the metal, a blood level of greater than 1.5μmol/l should be followed up. In lead workers, the presently acceptable upper limit for blood lead is 3.9μmol/l. Symptomatic lead poisoning is usually associated with levels in excess of 5μmol/l, but in children symptoms may be present at lower levels.

Tests

Although blood lead measurement is the screening method of choice for excessive exposure, other tests may sometimes be useful. Lead interferes with several steps in porphyrin synthesis (see page 261) and porphyrinuria (due mainly to coproporphyrin III) may be present in lead poisoning although this is not a very sensitive test. An excess of δ-aminolaevulinic acid in the urine is also characteristic but not specific. An excess of protoporphyrin in erythrocytes is a more sensitive indicator of excessive exposure to lead but again is not specific, also being a feature of iron deficiency. However, the results of these tests may indicate a need to measure blood lead concentration itself, which may involve sending the blood sample to a specialized laboratory if the assay is not available locally.

Chemical pathology laboratories are becoming increasingly involved in screening for other industrial toxins, and in particular heavy metals. Specialized laboratories should be able to provide an analytical service for cadmium and mercury as well as lead.

Clinical features and management

Lead poisoning causes nausea, vomiting and severe abdominal colic. In the nervous system, encephalopathy with convulsions and impairment of consciousness may lead to coma and death. In severe cases, usually due to acute lead poisoning, active treatment with the chelating agents calcium versenate and dimercaprol given parenterally is required. In milder cases, oral penicillamine is useful, while for asymptomatic persons whose blood level indicates excessive exposure to lead, the source of lead should be identified and removed or the exposed person removed from the source.

Alcohol

Although there is no specific antidote to ethanol, drug overdose is often complicated by the simultaneous ingestion of alcohol. It potentiates the action of many drugs and measurement of blood alcohol concentration may provide the explanation for an unexpected delay in a patient's recovery from drug overdose.

Blood alcohol measurements may also be of value in the management of patients with head injuries, when the effects of alcohol may make it difficult to assess the severity of any brain damage due to the injury itself.

Clinical features and effects

Chronic alcoholism is now a major health problem in many areas of the world. In addition to its well-known harmful effects on the liver, chronic alcohol ingestion can damage many organs and tissues in the body. Metabolic sequelae include hypertriglyceridaemia, hypoglycaemia, hypogonadism, hyperuricaemia, a form of Cushing's syndrome and cutaneous hepatic porphyria. Blood alcohol levels may be of value in establishing a diagnosis of alcoholism.

It has been suggested that the finding of an ethanol concentration of greater than 65mmol/l (300mg/100ml) at any time is diagnostic; in a patient who is asymptomatic, the level suggested is 33mmol/l (150mg/100ml). The combination of a raised serum γ-glutamyl transpeptidase and increased mean red cell volume is a characteristic and sensitive index of excessive alcohol intake, although not an entirely specific one. Better laboratory tests are urgently required; red cell aldehyde dehydrogenase, serum desialotransferrin and serum mitochondrial aspartate transaminase offer some promise for the future.

SCREENING FOR DRUGS

It has been pointed out that when a toxin does not have a specific antidote, precise knowledge of its serum concentration does not contribute to patient management. Nevertheless, qualitative rather than quantitative measurement of a toxic agent may be desirable. If a patient is admitted to hospital unconscious for no readily discernable reason, identification of the drug may be of prognostic value or suggest the need to attempt to remove it from the body, for example, by haemodialysis or haemofiltration.

Screening for drugs is also necessary in cases of suspected brain death. Symptoms of apparent brain death may be due to the presence of CNS-depressant drugs and it is vital that this possibility is considered and eliminated before true brain death is diagnosed.

In cases of suspected homicide, the identification of any poisons present is vital and must be carried out by suitably qualified personnel whose testimony would be accepted by the court if they were to be called upon as witnesses.

It is not practical for all laboratories to provide facilities for screening for all possible toxins. In the United Kingdom, there is a network of poisons reference laboratories which provide advice and an analytical service for such purposes. Samples for drug screening should be collected after consultation with the nearest laboratory.

Urine is in general more useful for screening than blood, as many drugs and their metabolites are cleared rapidly from the blood but are present in high concentration in the urine. Urine and blood samples, stomach contents and any tablets or material that may have been ingested should be collected, carefully labelled and, if analysis is not going to be performed immediately, stored in a refrigerator or deep-frozen.

SUMMARY

A knowledge of the concentration of a drug in the serum can be of considerable assistance when deciding the appropriate drug dose to prescribe. This is particularly likely when the drug has a narrow therapeutic ratio (i.e. when the range of serum concentrations over which the maximum beneficial effect is seen is only a little less than that at which it becomes toxic) and when it is difficult to assess the effect clinically. For drug concentrations to be used rationally for this purpose, it is essential that there is a predictable and defined relationship between concentration and effect. For many drugs, however, such therapeutic monitoring is unnecessary, as for example when the effect can readily be assessed clinically or by clinical or laboratory measurements and when a drug of low toxicity has a virtually guaranteed effect when given in a standard dose. It is of no value when the effect of a drug is due to a metabolite, unless the concentration of the metabolite can be measured.

Therapeutic drug monitoring is an established technique in relation to the use of phenytoin, lithium, digoxin, aminoglycoside antibiotics, aminophylline and cyclosporin A. It is also used for certain anti-disrhythmic drugs and for anticonvulsants other than phenytoin, but the rationale for therapeutic drug monitoring in some of these cases is open to argument and results should be interpreted and used with caution.

Measurements of drug concentrations in body fluids are also valuable in the investigation and management of patients who have taken overdoses of drugs or have been poisoned. Whilst the management of many forms of drug overdosage and poisoning is essentially conservative so that identification of the substance is of little direct use in management, for those drugs for which specific antidotes exist, or for which it is possible to take measures to promote their excretion, measurement of serum concentrations can be very helpful. Thus the decision to treat paracetamol overdosage with N-acetyl cysteine is based on whether the untreated patient would develop hepatic failure; this can be predicted from the serum paracetamol concentration provided that the time of the drug overdose is known. With regard to salicylate overdosage, although there is no specific antidote, the excretion of the drug can be accelerated by alkaline diuresis and the serum concentration provides a guide to whether this is necessary.

Specific measures are also available to increase the rate of elimination of iron and lead from the body, and the concentrations of these substances in serum (iron) or blood (lead) can help decide whether or not to use such measures. Exposure to lead is an occupational hazard in certain industries and measurement of either blood lead concentration or erythrocyte protoporphyrin concentration can be used to detect those who have had excessive exposure.

Patients who have taken drug overdoses or been poisoned frequently develop metabolic problems, the management of which requires the close cooperation of the clinical chemistry laboratory. Salicylate overdosage, for example, can result in profound disturbances of acid–base, glucose and electrolyte metabolism. Serial measurements of the serum concentration of a drug will indicate the efficacy of any treatment designed to increase its excretion.

FURTHER READING

Anon (1985) What therapeutic drugs should be monitored? *Lancet*, **1**, 309–310.

Vale JA & Meredith TJ (1981) *Poisoning – Diagnosis and Treatment*. London: Update Books.

Widdop B (ed) (1985) *Therapeutic Drug Monitoring*. Edinburgh: Churchill Livingstone.

23. Clinical Nutrition

INTRODUCTION

An adequate intake of nutrients is essential for normal growth and development and for the maintenance of health. These nutrients include proteins, to supply amino acids, energy substrates (carbohydrates and fat), inorganic salts, vitamins and other essential nutrients, such as essential fatty acids. The daily requirements for these nutrients are determined by many factors, including age, sex, physical activity and the presence of disease; if an individual's requirements are not met, he is at risk of developing a clinical deficiency syndrome. Excessive intake of nutrients can also be harmful. Obesity is a common condition in the developed world and is related to an intake of energy substrates in excess of the body's requirements. There is much evidence linking several common diseases, including coronary heart disease and hypertension, with a relative excess or insufficiency of one or more components of the diet.

This chapter discusses the pathology of some specific deficiency syndromes, with particular reference to the role of the laboratory in their diagnosis and management. This role is also discussed in relation to patients suffering from, or at the risk of, generalized malnutrition. Nutritional support for these patients may be provided enterally (that is, into the alimentary tract, either by mouth or through a feeding tube), or parenterally (intravenously, bypassing the gut). Such treatment requires close cooperation between the clinician and the laboratory.

VITAMIN DEFICIENCIES

The biochemical functions of the vitamins are in most cases well understood: for example, many B vitamins are essential precursors or components of coenzymes; vitamin D is the precursor of the hormone calcitriol, vital to calcium homoeostasis; and vitamin K is required for the γ-carboxylation of glutamic acid residues in certain blood coagulation factors which permits them to bind calcium ions, an essential step in their activation.

These functions are almost entirely intracellular. However, the plasma concentration of a vitamin is not necessarily related to its tissue concentration, in other words, availability for normal function. It follows that plasma concentrations of vitamins may be unreliable as indicators of the body's vitamin status. In deficiency states, plasma levels tend to fall before tissue levels. On the other hand, if a vitamin is administered to a deficient patient, a rise in plasma concentration to normal is not necessarily indicative of adequate replacement.

In practice, the best means of assessing a patient's vitamin status depends upon the vitamin in question. The range of techniques that can be used is illustrated by the following examples.

WATER-SOLUBLE VITAMINS

Vitamin B$_1$ (thiamin)

Thiamin pyrophosphate is a cofactor in the metabolism of pyruvate and 2-oxoglutarate to acetyl-CoA and succinyl-CoA respectively, and in a reaction of the pentose shunt pathway catalyzed by the enzyme transketolase. The body contains only about thirty times the daily requirement of this vitamin. Subclinical thiamin deficiency may be unmasked in malnourished patients given glucose intravenously, which increases the metabolic requirement for the vitamin.

Deficiency of vitamin B$_1$ causes beriberi; one of the manifestations of this is Wernicke's encephalopathy, characterized by memory loss and nystagmus, and seen in the United Kingdom chiefly in chronic alcoholics whose diet is poor.

Other features of thiamin deficiency include peripheral neuropathy, muscle weakness, dementia and cardiac failure. Wernicke's encephalopathy responds rapidly to thiamin and since the vitamin is cheap and non-toxic this therapeutic response can be used to make the diagnosis. Laboratory tests for deficiency are seldom necessary.

It may, however, be necessary formally to document deficiency in nutritional research. One method involves administration of a glucose load and measurement of the plasma pyruvate concentration. An excessive rise is seen in thiamin deficiency because the vitamin is a cofactor for the conversion of pyruvate to acetyl-CoA. However, the most sensitive method, which will detect subclinical deficiency, is measurement of transketolase in a red cell haemolysate, the enzyme activity being measured both with and without the addition of thiamin pyrophosphate to the reaction mixture. Enzyme activity may be normal in subclinical deficiency but is increased by the addition of the coenzyme. If the deficiency is clinically obvious, the basal enzyme activity will be low.

CASE HISTORY 23.1

An elderly lady, resident in a private nursing home, complained of difficulty in walking, with paraesthesiae and numbness in her legs. The physical signs were consistent with a peripheral neuropathy.

There had been suggestions that residents were not fed adequately and the doctor took a blood sample for measurement of transketolase before giving his patient vitamin supplements.

Investigations

red cell transketolase activity:
  without added thiamin pyrophosphate
    $2.0 \mu mol/h/10^9$ red cells
  with added thiamin pyrophosphate
    $2.4 \mu mol/h/10^9$ red cells

Comment

The patient's symptoms showed some improvement with the vitamin supplements. Red cell transketolase activity (measured by the decrease in substrate concentration as it is metabolized) was at the lower limit of normal and increased by twenty percent in the presence of thiamin pyrophosphate. This is consistent with mild thiamin deficiency; an increase of up to fourteen percent is considered normal, while an increase of greater than twenty-five percent is clear evidence of deficiency. Peripheral

neuropathy is a common clinical problem; nutritional deficiency is but one of many causes.

An analagous technique can be used for assessing riboflavin status (by measurement of the red cell enzyme glutathione reductase with and without the vitamin) and pyridoxine (using red cell alanine or aspartate transaminases).

Nicotinic acid

Nicotinic acid is the precursor of nicotinamide. This is a constituent of the coenzymes nicotinamide adenine dinucleotide (NAD) and its phosphate (NADP) which are essential to glycolysis and oxidative phosphorylation.

Part of the body's nicotinic acid requirement is met by endogenous synthesis from tryptophan. The deficiency syndrome, pellagra, develops because of an inadequate dietary intake of nicotinic acid or because of decreased synthesis. The latter may be a feature of the carcinoid syndrome, in which there is increased metabolism of tryptophan to hydroxyindoles so that less is available for nicotinic acid synthesis, and in Hartnup disease, a rare inherited disorder of the epithelial transport of neutral amino acids, due to decreased intestinal absorption of tryptophan from the gut.

Nicotinic acid status can be assessed either by a microbiological assay of the vitamin in serum or by measurement of its urinary metabolites.

Folic acid

A derivative of folic acid is vital to purine and pyrimidine (and hence nucleic acid) synthesis. Folic acid deficiency is relatively common; its most usual manifestation is as a megaloblastic anaemia. This vitamin is now usually measured in haematology departments by immunoassay, although microbiological assays were widely used in the past. The concentration in red cells reflects the body's folate status more accurately than that in serum.

Tests involving the chemical measurement of formimino-glutamate, an intermediate in the degradation of histidine which requires tetrahydrofolate for its further metabolism, are now obsolete.

Vitamin B$_{12}$

Vitamin B$_{12}$ comprises a number of closely related substances called cobalamins which are essential to nucleic acid synthesis. Deficiency of the vitamin causes a megaloblastic anaemia and, in severe cases, subacute combined degeneration of the spinal cord.

Dietary deficiency of this vitamin is rare except in strict vegetarians (vegans) and considerable amounts are stored in the liver, so that deficiency is not common even with severe malabsorption (unless very long-standing). Vitamin B$_{12}$ deficiency is most commonly seen in pernicious anaemia. This is an autoimmune disease, in most cases of which there is a lack of intrinsic factor essential for the absorption of the vitamin from the gut.

Vitamin B$_{12}$ is measured in serum by immunoassay, usually in departments of haematology. Tests of vitamin B$_{12}$ absorption are discussed in *Chapter 7*.

Vitamin C (ascorbic acid)

The precise mode of action of ascorbic acid is uncertain but it is known to be essential for normal collagen synthesis. Subclinical deficiency of ascorbic acid is quite often present in elderly housebound people. The concentration of ascorbate in serum reflects dietary intake and is a poor index of tissue stores of the vitamin. These are better assessed by determination of ascorbate concentration in leucocytes. In practice, this is seldom necessary, since ascorbic acid is cheap and non-toxic, so a therapeutic trial of vitamin supplementation is the simplest procedure in suspected vitamin C deficiency.

CASE HISTORY 23.2

An eighty-year-old widow was admitted to hospital with bronchopneumonia and obvious self-neglect. She lived alone but had several cats and a neighbour who had called the doctor said that most of the woman's pension was spent on her pets. On examination, she was seen to have widespread perifollicular haemorrhages and a clinical diagnosis of scurvy was made. She was given ascorbic acid (11mg/kg body weight/day) and her urinary ascorbate excretion measured.

Only after eight days of treatment was there any increase from the initial very low level.

Comment

In a person with normal tissue ascorbate stores, ascorbate ingested in excess of requirements is rapidly excreted in the urine. In a patient with ascorbate deficiency, the vitamin is retained until tissue stores are replenished; in severe deficiency this can take more than a week. The test used to diagnose the deficiency only provides confirmation of the diagnosis retrospectively and is rarely performed except for research purposes.

FAT-SOLUBLE VITAMINS

Vitamin A

This vitamin is a constituent of the retinal pigment rhodopsin. It is also essential for the normal synthesis of mucopolysaccharides and growth of epithelial tissue. Mild deficiency causes night blindness while in more severe cases degenerative changes in the eye may lead to complete loss of vision. The normal liver contains considerable stores of the vitamin and deficiency is rarely seen in affluent societies. It is, however, an important cause of blindness in many areas of the world.

Vitamin A is present in the diet and can also be synthesized from dietary carotenes. In the circulation it is transported by prealbumin and a specific retinol-binding protein.

Vitamin A can be measured in serum but this assay is rarely required in the United Kingdom. In areas where deficiency of the vitamin is endemic the diagnosis is usually obvious clinically, and the facilities required to provide laboratory confirmation of the diagnosis are often not available.

Vitamin D

Vitamin D is obtained from endogenous synthesis, by the action of ultraviolet light on 7-dehydrocholesterol in the skin to form cholecalciferol (vitamin D$_3$), and from the diet. Dietary vitamin D is largely vitamin D$_2$ (ergocalciferol); the only important dietary sources are fish and margarine, which is artificially fortified with vitamin D. Vitamins D_2 and D_3 appear to undergo the same metabolic changes in the body and have identical physiological actions. For this reason, the terms cholecalciferol or vitamin D are frequently used to refer to both forms of the vitamin.

In most individuals endogenous synthesis is the major source of vitamin D. Privational (dietary) vitamin D deficiency does occur, most commonly in people who have decreased endogenous synthesis, such as the elderly housebound. It is also seen in immigrants from the İndian subcontinent, particularly women, in whom the effects of poor intake may be exacerbated by decreased exposure to sunlight due to their traditional clothing. Infants are at risk of vitamin D deficiency particularly if premature (the vitamin is transported across the placenta mainly in the last trimester of pregnancy) or if the mother is vitamin D-deficient.

Cholecalciferol itself has little physiological activity. It is hydroxylated first in the liver to 25-hydroxycholecalciferol (25-HCC, calcidiol) and then in the kidney to 1,25-dihydroxycholecalciferol (1,25-DHCC, calcitriol) and to various other metabolites (see Fig.14.6). These metabolites are transported in the circulation by a specific binding protein. Calcitriol is a hormone of vital importance in calcium homoeostasis; its actions and the control of its production are discussed in *Chapter 14*.

Cholecalciferol is stored in the liver and in adipose tissue. Vitamin D status can be assessed in the laboratory by measurement of the serum concentration of calcidiol, the major circulating metabolite. This undergoes seasonal variation, being higher in the summer than in the winter.

Decreased synthesis or dietary deficiency of vitamin D causes rickets in children and osteomalacia in adults. Other causes include disordered metabolism of cholecalciferol and malabsorption. Calcidiol undergoes enterohepatic circulation and with malabsorption the uptake of calcidiol as well as that of dietary vitamin D is affected. The chemical pathology of rickets and osteomalacia is considered in more detail in *Chapter 14*.

Vitamin K

Vitamin K has an essential role in the synthesis of certain blood-clotting factors and thus its status can be assessed by a functional assay of relevant clotting factor activity, the prothrombin time. This is also dependent on the synthetic capacity of the liver. In practice, the prothrombin time is most commonly used as a test of liver function and in the monitoring of patients treated with anticoagulants, for example, warfarin, which are antagonists of vitamin K.

Vitamin E

The exact function of this vitamin, a physiological antioxidant, remains uncertain at present. Clinical deficiency in man has only been found in premature infants in whom it may be associated with a haemolytic anaemia.

TRACE ELEMENTS

The maintenance of normal health requires provision in the diet not only of adequate protein, energy substrates and vitamins, but also of various inorganic salts and trace elements. Trace elements in the body are by definition present in concentrations less than 100 parts per million (ppm); these are shown in Fig.23.1. None is required in more than milligram quantities per day while the daily requirement for some is measurable in micrograms. Consequently the 'essential' status of some of these trace elements is difficult to confirm.

Trace element deficiency

The commonest trace element deficiency is that of iron; it is common even in affluent societies, particularly in women during the reproductive years. Iodine deficiency causes goitre and, if severe, hypothyroidism; it is now uncommon in the developed world but is still a problem in some areas. Deficiency of other trace elements is uncommon except under special circumstances. These include severe malnutrition, artificial feeding (especially if prolonged), prematurity and the presence of excessive losses (such as with enterocutaneous fistulae and severe diarrhoea). Multiple deficiencies may occur in these conditions, confusing the clinical picture and making diagnosis difficult.

Essential trace elements in the body	
Element	**Function**
chromium	deficiency causes glucose intolerance
cobalt	component of vitamin B_{12}
copper	cofactor for cytochrome oxidase
fluorine	present in bone and teeth
iodine	component of thyroid hormones
iron	component of haem pigments
manganese	cofactor for several enzymes
molybdenum	cofactor for xanthine oxidase
nickel	important in maintenance of membrane structures
selenium	cofactor for glutathione peroxidase
silicon	present in cartilage (but essential?)
tin	?
zinc	cofactor for various enzymes

Fig.23.1   Essential trace elements in man.

Laboratory assessment

Unfortunately, the laboratory assessment of the body's trace element status is difficult, as specialized equipment and considerable technical expertise are required. Measurements are often made in plasma, but these may not accurately reflect the concentration of a trace element at its site of action. Although a low plasma level may not indicate deficiency in the tissues, such deficiency is usually accompanied by a low plasma level, so that if a low level is found, it is reasonable to provide appropriate supplementation. Trace element deficiency should be anticipated in patients at risk and steps taken to prevent the occurrence of a deficiency syndrome.

Zinc

Zinc is a trace element of particular importance. It is essential for the activity of a number of enzyme systems, including carbonate dehydratase and enzymes involved in nucleic acid and protein synthesis. The clinical manifestations of zinc deficiency include dermatitis and delayed wound healing; there is, however, no evidence that zinc supplementation accelerates wound healing in patients who are not deficient. Zinc deficiency is a well-recognized potential complication of artificial, particularly parenteral, nutrition if insufficient supplementation is provided. Patients who are catabolic, for example following trauma or major surgery, lose large amounts of zinc in the urine and are at risk of becoming zinc-depleted. Severe deficiency is seen in the condition acrodermatitis enteropathica, in which there is a defect in intestinal zinc absorption.

Serum zinc concentrations must be interpreted with caution; blood should be collected in the fasting state since zinc levels may fall by up to twenty percent following a meal. Low serum levels are not exclusive to zinc deficiency; they are also seen following myocardial infarction and in other conditions such as malignant disease and chronic liver disease without associated clinical evidence of tissue deficiency.

Copper

Copper is also essential for the activity of certain enzymes, notably cytochrome oxidase and superoxide dismutase. Copper deficiency is uncommon; manifestations include anaemia and leucopenia. Wilson's disease, a disorder characterized by excessive tissue deposition of copper is discussed in *Chapter 6*.

PROVISION OF NUTRITIONAL SUPPORT

Patients requiring nutritional support should be fed enterally wherever possible; as this is more natural, cheaper and less hazardous than parenteral feeding. Parenteral nutrition is, however, required in patients with intestinal failure, pancreatic and high intestinal fistulae, and sometimes in those with a very high energy requirement (such as following severe burns) when it may be impossible to provide adequate nutrition by the enteral route alone.

Nutritional support may be required for patients who are already malnourished, as a result of, for example, severe small bowel disease or a stenosing carcinoma of the oesophagus. In such patients it is necessary both to correct existing deficits and to supply continuing needs. It is far better, when possible, to introduce nutritional support before serious deficiencies have developed and need should be anticipated by identifying patients at risk.

The diagnosis of severe malnutrition does not require the aid of the laboratory since it is clinically obvious. A serum albumin concentration of below 30g/l is often held to be an index of malnutrition, but the serum albumin can be low for many other reasons (see pages 210 & 211) and may be higher than this in a dehydrated patient with malnutrition. Other biochemical tests offer no advantages over clinical examination. However, the laboratory will be required to provide data on serum and urine for monitoring malnourished patients and it is important to be aware of both the value and the limitations of such data.

Laboratory monitoring of parenteral nutrition

Patients fed parenterally receive a (usually continuous) intravenous infusion of glucose, amino acids, vitamins and inorganic salts in solution in water. Fat emulsions are also used as an energy source and there is an increasing tendency to mix all the patient's nutritional requirements for a twenty-four-hour period in a single container. This is done under sterile conditions and the patient's intravenous catheter only needs to be handled once in twenty-four hours when the container is changed.

Patients receiving parenteral nutrition require careful clinical monitoring. Fluid status should be assessed both clinically and by means of fluid balance charts to ensure that the patient is not overhydrated or underhydrated, especially if there are abnormal losses. Short-term changes in weight reflect changes in fluid status but weighing the patient may be impracticable.

Plasma potassium and glucose

Plasma potassium and glucose must be measured daily in patients receiving parenteral nutrition. The urine should be checked for glucose every six hours once parenteral nutrition is started; glucose intolerance is common and glycosuria will then usually occur. When glucose intolerance does occur, blood or plasma measurements should be made more frequently to monitor the effect of any changes in glucose or insulin input. Hypoglycaemia is usually only a problem if parenteral feeding is stopped suddenly.

Sodium

Mild hyponatraemia (sodium concentration 125–135mmol/l) is common in patients receiving parenteral nutrition. It may be multifactorial in origin and is not an indication for increasing the sodium input. Measurement of urinary sodium excretion is useful if there is hyponatraemia; when it is due to sodium depletion and renal function is normal, the urine will contain little sodium. Spurious hyponatraemia (see page 20) due to the infusion of lipid emulsions should not be a problem in practice. If during continuous lipid infusion the plasma is more than faintly opalescent to the naked eye, the lipid is not being cleared adequately and the rate of administration should be decreased. Significant spurious hyponatraemia only occurs if the plasma is frankly lipaemic. Hypernatraemia is much less common than hyponatraemia and is usually due to lack of water rather than excess of sodium. The cause of hypernatraemia should always be determined and treated appropriately.

Serum creatinine and urea

If the patient has a good urinary output it is unnecessary to measure the serum creatinine concentration more than twice a week. However, the urea should be measured more frequently, especially after starting a patient on parenteral nutrition. If the input of amino acids exceeds the body's ability to utilize them for protein synthesis, there will be an increase in urea formation and the serum urea concentration may increase.

Plasma bicarbonate

The plasma bicarbonate is of value as an indicator of potassium status. Also patients receiving parenteral nutrition occasionally develop a metabolic acidosis, accompanied by a fall in plasma bicarbonate concentration. In practice, given the availability of multichannel analyzers, the tendency is to measure all these analytes daily, even though this may not be strictly necessary. Many patients receiving nutritional support are seriously ill or have existing deficiencies or increased losses of nutrients, factors that may necessitate more frequent monitoring. However, in patients on parenteral nutrition who are otherwise well, biochemical measurements may be made less frequently.

Plasma albumin

Plasma albumin concentration is frequently decreased in patients requiring nutritional support. Because the plasma half-life of albumin is long, and many factors influence its concentration, it is of little value in monitoring patients except that in the short-term it provides information on fluid balance. Transferrin has a shorter half-life, but its usefulness in this context is limited because its concentration increases in iron deficiency and because it is an acute phase protein. The plasma levels of prealbumin (half-life two days) and retinol-binding protein (half-life 12h) are reduced in malnutrition and increase rapidly when adequate nutritional support is provided. In practice, however, such measurements add little to what is often obvious clinically.

Other measurements in serum

The frequency of monitoring of other analytes will be determined by the clinical circumstances. In general, it is advisable to measure serum calcium and phosphate twice weekly (hypophosphataemia is a serious potential complication of parenteral nutrition). Liver function tests are often performed. Mild cholestasis (probably due to the deposition of fat in the liver) can complicate parenteral nutrition but other causes, for example, septicaemia, should always be considered.

Measurements of other elements, for example, magnesium and zinc, are required less frequently. However, these measurements should be interpreted with caution since plasma levels may be misleading if an infusion was in progress when the blood sample was obtained. A low urinary excretion of the element in question may more accurately represent a decreased tissue level of the trace element.

Haematological measurements

Regular haematological measurements must be performed and will give an indication of haematinic deficiencies.

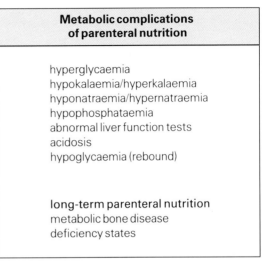

Metabolic complications of parenteral nutrition
hyperglycaemia
hypokalaemia/hyperkalaemia
hyponatraemia/hypernatraemia
hypophosphataemia
abnormal liver function tests
acidosis
hypoglycaemia (rebound)
long-term parenteral nutrition
metabolic bone disease
deficiency states

Fig.23.2 Metabolic complications of parenteral nutrition, listed in approximate order of frequency.

Urine analysis

Nitrogen balance can be assessed from a comparison of known nitrogen input with nitrogen excretion. A crude estimate of nitrogen excretion is provided by urinary urea (this being the major route of nitrogen excretion) provided that the plasma urea is constant and that there are no unusual losses. Measurements of urinary sodium and potassium output may be misleading unless considered together with the plasma concentrations, input, and the patient's renal function. A high urinary sodium excretion most commonly reflects the normal response to excessive sodium administration and so is not on its own an indication for increasing the input further. However, a low urinary sodium excretion is usually indicative of sodium depletion except in patients who are stressed. Urinary potassium excretion must also be interpreted with regard to intake. Potassium balance is negative in patients who are catabolic, but becomes positive when new tissue is being laid down.

Some of the commoner metabolic complications of parenteral nutrition are summarized in Fig.23.2.

SUMMARY

Nutritional disorders can be due to a deficiency or an excess of nutrients. Deficiency syndromes include those due to the lack of a single nutrient and those in which there is generalized deficiency; some essential nutrients can be harmful if taken in excess and if the total energy intake is greater than an individual's requirements, obesity will develop.

Specific laboratory methods are available for the diagnosis of deficiencies of individual water soluble vitamins but, with the exception of those for folic acid and vitamin B_{12}, they are rarely required in clinical practice. Among the fat-soluble vitamins, Vitamin A deficiency is rare in the developed world but Vitamin D deficiency, leading to rickets and osteomalacia occurs relatively frequently, particularly in the elderly, premature infants, patients with malabsorption and in certain racial groups. The diagnosis can be confirmed by demonstrating a low serum concentration of 25-hydroxycholecalciferol. Vitamin K deficiency leads to impairment of blood clotting with prolongation of the prothrombin time; serum measurements of the vitamin itself are not required for diagnosis.

Deficiencies of minerals required in large amounts by the body (for example, sodium, potassium, calcium, magnesium) can usually be inferred from clinical observation and measurement of their serum concentrations. It is more difficult to diagnose deficiencies of trace elements, such as zinc, manganese and copper since serum levels may not accurately reflect the body's status with regard to these elements.

Patients with generalized malnutrition show characteristic, though not specific, biochemical abnormalities, for example, low serum concentrations of albumin, transferrin and certain other proteins and decreased urinary creatinine excretion. There may also be evidence of specific deficiencies of vitamins or minerals. These patients require nutritional support. This should be enteral wherever possible, that is, using the gut either by supplementation of the diet or by tube feeding. In patients with intestinal failure, however, feeding must be parenteral. This entails the infusion of nutrients intravenously and is a potentially hazardous procedure. There is a risk of metabolic complications, for example, hyperglycaemia, hypophosphataemia and hypo- and hyperkalaemia, but these should be preventable by frequent biochemical monitoring. Biochemical and clinical monitoring are also necessary to assess nitrogen balance and to follow the patient's response to treatment.

FURTHER READING

Bender AE & Bender DA (1982) *Nutrition for Medical Students*. Chichester: John Wiley & Sons.

Marshall WJ & Mitchell PEG (1987) Total parenteral nutrition and the clinical chemistry laboratory. *Annals of Clinical Biochemistry*, **24**, 327–336.

Silk DBA (1983) *Nutritional Support in Hospital Practice*. Oxford: Blackwell Scientific Publications.

24. Clinical Chemistry at the Extremes of Age

OLD AGE: INTRODUCTION

The investigation and management of illness in the elderly poses a number of special problems, for both the physician and the chemical pathologist. These include:

- Different patterns of disease
- Different presentation of disease
- Decline in normal functions with age
- Different reference ranges

Many conditions are more common in the elderly than in younger adults; examples of such conditions of particular interest to the chemical pathologist include diabetes mellitus (see *Chapter 12*), Paget's disease of bone (see page 235) and thyroid disease (see pages 147–154). Further, the presentation of diseases in the elderly may be different from that normally seen in younger people. Thus myocardial infarction may present with confusion, consequent on a reduction in cerebral blood flow, rather than chest pain; the presenting feature of diabetes mellitus may be one of its complications, for example ischaemic ulceration, rather than polyuria and thirst. *Case histories 24.1 to 24.4* provide more detailed examples of these problems.

The functioning of some organs declines with age and this decline may be accelerated by even mild disease. The glomerular filtration rate decreases with age as does the creatinine clearance. However, the serum creatinine concentration changes little, because creatinine production also falls with age; this reflects a decrease in muscle mass and often also in meat consumption. Despite the fall in the glomerular filtration rate, renal function remains sufficient for normal homoeostasis although it may not be adequate to allow complete excretion of a drug or to sustain any further decrease in glomerular filtration without a failure of homoeostasis.

REFERENCE RANGES

Such changes in normal function mean that the reference ranges applicable to healthy adults may not

Serum constituents showing age-dependent changes in concentration	
cholesterol	increases progressively during adult life
glucose	increases (glucose tolerance decreases with age)
alkaline phosphatase	increases
urate	increases
total protein	decreases (slight decrease probably related to decreased protein intake)
albumin	decreases (as total protein)

Fig.24.1 Serum constituents showing age-dependent changes in concentration.

be applicable to the elderly, while the increased incidence of many diseases with increasing age makes it difficult to obtain data on normal people. Ideally, laboratories should construct age-related reference ranges for age-dependent analytes (Fig.24.1), but in practice this is rarely done.

This problem is exemplified by the enzyme alkaline phosphatase. Common causes of raised levels of this enzyme in the serum of the elderly include malignant disease with metastasis to bone or liver; osteomalacia; and Paget's disease of bone. In the United Kingdom, the incidence of Paget's disease exceeds five percent in people over sixty. Many cases are mild and clinically silent, being discovered only after a raised serum alkaline phosphatase has been found, often as part of a biochemical screening test. Asymptomatic patients with Paget's disease do not require treatment, but theoretically screening programmes are only worthwhile if abnormal results are followed up. How extensively this can be done is governed by economic factors. The practice in many laboratories is to assume that, in the absence of any clinical or other laboratory evidence of disease, an alkaline phosphatase of up to one and a half times the upper limit of normal for young adults does not justify further investigation in an elderly subject.

SCREENING

The higher prevalence of many diseases in the elderly provides some of the justification for screening programmes. If a condition has a high prevalence in a population, the predictive value of a positive test is much higher than if it is low (see *Chapter 1*). Such screening may be carried out in general practice, at over-sixties clinics, in geriatric assessment centres or in geriatric hospitals. The biochemical tests which should form part of such a screen (Fig.24.2) reflect the diseases that are of particular concern in this age group, some of which have been mentioned above. The serum potassium is included since diuretics are commonly prescribed for the elderly and, according to the type used, may cause hypokalaemia or hyperkalaemia. The possible influence of inter-current disease on tests of thyroid status must be borne in mind. The results of such tests may erroneously suggest thyroid disease in a patient who is ill for some other reason (sick euthyroid disease) and it is best to avoid doing these tests at such a time.

Biochemical tests used to screen for disease in the elderly	
Analyte	**Common abnormalities**
serum potassium	hypokalaemia (diuretic and purgative-induced) hyperkalaemia (potassium-sparing diuretic with poor renal function
serum creatinine	increased (renal impairment)
serum calcium	hypercalaemia (hyperparathyroidism) hypocalcaemia (osteomalacia)
serum alkaline phosphatase	increased (osteomalacia, Paget's disease and malignancy)
serum glucose	increased (diabetes mellitus)
serum TSH (or thyroxine)	hypothyroidism and hyperthyroidism

Fig.24.2 Biochemical tests used to screen for disease in the elderly.

CASE HISTORY 24.1

A general practitioner was called to see a previously fit man in an old people's home. The patient had become acutely short of breath two hours before, shortly after his breakfast, and developed a cough with frothy white sputum. He also complained of dizziness, but denied chest pain.

On examination, he had widespread crepitations throughout his lung fields; his blood pressure was 120/70mmHg but had been 150/ 90mmHg when checked by the doctor two months previously.

He was given a diuretic, with considerable symptomatic relief. An ECG showed changes consistent with a very recent myocardial infarct. The doctor took a blood sample for measurement of serum creatine kinase and was surprised when the laboratory telephoned him the next day to say that this was normal.

Comment

The breathlessness, cough and crepitations are classical features of left ventricular failure. A likely cause for this and the fall in blood pressure, was myocardial infarction; chest pain does not always occur, particularly in the elderly. The general practitioner should not have been surprised that the creatine kinase was normal – the blood had been taken too soon after the presumed infarction. He was advised by the chemical pathologist to take a further blood sample; this was timed at twenty-six hours after the onset of symptoms and the creatine kinase was clearly raised at 280iu/l.

CASE HISTORY 24.2

An elderly lady presented with an exacerbation of congestive cardiac failure. She was being treated with digoxin and a thiazide diuretic.

Investigations

serum:	digoxin	3.2nmol/l
	(twelve hours after previous dose)	
	potassium	3.0mmol/l
	urea	11.2mmol/l
	creatinine	160μmol/l

Comment

Drug interactions are an important cause of ill-health at all ages, but particularly in the elderly. An exacerbation of cardiac failure in a patient treated with digoxin should raise the suspicion of digoxin toxicity. The serum level here is compatible with this and digoxin toxicity is enhanced by hypokalaemia; thiazide diuretics are an important cause of this. This patient also has evidence of impaired renal function, which can impair digoxin excretion (see also *Case history 3.1*).

CASE HISTORY 24.3

A seventy-year-old woman presented with a painful ulcer on the sole of her left foot. On examination her foot felt cold and appeared

ischaemic; no pulses were palpable below the femorals on either side.

Her urine contained a trace of sugar and a biochemical screen revealed a random serum glucose level of 15mmol/l but she denied any thirst or polyuria.

Comment

This patient's random serum glucose is diagnostic of diabetes mellitus. The classic thirst and polyuria of diabetes may not always be present, particularly in the elderly, in whom the renal threshold for glucose is often elevated as a result of a decreased glomerular filtration rate. This may just be a feature of declining renal function with age, but can be exacerbated by renal disease which can develop as a complication of diabetes.

Many factors are involved in the pathogenesis of occlusive arterial disease, of which diabetes is only one. In diabetes, the presence of arterial disease appears to be related more to the patient's age than to the duration of their diabetes.

CASE HISTORY 24.4

An elderly lady was admitted to hospital after she had fallen at home and fractured her femur. She was a recluse and rarely went out, depending on a home help to do her shopping.

In addition to the fracture a radiograph showed classic features of osteomalacia.

Investigations

serum:	calcium	1.75mmol/l
	phosphate	0.70mmol/l
	alkaline phosphatase	440iu/l
	albumin	30g/l

Her fracture was treated by replacement arthroplasty. After her operation, a medical student took a detailed history from the patient and discovered that she had recently developed constipation and had passed some fresh blood per rectum. He found her liver to be enlarged and a barium enema revealed a stenosing carcinoma of the sigmoid colon. A laparotomy was performed and the tumour was resected, but the liver was seen to contain several metastatic tumour deposits. Measurement of alkaline phosphatase isoenzymes showed an increase in both the bone and the liver isoenzyme.

Comment

The low serum calcium (even when the low albumin is taken into account), slightly reduced phosphate (a reflection of secondary hyperparathyroidism) and raised alkaline phosphatase (reflecting increased osteoblastic activity) are typical of osteomalacia. This is more common in the elderly and both poor nutrition (the low albumin would be consistent with this) and decreased endogenous synthesis of vitamin D (due to lack of exposure to sunlight) may be important in its pathogenesis. The serum 25-hydroxycholecalciferol concentration is usually low. Typical radiological features are not always present; the definitive technique for making the diagnosis is histological examination of a bone biopsy, but this is a specialized, invasive procedure and the diagnosis is often confirmed by the response to a therapeutic trial of vitamin D.

This case illustrates another important aspect of geriatric medicine – that patients may have more than one disease. Although this is true at any age, it is more likely in the patient of advanced years.

Other case histories of particular relevance may be found on pages 22, 51 and 152.

CHILDHOOD: INTRODUCTION

Just as the elderly provide particular problems for the chemical pathologist, so too do children. The most obvious of these relates to the size of the blood sample. For the very young it is essential to employ analytical methods that will use the smallest possible amount of serum and this usually means providing special equipment. Small quantities of blood can be conveniently collected by pricking the heel, but this should be done by experienced personnel and the

Common analytes having different reference ranges in children	
Analyte	**Difference**
serum potassium	mean and upper limit higher in neonates
serum calcium	higher at birth; normal adult levels by 72h
serum phosphate	higher at birth, then falls but remains higher than adult levels throughout childhood; rises at puberty then falls to adult levels
serum alkaline phosphatase	as phosphate
serum glucose	lower low limit for first 24h of life

Fig.24.3 Common analytes with different reference ranges in children.

results obtained may be affected by haemolysis or by contamination with tissue fluid.

Complete, accurately timed collections of urine are very difficult to obtain in children. It is usually more reliable to relate the concentrations of urine constituents to urine creatinine concentration.

Many conditions present exclusively, or predominantly, in the neonatal period, examples include congenital diseases and inherited metabolic disorders (see *Chapter 18*). Other disorders may be recognized at any time throughout childhood, particularly disorders of growth and of sexual differentiation and development.

REFERENCE RANGES

The reference ranges for certain analytes are different in the newborn from the adult (Fig.24.3) and may vary through childhood; the concentrations of some analytes, in particular phosphate and calcium, are affected by the diet. A result should always be interpreted in the light of the reference range appropriate to the child's age. The age-related changes in serum alkaline phosphatase are discussed in *Chapter 17*; and in immunoglobulins in *Chapter 15*. Creatinine clearance must be corrected for surface area in a child, since it increases as the child grows.

SCREENING

Paediatric medicine no longer starts with the birth of the child. Antenatal screening for inherited metabolic disease is discussed in *Chapter 18*, but it is now becoming possible to carry out diagnostic and therapeutic procedures on the fetus *in utero* and the chemical pathologist will be required to provide a service in support of this fetal medicine.

In a book of this size, it is possible only to outline some of the more important areas where paediatric medicine impinges upon chemical pathology. The reader seeking more detailed information is referred to the *Further reading* section.

CHILDHOOD DISORDERS

Neonatal hypoglycaemia

This important condition is discussed in *Chapter 13*. It is particularly likely to occur in low birth weight infants, both premature and 'small-for-dates'; babies born to diabetic mothers; and babies who are ill or who have feeding problems. In such babies, blood glucose measurements should be made every four hours for the first forty-eight hours and at appropriate intervals thereafter to monitor treatment if hypoglycaemia has occurred.

Neonatal hypocalcaemia and hypomagnesaemia

The clinical signs of hypoglycaemia include irritability, twitching and convulsions. If the baby's blood glucose is not low, hypocalcaemia, which presents with similar signs, should be suspected.

Hypocalcaemia occurring within forty-eight hours of birth is seen particularly with very small babies. It can be prevented by giving adequate calcium; if the baby is not feeding normally, calcium must be given intravenously. Neonatal hypocalcaemia is also seen following hypoxia and sometimes in the infants of diabetic mothers.

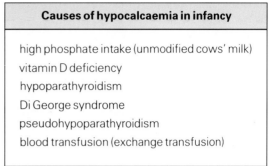

Causes of hypocalcaemia in infancy

high phosphate intake (unmodified cows' milk)

vitamin D deficiency

hypoparathyroidism

Di George syndrome

pseudohypoparathyroidism

blood transfusion (exchange transfusion)

Fig.24.4   Causes of hypocalcaemia in infancy excluding transient neonatal hypocalcaemia.

This early neonatal hypocalcaemia is usually only transient. Causes of hypocalcaemia occurring after this, and up to about ten days of life, are shown in Fig.24.4. Most of these conditions are discussed in detail in *Chapter 14*. Hypocalcaemia is a potential complication of exchange blood transfusion (clotting of donor blood is prevented by chelating the calcium) and can be prevented by giving calcium during a transfusion.

It should be noted that the reference ranges for both serum calcium (2.20–3.00mmol/l) and phosphate (1.00–2.60mmol/l) are higher in the newborn infant than in the adult. Serum phosphate levels rise further in the first two to three weeks of life.

Hypocalcaemia is often accompanied by hypomagnesaemia, although isolated hypomagnesaemia is uncommon. For this reason, magnesium supplements should be given together with calcium in treating hypocalcaemia. If magnesium is not given, hypocalcaemia is often resistant to treatment.

Jaundice

Most babies become mildly jaundiced shortly after birth. This 'physiological' jaundice is due to immaturity of the hepatic conjugating enzymes, to normal postnatal haemolysis and to enterohepatic circulation of bilirubin (conversion of bilirubin to urobilinogen in the gut cannot occur until the gut becomes colonized with bacteria). In physiological jaundice the bilirubin is primarily unconjugated and its serum concentration rarely exceeds 100μmol/l. At high levels of unconjugated bilirubin (>350μmol/l) there is a risk of kernicterus developing. Since unconjugated bilirubin is bound to albumin the risk is greater if the serum albumin is decreased or bilirubin is displaced from albumin, for example, by hydrogen ions in acidosis, by certain drugs or by high levels of free fatty acids. Unconjugated hyperbilirubinaemia can be treated by increasing water intake, phototherapy or exchange transfusion as appropriate, and of course by treatment of the underlying cause if this can be ascertained and treatment is feasible. Circumstances which should prompt investigation of neonatal jaundice are given in Fig.24.5.

Causes of unconjugated hyperbilirubinaemia in the newborn are given in Fig.24.6.

There are also many causes of conjugated hyperbilirubinaemia in infants; some of these are listed in Fig.24.7.

Metabolic disorders

The clinical features of metabolic disorders are rarely specific to any one condition; the salt loss and virilization of female infants with steroid 21-hydroxylase deficiency (see page 250) are exceptional in this respect. Some clinical features, such as severe acidosis and coma, suggest that a metabolic disorder may be present, but in the majority of cases they are nonspecific, babies afflicted by such disorders presenting, for example, with vomiting or 'failure to thrive'.

The determination of the precise diagnosis of a metabolic disorder may require complex and lengthy investigation, so it is important to be able to carry out some simple screening tests to indicate whether a metabolic disorder may be the cause of a baby's illness. An appropriate battery of tests is shown in Fig.24.8. If the results of these are all normal, a metabolic disorder is unlikely; if there are abnormalities, the pattern of these may suggest a possible diagnosis or indicate what further investigations would be appropriate. It is important that the child should, if at all possible, be on a normal diet when these tests are done; potential abnormalities may otherwise be masked. Thus disorders associated with an abnormal pattern of amino acid secretion may be missed if the infant does not have a normal protein intake.

When to investigate neonatal jaundice
presence at birth or appearance during first 24h of life
persistence beyond 14 days of life
total serum bilirubin concentration >250μmol/l
conjugated hyperbilirubinaemia
jaundice associated with other signs or symptoms of disease

Fig.24.5 Circumstances in which neonatal jaundice should be investigated.

CASE HISTORY 24.5

Thirty-six hours after birth a male infant started vomiting, developed grunting respiration and rapidly became lethargic and unresponsive. He appeared physically normal and was born at term after a normal pregnancy. The parents were first cousins; it was the woman's first pregnancy. A metabolic screen revealed a very high plasma ammonia concentration (>1000μmol/l). The plasma urea was at the lower end of the normal range and plasma amino acid chromatography showed an increase in glutamine and alanine. Despite intensive treatment, including peritoneal dialysis, the baby died seventy-two hours after birth.

Comment

Hyperammonaemia is an important cause of both morbidity and mortality in infants, but ammonia is technically difficult to measure and the condition may not be diagnosed. This was a typical presentation of hyperammonaemia; toxic encephalopathy is usually a prominent feature. Although there are many causes of hyperammonaemia (Fig.24.9), a case as severe as this, without any suggestion of liver disease, and in a child born of a first cousin marriage, should raise the suspicion of an inherited metabolic disorder of the urea cycle. The excess plasma glutamine and alanine, with a low to normal urea, are consistent with this. This child's urine was found to contain a high concentration of orotic acid. This pattern of abnormalities suggests deficiency

Causes of unconjugated hyperbilirubinaemia in the newborn

Increased haemolysis

rhesus blood group incompatibility
ABO blood group incompatibility
red cell enzyme defects:
    glucose-6-phosphate dehydrogenase
        deficiency
    pyruvate kinase deficiency

Decreased conjugation

Crigler–Najjar syndrome
hypothyroidism
breast milk jaundice (a benign condition seen
    in some breast-fed infants and thought to be
    due to interference with bilirubin conjugation
    by free fatty acids)

Fig.24.6   Causes of unconjugated hyperbilirubinaemia in the neonate.

Causes of conjugated hyperbilirubinaemia in the newborn

haemolytic conditions (enterohepatic circulation
    of bilirubin)
hepatic dysfunction ('neonatal hepatitis') due to:
    infection:
        congenital, e.g. rubella, cytomegalovirus
            and syphilis
        acquired, e.g. urinary tract infection,
            septicaemia and hepatitis
    metabolic disorder:
        α_1-antitrypsin deficiency
        galactosaemia
        tyrosinaemia
    congenital abnormality:
        biliary atresia

Fig.24.7   Causes of conjugated hyperbilirubinaemia in the neonate.

of ornithine carbamoyl transferase and this was confirmed on post-mortem biopsy of the liver.

Consanguineous parents, or a history of a previous neonatal death, should increase one's suspicion that an inherited metabolic disease may be responsible for a child's illness.

Screening tests for metabolic causes of illness in the newborn

Urine

reducing	bilirubin
substances	sugar and amino acid
glucose	chromatography
ketones	

Blood

glucose	hydrogen ion

Serum

sodium	magnesium
potassium	conjugated bilirubin
urea	ammonia
creatinine	chromatography for
calcium	amino acids
phosphate	

Fig.24.8   Screening tests for metabolic causes of illness in the neonate.

Some causes of hyperammonaemia in infancy

transient neonatal hyperammonaemia*
inherited disorders of the urea cycle*
other inherited metabolic disorders*,
    such as organic acidaemias
liver disease (including Reye's syndrome)
severe systemic illness* (asphyxia, infection,
    sepsis)
parenteral nutrition (excessive amino acid
    input)
sodium valproate therapy

*important causes in the neonate

Fig.24.9   Some causes of hyperammonaemia in infancy. Reye's syndrome is a cause of encephalopathy in children, associated with fatty infiltration of the liver and hyperammonaemia; the cause is unknown but it is most commonly seen following mild infection.

Some causes of failure to thrive
malnutrition
malabsorption
inherited metabolic diseases
infection
chronic disease:
renal
hepatic
pulmonary
cardiac
psychosocial deprivation
hypothyroidism
hypopituitarism

Fig.24.10   Some causes of failure to thrive.

Causes of precocious puberty and pseudoprecocious puberty
Precocious puberty
idiopathic
pineal tumours and hypo-thalamic hamartomas
post meningitis and encephalitis
hypothyroidism
Pseudoprecocious puberty
gonadotrophin-secreting tumours
congenital adrenal hyperplasia
adrenal tumours
ovarian and testicular tumours

Fig.24.11   Causes of precocious puberty and pseudoprecocious puberty.

Causes of delayed puberty
constitutional (idiopathic)
chronic systemic illness,
e.g. renal failure and hypothyroidism
undernutrition
chronic administration of
corticosteroids
hypothalamic or pituitary
insufficiency
primary gonadal failure

Fig.24.12   Causes of delayed puberty. Constitutional delayed puberty is by far the commonest cause, reflecting one extreme of the normal range and affecting some 2.5% of all children.

Failure to thrive

This is a common paediatric problem and some of the causes are shown in Fig.24.10. Where there are no suggestive clinical features, either in the history or on examination, the results of tests listed in Fig.24.8, together with simple haematological tests and a screen for infectious disease, will in many cases provide a starting point for definitive investigation.

Disorders of sexual differentiation and abnormal puberty

Precocious sexual development, which may become apparent shortly after birth, is rare; some causes are given in Fig.24.11. It is important to distinguish between true precocious puberty, in which the gonads are fully developed and contain gametes, and pseudo-precocious puberty (Fig.24.11) in which they are not. Pseudoprecocious puberty is often amenable to treatment, albeit palliative, whereas true precocity is usually not. Delayed puberty is much more common; causes are given in Fig.24.12. Causes of virilization during childhood and adolescence are summarized in Fig.24.13. Many of the conditions listed are rare, but the results of relatively simple tests, for example, the measurement of adrenal and gonadal steroids, and of gonadotrophins, are invaluable in formulating a differential diagnosis. The same holds true for disorders of sexual differentiation, examples of which are shown in Fig.24.14. Although also rare, all these conditions are of immense importance to the patient and their parents, and laboratory investigations are vital in their diagnosis and management.

Causes of virilization in girls
Adrenal
congenital adrenal hyperplasia
Cushing's syndrome
adrenal tumours
premature adrenarche
Ovarian
ovarian tumours
polycystic ovary syndrome

Fig.24.13 Causes of virilization in girls. In boys these adrenal conditions, testicular tumours and ectopic gonadotrophin secretion may cause pseudoprecocious puberty.

Causes of abnormal sexual differentiation
Male pseudohermaphroditism (genotypic males with incomplete masculinization)
decreased testosterone production: various inherited enzyme abnormalities impaired testosterone metabolism: 5α-reductase deficiency androgen insensitivity syndromes congenital anomalies
Female pseudohermaphroditism (genotypic female with virilization)
congenital adrenal hyperplasia adrenal tumours high maternal androgens
Syndromes of abnormal gonadal differentiation
Turner's syndrome (45X0 karyotype) Klinefelter's syndrome (47XXY karyotype) other chromosomal abnormalities true hermaphroditism

Fig.24.14 Causes of abnormal sexual differentiation.

Disorders of growth

Many disorders can cause retardation of growth including most of the causes of delayed puberty indicated in Fig.24.12. Simple laboratory tests can provide important diagnostic information in such cases, but do not obviate the need for accurate clinical and anthropometric assessment. Growth hormone deficiency is rare; its diagnosis is discussed in *Chapter 8*. It can be treated by hormone replacement. The effects and diagnosis of growth hormone excess are also considered in *Chapter 8*.

SUMMARY

Many biochemical and physiological functions change with age; some of these are related to specific events, in particular puberty and the menopause, but for others the change is more gradual, for example, a decrease in the glomerular filtration rate in the elderly. This must be borne in mind when interpreting the results of biochemical tests in the elderly and ideally such results should be compared with age-related reference ranges. Thus the serum cholesterol concentration increases throughout adult life as does the serum urate; and glucose tolerance decreases in the elderly. The presentation of certain diseases may be different in the elderly and biochemical tests assume a greater importance in diagnosis; furthermore, many diseases occur more frequently in the elderly and it may be justified to screen for thyroid disease, diabetes mellitus and osteomalacia, among others, in elderly people.

In children too, the reference ranges for some biochemical variables is different from that in adults. Examples include serum phosphate concentration and alkaline phosphatase activity (both higher) and cholesterol and urate (both lower). Many conditions present most frequently, or even exclusively, in childhood; thus many inherited metabolic diseases characteristically present at or soon after birth.

Metabolic problems which occur particularly frequently in the newborn include hypoglycaemia, hypocalcaemia and hypomagnesaemia. Many infants become jaundiced in the first few days of life but in the majority of cases this is entirely benign. This 'physiological' jaundice is due to an increase in unconjugated bilirubin. Conjugated hyperbilirubinaemia is always pathological.

The clinical features of inherited metabolic disorders presenting in infancy and childhood are often non-specific. In children, who for example fail to thrive or show unusual irritability or lethargy, simple screening tests on urine and plasma should be performed to identify any abnormality which may be due to an inherited metabolic disease.

Disorders of sexual differentiation are uncommon but following clinical assessment the results of simple laboratory tests (for example the levels of adrenal and gonadal hormones, gonadotrophins) are often of vital importance in formulating a differential diagnosis, and indicating the course of further investigations. This is also true of delayed puberty, a much more common complaint. There are many causes of growth failure, including systemic disease, social deprivation and malabsorption; few are due to growth hormone deficiency. Again, the results of accurate clinical assessment, combined with simple laboratory tests will often indicate the diagnosis and thus the appropriate mode of treatment.

FURTHER READING

Clayton BE & Round JM (eds) (1984) *Chemical Pathology and the Sick Child.* Oxford: Blackwell Scientific Publications.

Hodkinson M (ed) (1984) *Clinical Biochemistry of the Elderly.* Edinburgh: Churchill Livingstone.

Adult Reference Ranges

These reference ranges, from the author's laboratory, are provided for the interpretation of data presented in the case histories. Readers should note that reference ranges may differ between different laboratories; this applies particularly to hormones and enzymes. All values are for concentrations (activities in the case of enzymes) in serum or plasma, except where indicated otherwise.

acid phosphatase: total	4–11iu/l		creatinine	60–110μmol/l
prostatic	<4iu/l		follicle-stimulating hormone (FSH):	
adrenocorticotrophic hormone (ACTH): at 0900h	10–90ng/l		adult males	1.5–9.0u/l
			females:	
albumin	35–50g/l		follicular phase	3.0–8.0u/l
			post-menopausal	>15u/l
aldosterone: recumbent	100–500pmol/l		glucose: fasting	2.8–6.0mmol/l
alkaline phosphatase	30–90iu/l		γ-glutamyl transpeptidase (γGT)	<60iu/l
alphafetoprotein (AFP)	<10ku/l			
ammonia	10–47μmol/l		growth hormone:	
			following glucose load	<2mu/l
amylase	<300iu/l		following stress	>20mu/l
aspartate transaminase (AST)	10–50iu/l		haemoglobin: males	13–18g/dl
			females	12–16g/dl
bicarbonate (total CO_2)	22–30mmol/l		hydrogen ion: arterial blood	35–46nmol/l (pH 7.36–7.44)
bilirubin: total	3–20μmol/l			
calcium	2.2–2.6mmol/l		hydroxybutyrate dehydrogenase (HBD)	<250iu/l
carbon dioxide ($P\text{co}_2$): arterial blood	4.5–6.0kPa (35–46mmHg)		insulin: fasting	<15mu/l
			in hypoglycaemia	<2mu/l
cholesterol: total	3.0–6.7mmol/l		luteinizing hormone (LH):	
high density lipoprotein (HDL)	1.05–1.45mmol/l		adult males	2.0–10u/l
			adult females:	
copper	12–19μmol/l		follicular phase	3.0–12u/l
			post-menopausal	>20u/l
cortisol: at 0900h	280–700nmol/l			
at 2400h	<220nmol/l		magnesium	0.7–1.0mmol/l
creatine kinase	<90iu/l		osmolality	280–295mmol/kg

oxygen (P_{O_2}): arterial blood	11–15kPa (85–105mmHg)	testosterone: adult males adult females	9–30nmol/l 0.5–2.5nmol/l
parathyroid hormone (N-terminus)	<120pg/ml	thyroid-stimulating hormone (TSH, thyrotrophin)	0.4–5.0mu/l
phosphate	0.8–1.4mmol/l	thyroxine (T_4): total free	60–150nmol/l 9–26pmol/l
potassium	3.6–5.0mmol/l	triglyceride: fasting	0.4–1.8mmol/l
prolactin	50–450mu/l	triiodothyronine (T_3)	1.2–2.9nmol/l
protein: total	60–80g/l	urea	3.3–6.7mmol/l
renin (plasma renin activity, PRA): recumbent	1.2–2.4pmol/h/ml	uric acid	0.1–0.4mmol/l
sodium	135–145mmol/l	zinc	12–20μmol/l

Case History Index

Index